Multitemporal Earth Observation Image Analysis

SCIENCES

Image, Field Director – Laure Blanc-Féraud

Information Seeking in Images and Videos,
Subject Head – Hichem Sahbi

Multitemporal Earth Observation Image Analysis

Remote Sensing Image Sequences

Coordinated by
Clément Mallet
Nesrine Chehata

iSTE

WILEY

First published 2024 in Great Britain and the United States by ISTE Ltd and John Wiley & Sons, Inc.

ISTE Ltd
27-37 St George's Road
London SW19 4EU
UK

www.iste.co.uk

John Wiley & Sons, Inc.
111 River Street
Hoboken, NJ 07030
USA

www.wiley.com

Library of Congress Control Number: 2023944801

British Library Cataloguing-in-Publication Data
A CIP record for this book is available from the British Library
ISBN 978-1-78945-176-4

ERC code:
PE6 Computer Science and Informatics
PE6_11 Machine learning, statistical data processing and applications using signal processing (e.g. speech, image, video)
PE6_12 Scientific computing, simulation and modelling tools
PE10 Earth System Science
PE10_14 Earth observations from space/remote sensing

Contents

Foreword

Francesca BOVOLO
Remote Sensing for Digital Earth, Fondazione Bruno Kessler, Trento, Italy

Satellite image time series was generated since the beginning of the satellite-borne remote sensing era. At first, time series was sparse in time and had limited trade-offs in terms of spatial, spectral and radiometric resolution. This was related to technological limits of acquisition sensors, on-board and ground data storage capabilities, acquisition as well as data distribution policies. As the technology evolved and sensors could allow for a better trade-off in spatial/spectral/radiometric resolutions, the advent of satellite constellations gave an impulse to the increment of temporal resolution, data storage capabilities increased and the data providers started to move toward an open access data policy. Accordingly, time series showing better trade-offs in terms of spatial, spectral, radiometric and temporal resolutions became more and more accessible. Among the other freely available data, let us mention here the NASA Landsat family, the NASA MODIS data collections and the ESA Sentinel missions data. Furthermore, we observe that some recent mission sensors have been designed to maintain a synergy with earlier ones (e.g. Sentinel-2 and Landsat, Sentinel-3 and MODIS). This allows for constructing longer time series with similar data and denser time series (with respect to the use of a single sensor) in the period of cooperation. Yet the sensors are not identical, thus their joint use requires adequate data management. We also observe the increment of satellite constellations. We have small constellations of few satellites (e.g. Sentinel-2 with 2 elements, RapidEye with 5) to very large constellations with hundreds of elements (e.g. PlanetScope with 180+).

Such an amount of data unlocks a wide range of possible uses and applications (e.g. denser time series samples faster temporal behaviours being invisible before), raises up novel methodological challenges (e.g. multisensor multitemporal data fusion), provides possible novel solutions for applications, introduces multisensor time series of data with strong variability and irregularity, and sets down the issue of handling large scale information in both space and time. The latter one has for long been a critical bottleneck due to the limitations in both the availability and the access to computational power and data storage. However, nowadays we are experiencing a rapid technological evolution and broadening of access to high-power processing tools and data archiving facilities with lower costs. Accordingly, we have the chance to analyze time series and model long-time and large-space span phenomena back to the past as well as predicting future behaviors. This pushes us to rethink and revise concepts such as problem formulations in terms of land-cover classes (e.g. seasonal classes against permanent ones, class variability in space and time) and land-cover change classes, data harmonization, temporal consistency, scalability of methodologies, reusability of (intermediate) products, etc.

The chapters in this book touch some of the above-mentioned issues. Chapter 1 approaches the issue of aligning time series of stereoscopic images relaying on the scale-invariant future transform (SIFT) concept. The widely known concept is redefied by accounting for the temporal dimension and becomes therefore effective over multiple scales in space and time. The mechanism unlocks the possibility to model 3-D description of changes. The next three chapters focus attention on vegetation time series analysis and in particular on crop monitoring. Several methodological and application aspects are investigated. Among them, Chapter 1 deals with the issues of species spectral behaviors in time, space and with the direct impact on the legend definition, the issue of imbalanced classes (in space and time) and the design of the methodological approach for effective classification. Chapter 2 contributes to the analysis of scalability and geographic generalization aspects (regional and global) of crop type mapping in time series. It accounts for space and time variability with a focus on pre-processing aspects. Chapter 3 adds a level of complexity by introducing multisensor time series data processing and the challenging task of locating in time-specific events like the irrigation ones, as emphasized in Chapter 4. Chapters 5 and 6 deal with data management in terms of the need to fill the gaps (e.g. cloud impacts) and/or interpolating irregular image time series to obtain complete and regularly gridded products. The former focuses on phenology analysis, whereas the latter focuses on ocean surface topology retrieval. Chapter 7 focuses on spatially and temporally rare phenomena such as tropical cyclones where spatiotemporal forecasting precision requires accurate and hyperparametric dynamic models. The chapter is an excellent example of how the increase in computational power, multisource data access and the proper management historical spatiotemporal data may improve product accuracy.

CD and TS Book 1	CD and TS Book 2	This book
1. Hierarchical Markov Random Fields for High Resolution Land Cover *Classification of Multisensor and Multiresolution* Image Time Series 2. Pixel-based *Classification* Techniques for Satellite Image Time Series 3. *Semantic Analysis* of Satellite Image Time Series 4. Optical Satellite Image Time Series Analysis for Environment Applications: From Classical Methods to *Deep Learning* and Beyond 5. A Review on Multi-temporal Earthquake *Damage Assessment* Using Satellite Images 6. Multiclass *Multilabel Change* of State Transfer Learning from Image Time Series	1. Unsupervised *Change Detection* in Multitemporal Remote Sensing Images 2. *Change Detection* in Time Series of Polarimetric SAR Images 3. An Overview of Covariance-based *Change Detection* Methodologies in Multivariate SAR Image Time Series 4. Unsupervised Functional Information Clustering in Extreme Environments from Filter Banks and Relative Entropy 5. Thresholds and Distances to Better Detect *Wet Snow* Over Mountains with Sentinel-1 Image Time Series 6. Fractional Field Image Time Series Modeling and Application to *Cyclone Tracking* 7. Graph of Characteristic Points for Texture Tracking: Application to *Change Detection and Glacier Flow* Measurement from SAR Images 8. *Multitemporal Analysis* of Sentinel-1/2 Images for Land Use Monitoring at Regional Scale 9. Statistical Difference Models for *Change Detection* in Multispectral Images	1. Broader Application of the *Time-SIFT Method*: Proof-of-Concept of 3-D Monitoring Study Cases with Various Spatiotemporal Scales 2. Hierarchical *Crop Mapping* from Satellite Image Sequences with Recurrent Neural Networks 3. Exploiting Multitemporal Multispectral High-resolution Satellite Data Toward Annual Land *Cover and Crop Type* Mapping 4. *Irrigation Monitoring* Using High Spatial and Temporal Resolutions Remote Sensing Time Series 5. *Trends in Satellite Time Series* Processing for Vegetation Phenology Monitoring 6. Data-Driven *Spatiotemporal Interpolation* for Satellite-Derived Geophysical Tracers 7. Recent Advances in *Tropical Cyclone Forecasting* using Machine Learning on Reanalysis and Remote Sensing

Table F.1. *Book comparisons between Change Detection and Image Time Series Analysis Volumes 1 and 2 and Multitemporal Earth Observation Image Analysis* (numbers in the list refer to chapter numbers)

In this book, methodologies have been employed coming from the statistical and pattern recognition field, as well as from the machine and deep learning family. Chapters include test cases and practical examples approaching several applications by means of time series of data acquired by different sources, with different space and time scales, and over different geographical areas.

In the future, we expect to observe a further increment of data accessibility, with the chance to have free-access georeferenced databases with time series of multi-source/sensor data. The simplified access to processing facilities will allow the exercise of methodologies, speed up theoretical development and make available spatiotemporal data becoming pervasive and wide-spread use in space, time and in multiple applications.

This book comes as a complementary contribution on the topic of time series analysis with respect to two recently published books on Change Detection and Image Time Series Analysis (Atto et al. 2022a, 2022b). These two books focus mainly on information extraction with a significant attention to change detection by both supervised and unsupervised methods. Whereas the current book devotes more of its attention to some pre-processing aspects such as time series alignment and interpolation, and time series analysis for the solution of agricultural-related challenges.

References

Atto, A., Bovolo, F., Bruzzone, L. (2022a). *Change Detection and Image Time Series Analysis 1: Unsupervised Methods*. ISTE Ltd, London and John Wiley & Sons, New York.

Atto, A., Bovolo, F., Bruzzone, L. (2022b). *Change Detection and Image Time Series Analysis 2: Supervised Methods*. ISTE Ltd, London and John Wiley & Sons, New York.

1

Broader Application of the Time-SIFT Method: Proof-of-Concept of 3-D-Monitoring Study Cases with Various Spatiotemporal Scales

Denis FEURER[1], Sean BEMIS[2], Guillaume COULOUMA[1], Hatem MABROUK[3], Sylvain MASSUEL[4], Romina Vanessa BARBOSA[5], Yoann THOMAS[5], Jérôme AMMANN[6] and Fabrice VINATIER[1]

[1]LISAH, INRAE, IRD, Institut Agro, University of Montpellier, France
[2]Department of Geosciences, Virginia Tech, Blacksburg, USA
[3]Institut National Agronomique de Tunisie (INAT), Tunis, Tunisia
[4]G-EAU, AgroParisTech, Cirad, INRAE, IRD, Institut Agro, University of Montpellier, France
[5]CNRS UMR 6539 LEMAR, UBO, IRD, Ifremer, IUEM Plouzané, France
[6]CNRS UMR 6538 LGO, IUEM Plouzané, Université de Bretagne Occidentale, France

1.1. Introduction

Photography, aerial photography and satellite earth observation, along with the rapidly expanding accessibility of uncrewed aerial vehicle (UAV)-based remote sensing and smartphones, are a rich source of potential image time series that allow the monitoring of a wide range of phenomena across a broad scope of scientific domains. Some recent advances in remote sensing hence renewed the tools for the monitoring of 3-D change. This is particularly notable for the monitoring of sub-metric 3-D changes such as geomorphic studies, vegetation or erosion monitoring (e.g. Marcus and Fonstad 2008; Brodu and Lague 2012; Eltner et al. 2016). For a long time, studies focusing on these scales relied on topographic ground surveys, which are very cumbersome and costly. Terrestrial laser scanning provides denser topographic data than can be acquired from typical ground survey methods, such as a total station, but still has the drawback of being a ground-based technique with the corresponding limitations (e.g. occlusions, implementations, difficult coverage of large areas). Dense airborne laser scanning allows the collection of such topographic data, but the relative novelty of this technique results in a lack of historical depth. Meanwhile, the last decade saw the rapid expansion of methods from the field of computer vision adopted into the earth sciences community (Westoby et al. 2012; Fonstad et al. 2013; Bemis et al. 2014). Structure from Motion (SfM) photogrammetry is now widely used to provide detailed 3-D descriptions of various natural and anthropogenic objects of interest or systems (see, for example, reviews from Carrivick et al. 2016; Eltner et al. 2016; Smith et al. 2016). Moreover, SfM has allowed new insights to be extracted from archival aerial imagery (e.g. Verhoeven 2011; Salach 2017; Sevara et al. 2017) and hence access to an unprecedented historical depth for the monitoring of 3-D change. Still, in order to compare data acquired at different epochs, we must ensure that multi-temporal data share a common geometric reference. Co-registration of data of different epochs is indeed – by nature – the prerequisite of diachronic studies. In some experimental designs, permanent reference points can be established to provide common registration points for a multi-temporal study. However, there are a wide range of examinations for which it is not possible to establish reference points prior to imagery collection. For example, study areas with difficult access or environmental sensitivity may preclude the establishment of semi-permanent reference features. Furthermore, a wealth of temporal data already exists in archival imagery where it would be impossible for new studies to establish ground control references prior to imagery collection.

Archival aerial imagery is the most ancient source of multi-temporal Earth observation data, with coherent archives that go back to the early years of the 20th century (Cowley and Stichelbaut 2012). Methods for processing time series of the Earth's surface were first focused on these kinds of data. These methods can differ according to the available data, but even for the most favorable cases – when all the camera calibration and ancillary data is available – we may still need to perform additional co-registration, as described by Fischer et al. (2011). Two main strategies exist for co-registration of multi-temporal 3-D datasets. One approach relies on a posteriori co-registration of the final 3-D models, whereas the other approach utilizes a suite of ground control points (GCPs) that can be found in all images. The first category of studies usually requires the user to have external fine topographic data available, such as a lidar-derived topographic dataset. This is the approach used by Bakker and Lane (2017), who noticed that the propagation of linear errors could not be limited during the SfM processing and that these errors resulted in spatial differences between the digital elevation model (DEM) of the different epochs. In such cases, authors perform co-registration with an a priori knowledge of stable zones. The "stable zones" approach has also been favorably used for UAV-based studies (e.g. Haas et al. 2016). The second category of studies seeks to have or create a set of GCPs that may be unambiguously found in all images of the multi-temporal dataset. This strategy seems to be the most used in the photogrammetric community, with remarkable recent works that proposed to automatically determine stable linear features (Nagarajan and Schenk 2016), or more recently, automatically derived multi-temporal GCPs (Giordano et al. 2018).

Despite the remarkable methods proposed by these previous works, a non-negligible amount of non-image information is needed, either in the form of external data or in the form of photogrammetric expertise, which significantly reduces the potential for a broader use of existing multi-temporal stereoscopic data. In this context, Feurer and Vinatier (2018a) and Feurer and Vinatier (2018b) proposed a new method, which allows a user to establish a unique geometric reference shared by all images of a multi-temporal stereoscopic image dataset. This method takes advantage of the invariance properties of feature detection algorithms such as the Scale Invariant Feature Transform (SIFT; Lowe 2004) to automatically link together images of different epochs and is hence called Time-SIFT. The idea of the Time-SIFT method has its roots in the breakthrough that SfM algorithms proposed in photogrammetry, with the provision of dense sets of matched feature points. Indeed, as noticed by several authors (Westoby et al. 2012; Fonstad et al. 2013), the fact that SfM processing of multi-view stereoscopic imagery relies first and foremost on image information only opens up new possibilities in photogrammetric processing. Among these new possibilities, archival aerial imagery was a compelling candidate because the "classical" photogrammetric workflows require a lot of ancillary data (calibration certificates) in addition to the images themselves, and images with a specific geometric processing if scanned from analogue photography. The study by Feurer and Vinatier (2018a, 2018b) showed that the Time-SIFT method allowed for the provision of stackable multi-temporal DEM from almost image information only. Using imagery that contains temporal change to establish a multi-temporal geometric reference seems counterintuitive as we may expect the change captured between epochs to contaminate the registration or reconstruction. Therefore, the Time-SIFT method requires further testing using test cases with a range of spatiotemporal scales. The initial paper (Feurer and Vinatier 2018b) focused on one particular spatiotemporal scale and phenomena, and the use of archival aerial imagery to examine prior 3-D changes linked to human activity. The broader suitability of the method for other spatiotemporal scales and phenomena depends on the ability of the invariance properties of the SIFT-like algorithm to provide consistent results across varied implementations. This possible invariance is expected to result in a trade-off between the spatiotemporal scale of the image data set and the spatiotemporal scale of the studied processes.

In this context, our study aims to check the potential of the Time-SIFT method on varied test cases with different spatiotemporal scales, from the millimeter/minute to the kilometer/decade. The expected success criteria are the ability of the Time-SIFT method to build a complete (with all images and all epochs) and consistent (all data representing the same 3-D scene at the global scale) multi-temporal 3-D block, as well as provide informative 3-D data allowing for the description and characterization of temporal change associated with the targeted phenomena.

Considering the relatively low effort required by the Time-SIFT method (which at a minimum requires only image data), a significant gain should be expected relatively to other existing methods, either in terms of working and/or processing time, when using GCPs or when external registration data are feasible, or in terms of achievable results when these data are not available, which implies that classical methods would not be applicable.

The chapter is divided into two sections. The first section provides a highlight and a detailed description of the Time-SIFT method. The second section is dedicated to the description of five different test cases, presented in five different subsections. The chapter ends with a summary of the findings of these different test cases and concluding remarks.

1.2. The Time-SIFT method

1.2.1. *General principle*

The underlying hypothesis of the Time-SIFT method is that, given an object or a scene, there may always be a spatial scale or a set of spatial scales for which different images taken at different times would show similar local features. A direct implication of this hypothesis is that SIFT-like algorithms may be able to find homologous points in images of the same scene taken at different times, hence enabling the construction of a multi-temporal image block with a single SfM processing of a whole multi-temporal photogrammetric dataset. The immediate effect of processing all images together in a single block is that they share the same unique geometric reference. Hence the co-registration of multi-temporal data is done from the beginning and only necessitates image information. This limits the need for external registration data such as GCPs. The principle of the proposed method is depicted in Figure 1.1.

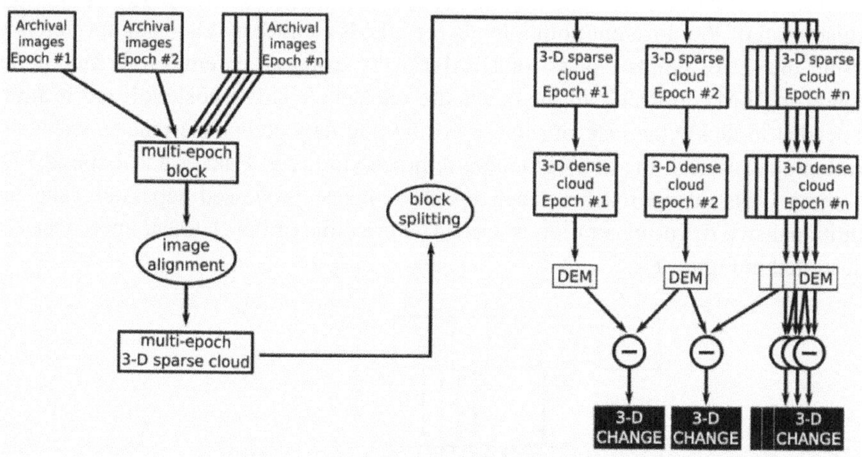

Figure 1.1. *Principle of the Time-SIFT method: use of a single block which joins multi-epoch images in the first step of the Structure from Motion processing. The whole dataset shares the same unique geometry because of a single multi-epoch block alignment. Each epoch is then separated while retaining orientation parameters so that subsequent dense 3-D point clouds and digital elevation models (DEM) maintain the same geometric reference*

The first step of the Time-SIFT method is to construct a single multi-temporal image block (or chunk). In this unique block, all images of all epochs are put together and image orientations are computed simultaneously for the whole dataset. As a result, all image orientations are estimated within only one single geometric reference and different epochs share exactly the same reference system by construction. This drastically simplifies the comparison of the multi-epoch images and may even eliminate the need for co-registration before comparing 3-D models of different epochs.

1.2.2. *Existing methods for multi-temporal photogrammetric processing*

To the best of our knowledge, with the notable and very specific exceptions of Chanut et al. (2017) and Vargo et al. (2017), prior multi-temporal studies utilizing SfM methods calculated 3-D models and image orientation estimates separately between epochs. There are then various strategies to ensure a good co-registration between the dense cloud (or the DEMs) of the different epochs, which can be generalized as two groups. The first group of strategies relies on external data that may be used to register the separate blocks. This usually implies the use of a significant number of GCPs. The second group of strategies relies on a posteriori co-registration of the different point clouds (or DEMs), either by choosing a reference within the set of points clouds (or DEMs) or by using an external reference point cloud (or DEM) on which all the others are registered. This a posteriori registration is needed to tackle the lack of precise positioning data and/or the inaccuracies and error propagation in the photogrammetric processing (e.g. Bakker and Lane 2017). In these classical approaches, each epoch is hence processed separately and an additional step of co-registration is used. The principle of the classical approaches is illustrated in Figure 1.2.

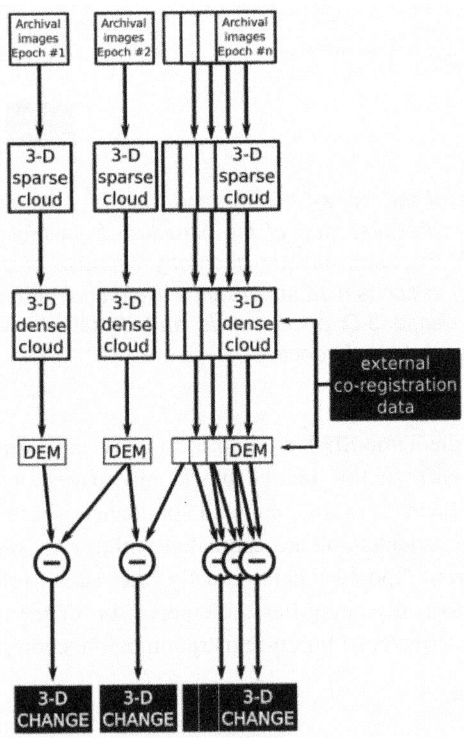

Figure 1.2. *Principle of "classical" methods. Each epoch has its own geometric reference and multi-epoch digital elevation models (DEMs) co-registration relies on external data, such as ground control points during alignment, an external lidar DEM or even external vector data*

1.2.3. *Step-by-step description of the Time-SIFT implementation*

The whole process of the Time-SIFT method can be summarized by the following three steps:

– adjust all image orientations in a single multi-temporal block;

– with image orientations fixed, split the block by epochs;

– for each epoch, compute dense cloud, DEM and/or orthoimage.

At the end of these three steps, the resulting 3-D data or orthoimages of the different epochs are perfectly stackable given that block geometry was estimated in a single step with all multi-temporal data and that this geometry was not changed during the following steps of dense image correlation and image mosaicking. The following sections detail the Time-SIFT process.

1.2.3.1. *Multi-epoch image orientations in a single multi-temporal block*

This first step is the core of the Time-SIFT method. All images of the same scene, taken at different epochs, are processed together to find homologous points and estimate cameras orientations. SfM algorithms first detect salient feature points, then match them and then eliminate outliers, using RANdom SAmple Consensus filtering (RANSAC; Fischler and Bolles 1981). With the image orientations, SfM methods estimate and iteratively refine camera positions and angles, resulting in a sparse 3-D point cloud. The underlying prerequisite of the Time-SIFT method is that there should be a majority of feature points that are stationary between two successive epochs, even if they are not found in other epochs of the study. There is probably a relationship between space and time scales for stationary features, as they would probably be kept for longer time spans when the imagery is of coarser resolution and covers larger spatial domains. When using the Time-SIFT method to monitor a changing phenomenon, we recommend that the whole image dataset covers an area containing enough stable zones and avoid a strong focus on the zone of change only (see, for example, the "soil bulk density" case study, where images were purposely taken to be processed with the Time-SIFT method).

As the final step, the image block geometry is refined (for instance, with the use of some GCPs in order to perform lens autocalibration) and – if not already done

– absolute orientation is estimated. At the end of this step, the final orientation of each image is obtained and will remain fixed for the whole workflow.

1.2.3.2. *Split of the multi-temporal block into individual blocks (one block per epoch, image orientations kept fixed)*

Proper implementation of this simple step is critical for the application of the Time-SIFT method. The multi-temporal block is split into different blocks, one for each epoch, which are then processed independently. In other words, the image block geometry must not be changed after this step, otherwise the datasets of different epochs would lose their intrinsic co-registration. In practice, this results in a critical bottleneck in the workflow: the estimation and refinement of the camera orientation and lens autocalibration must be done *before* the multi-temporal block split step and not changed afterwards.

1.2.3.3. *Dense cloud, DEM and/or orthoimage computing for each epoch*

Once the multi-temporal block is split, the workflow returns to a classical photogrammetric workflow: within each individual epoch, dense image correlation is computed. This step allows us to obtain a dense 3-D point cloud. The same process is repeated for each epoch. Depending on the requirements, these dense clouds, and also interpolated DEMs and/or orthomosaics, can be computed and exported. These output datasets, although coming from different epochs, share the same geometric reference and can thus be directly compared.

1.3. Case studies

To demonstrate the ability of the Time-SIFT method to produce coherent multi-temporal dataset in various situations, we present five distinct case studies. These studies differ in many ways, but are selected specifically for the following characteristics: (i) spatiotemporal scales of the image datasets (Figure 1.3); (ii) relative magnitude and spatial patterns of the expected changes; (iii) different data sources; and (iv) varied target phenomena.

For each case study, the following elements are given: the aim of the study, dataset description and results of the Time-SIFT method associated with its added value. For the sake of simplicity, we used a single software editor (Agisoft packages Photoscan Pro® version 1.2.6 and MetaShape Pro version 1.5.x) to apply the

Time-SIFT method to each case study. We hence computed the very same processing metrics for each dataset and were hence able to compare these metrics directly. This said, it must be noted that the method can be implemented in any SfM software, as demonstrated, for instance, through the recent application of the Time-SIFT method with MicMac by Filhol et al. (2019), PhotoModeler by Parente et al. (2021) or the work of Cook and Dietze (2019).

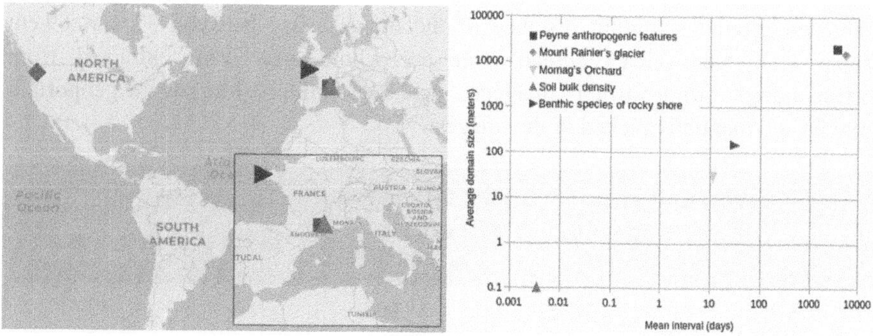

Figure 1.3. *Main characteristics of the case studies: (A) locations and (B) scales. For a color version of this figure, see www.iste.co.uk/mallet/multitemporal.zip*

1.3.1. *La Peyne*

1.3.1.1. *Objective*

This case study concerns the change in land management in a Mediterranean area due to anthropogenic activity, such as urban growth, civil engineering around highways, quarries and backfills. The process of land use change followed progressive loss in land and a growth of urbanization over the past 50 years (Vinatier and Arnaiz 2018). On a zone of several square kilometers, studied by Vinatier and Arnaiz (2018), 77% changed within 50 years. A 1-m resolution DEM is used to estimate changes in urban structures over this time span, and a call for archival aerial imagery to cover this long time period is needed. The process of attributing GCPs to sets of aerial images is very time consuming, making the use of Time-SIFT method appropriate for the documentation of past changes.

1.3.1.2. *Case description*

This case was the first test of the Time-SIFT method and is reported with more detail by Feurer and Vinatier (2018a, 2018b). On a 170 km² rectangle corresponding to the lower part of the Peyne catchment (Occitanie, southern France), investigated earlier in Vinatier and Arnaiz (2018), a part of the aerial imagery archive was downloaded (43° 35′N, 3° 19′E). The French geographic institute (IGN) provides free access to this archive, available at: https://remonterletemps.ign.fr/. Images are scanned with photogrammetric-grade scanners. The landscape is typical of Mediterranean agricultural areas dominated by vineyards and to a smaller extent, forests. Images of the 1971, 1981, 1990, and 2001 aerial campaigns were used. Estimated ground sampling distance of the images vary between 37 and 69 cm (Table 1.1). Nine GCPs spread over the whole area were extracted from corresponding orthorectified imagery available online (http://www.geoportail.gouv.fr) and manually picked in the images.

Epoch (date)	Images (#)	Focal length (mm)	Approximate scale	Scan resolution (μm)	Ground sampling distance (cm)
1971/06/21	61	152	1:18,000	21	37
1981/06/16	27	153	1:32,000	21	66
1990/06/25	31	153	1:32,000	21	69
2001/06/04	44	153	1:2,600	21	55

Table 1.1. *Parameters of scanned aerial images used in La Peyne case study*

1.3.1.3. *Application of the Time-SIFT method*

The Peyne case study was processed with Photoscan 1.2.6, for which the handling of fiducial marks was not yet possible. Archival aerial images were first resampled with the Micmac software (Rupnik et al. 2017) so that fiducial marks were positioned in the corners in the resulting image. More details can be found in Feurer and Vinatier (2018a, 2018b). As a result, pre-processed images share the same geometry and internal orientation has been taken into account before the SfM processing, which was realized with the parameters shown in Table 1.2.

Processing step	Property	Value
Alignment*	Accuracy	Highest
	Pair preselection	Disabled
	Key point limit	100,000
	Tie point limit	50,000
	Adaptive camera model fitting	No
Camera calibration	Camera groups	Split by epoch
Optimization	Lens parameters	f, b1, b2, cex, cy, k1, k2, k3, k4, p1, p2
	Marker accuracy (pix)	5
	Marker accuracy (m)	20
	Tie point accuracy (pix)	1
Dense cloud	Quality	High
	Depth filtering	Moderate
Mesh	Surface type	Arbitrary
	Interpolation	Enabled
	Quality	Medium
	Depth filtering	Moderate
DEM	Pixel size (m)	1
	Interpolation	Enabled

Table 1.2. *Time-SIFT parameters in La Peyne case study.*
**The sparse cloud is obtained after the "alignment" step*

1.3.1.4. *Results*

Application of the Time-SIFT method resulted in four stackable DEMs and orthophotos, as shown in Figure 1.4 (right). The quality of image orientations can be assessed through the root mean square (RMS) reprojection error values, which fall within the pixel size for all four epochs (Table 1.3). The quality of DEM co-registration can be investigated by comparing topographic profiles (Figure 1.4, bottom).

Epoch	Dense cloud (points)	Dense cloud (pts/m^2)	DEM (m/pixels)	RMS reprojection error (pixels)
1971	4.26×10^8	2.08	0.69	0.82
1981	2.20×10^8	0.61	1.28	0.59
1990	2.30×10^8	0.57	1.33	0.63
2001	2.01×10^8	0.87	1.07	0.68

Table 1.3. *Digital elevation models obtained with Time-SIFT for La Peyne*

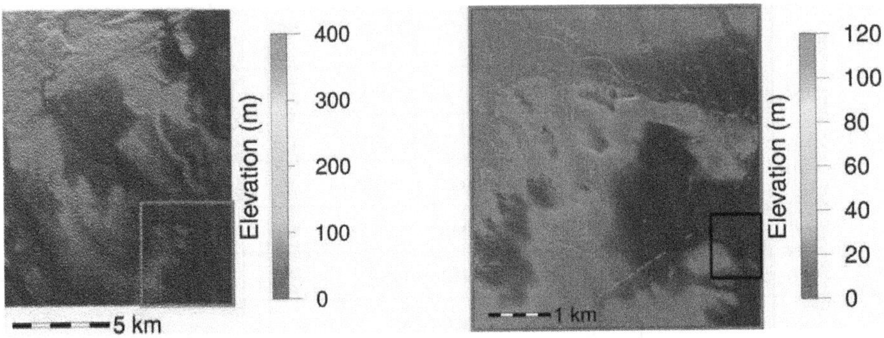

Figure 1.4. *Full digital elevation model of the La Peyne test site (left) and zoomed-in view (right). The black box on the right corresponds to the zone described in Figure 1.5. For a color version of this figure, see www.iste.co.uk/mallet/multitemporal.zip*

Several features are apparent on the topographic profiles (Figure 1.4). The left part corresponds to an area where no civil engineering took place. It also seems, from orthoimage analysis, that there were no major vegetation changes. This part can then be considered as a stable zone, with possible minor changes due to limited vegetation variation. This part of the profile allows us to estimate the remaining noise level: this noise is of several meters. Moreover, as already stated in Feurer and Vinatier (2018a, 2018b), this level of noise is confirmed by the fact that new buildings are clearly visible in the successive DEMs, some of them being single-story. The right part of the profile focuses on an area that was strongly modified through time, with the construction of a highway and an interchange. Coincidentally, 1,990 images were taken when the highway construction was ongoing and the north/south bridge was not present yet, which is clearly visible both on the 1990 DEM and orthoimage. The 1990 DEM seems to be a little more noisy than others. This may be related to the fact that this DEM was built from relatively coarser images (Tables 1.1 and 1.3). Conversely, the part of the 2001 DEM profile that corresponds to the highway is relatively even, which indicates that the 2001 DEM is less noisy.

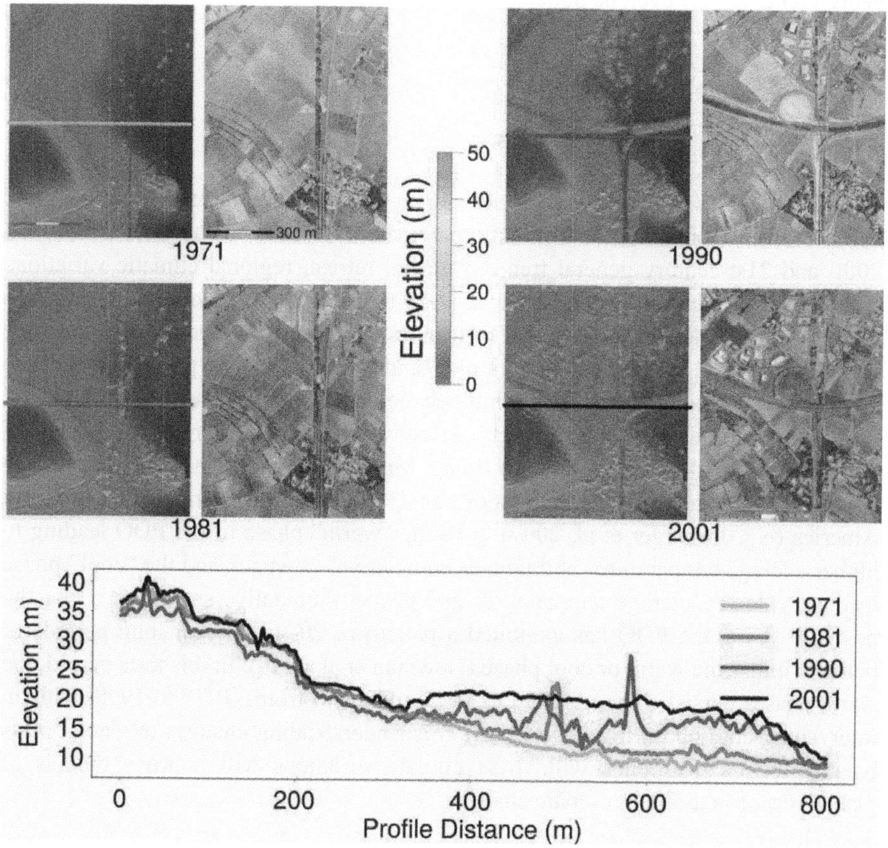

Figure 1.5. *La Peyne test site. Zoomed-in views of the four digital elevation models and orthophotos (top) with profiles extracted over the highway construction zone (bottom). The highway construction is visible on the right part of the profile. For a color version of this figure, see www.iste.co.uk/mallet/multitemporal.zip*

The most interesting finding of this test case is that, with almost only image information, we can virtually explore the whole archive of aerial imagery. In this study, only nine GCPs were used to estimate the absolute orientation and scale of the 3-D models produced from 27 to 61 images by epoch, but it demonstrates that this work may be extended to wider areas and longer time spans, provided that the required imagery is available. As shown in Feurer and Vinatier (2018a, 2018b), the only requirements are that images were scanned with photogrammetric-grade scanners and that fiducials marks are present in the images. All the remaining processing can be done with image information only. Absolute scaling and georeferencing may even be done afterward.

1.3.2. *Mount Rainier*

1.3.2.1. *Objective*

The processes and rates of glacier retreat and advance depend on a complex balance of environmental and climatic parameters. Even with the well-established 20th- and 21st-century general trend of glacial retreat, regional climatic variations and local site conditions have produced short-term glacial advances and changes in mass balance. The glaciated active volcanoes of the northwestern United States, including Mount Hood, Mount Rainier and Mount Baker, provide a unique opportunity to examine the relationship between decadal-scale climatic patterns and local site conditions, such as aspect and elevation. Climate variability associated with the Pacific Decadal Oscillation (PDO; Mantua and Hare 2002) exerts a strong control on temperature and precipitation trends in the northwestern portion of North America (e.g. Josberger et al. 2007), with the "warm" phase of the PDO leading to higher average temperatures and thinner cumulative snowpack and the "cool" phase leading to lower average temperatures and deeper cumulative snowpack. Over the past 100 years, the PDO has exhibited a pattern of 20- to 30-year long periods of being in either the warm or cool phase (Newman et al. 2016). In this time period, the warm phases persisted from ~1925 to 1947 and again from ~1977 to 1998, with an intervening cool phase from ~1947 to 1977. Understanding changes in glacial mass balance as it corresponds with these climate variations will improve models of glacial response to future climate changes.

1.3.2.2. *Case description*

To improve the temporal availability of high-resolution topographic data for use in measuring the mass balance of the glaciers across major shifts in the PDO, we tested the Time-SIFT methodology on archival aerial image collections from Mount Rainier, a 4,392-m-tall glaciated volcano located in the state of Washington in the United States (46.85° N, 121.75° W). Mount Rainier has 26 major glaciers occupying its flanks and 93 km^2 of permanent snowfield and glacier coverage. We selected Mount Rainier for this case study due to the availability of snowfall records extending back to 1920, previous studies of glacial mass balance, multiple generations of publicly available aerial images extending back to 1955 and full coverage by airborne lidar surveys in 2007–2008.

Using only aerial images that provide full coverage of Mount Rainier and are available at no cost (https://earthexplorer.usgs.gov/), we assembled image collections from 1955, 1970, 1983 and 1984. Adding in the 2007/2008 airborne lidar, topographic data from these time spans will provide several important qualitative and quantitative tests. For one case, Sisson et al. (2011) used DEMs based on topographic maps derived from the 1970 aerial images and differenced these with the 2007/2008 lidar data to measure the elevation change of glacier surfaces across the whole mountain. Measuring these elevation changes using the same datasets via the Time-SIFT method will provide a key reproducibility test. Another test follows the observed and expected trends of glacial advance and retreat, with 1955 and 1970 occurring in the middle of a cool PDO phase, and 1983 and 1984 following the transition from a cool to warm PDO phase. Accordingly, glacial volume increases from 1955 to 1970 followed by volume loss from 1970 to 1983/1984 and 2007/2008 would be consistent with the observations. Another test of reproducibility is the one-year interval between the 1983 and 1984 aerial image collections. If the Time-SIFT methods are producing scale and referenced models properly, the elevation differences between 1983 and 1984 should be proportionally small relative to the longer time intervals.

Epoch (date)	Images (#)	Focal length (mm)	Approximate scale	Scan resolution (μm)	Ground sampling distance (cm)
1955/09/02* 1955/09/18*	16 10	153.69 153.69	1:66,000 1:66,000	25 25	158 158
1970/09/09# 1970/09/26#	12 31	152.45 152.45	1:38,000 1:38,000	25 25	150 150
1983/08/21	18	208.401	1:58,000	25	142
1984/08/25	23	208.047	1:58,000	25	154

Table 1.4. *Parameters of scanned aerial images used in Mount Rainier case study. *Both sets of 1955 images were collected 16 days apart by the same camera and are processed in this study as a single epoch. #Both sets of 1970 images were collected 17 days apart by the same camera and are processed in this study as a single epoch*

1.3.2.3. *Application of the Time-SIFT method*

For Agisoft Metashape Pro, in order to recognize multiple images as being from the same camera, the digital images must have the same pixel dimensions. Our images were delivered with small variations in pixel dimensions within individual epochs. Prior to loading the images into Agisoft Metashape Pro (v. 1.5), we processed all images to trim excess pixels from the margins to ensure equal pixel dimensions for all images in each epoch. Within Metashape Pro, we defined camera groups, added calibrated focal length data from the respective calibration reports, masked the margins of the images and then implemented the built-in fiducial mark calibration features. For several 1955 images, we had to mask out large portions of the image due to small cracks in the source film that is apparent in the scanned image. To calibrate the photos, we manually marked four to eight fiducial marks on each image and calibrated each camera group individually. We did not perform fiducial mark-based calibrations during our initial trials of this technique, and this produced widespread topographic artifacts due to minor warping of the source film/images.

Processing step	Property	Value
Alignment	Accuracy	Medium
	Pair preselection	Yes
	Key point limit	100,000
	Tie point limit	50,000
	Adaptive camera model fitting	Yes
Camera calibration	Camera groups	Split by epoch
Optimization	Lens parameters	f, b1, b2, cx, cy, k1-k3, p1, p2
	Marker accuracy (pix)	5
	Marker accuracy (m)	4
	Tie point accuracy (pix)	2
Dense cloud	Quality	High
	Depth filtering	Moderate
DEM	Pixel size (m)	4
	Interpolation	Disabled

Table 1.5. *Time-SIFT parameters in Mount Rainier case study*

Following the Time-SIFT procedures, we performed the alignment and optimization of all images in a single multi-epoch block, using the parameters listed in Table 1.5. With the availability of airborne lidar, we chose to use that dataset as the ground control source in our study to facilitate direct comparison of the photogrammetric DEMs with the lidar DEMs. We established 25 GCPs, using features identifiable both on the lidar topography and the aerial imagery. These GCPs were marked manually on all images where they could be identified, and based on the size and type of features that we used as markers, we estimate that their locations were identified within 4 m. Following the optimization step, the initial multi-epoch block was duplicated to create separate image blocks for each epoch – 1955, 1970, 1983 and 1984. Both 1955 and 1970 epochs include images collected on two different days, but because the collection was only ~2 weeks apart and taken with the same camera and approximate scale, we combined both days as a single epoch. We then produced dense point clouds, orthomosaics and DEMs for each individual epoch, and calculated DEMs of Difference (DoDs, e.g. Chandler and Brunsden 1995) in ArcGIS (Table 1.6).

Epoch	Dense cloud (points)	Dense cloud (pts/m^2)	DEM (m/pixels)*	RMS reprojection error (pixels)
1955	1.26×10^8	0.1	4	1.08
1970	2.18×10^8	0.279	4	1.10
1983	1.26×10^8	0.124	4	1.05
1984	1.37×10^8	0.105	4	1.04

Table 1.6. *Digital elevation models obtained with Time-SIFT for Mount Rainier. *DEMs were initially built at the default resolution recommended by Agisoft Photoscan Pro and later resampled to equal resolutions in ArcGIS*

1.3.2.4. *Results*

The visual derivative products from our Time-SIFT analysis on Mount Rainier, including orthomosaics, shaded-relief maps and slope maps, provide a simple, first-order check on the quality of the topographic reconstructions. When viewed sequentially, the glaciers are the only features moving horizontally on the landscape, with the areas between glaciers remaining stationary through each time interval, as expected. For more quantitative assessments of reconstruction accuracy, we examined DoDs for different time intervals to compare our Time-SIFT derived measurements of ice thickness change to other studies of ice thickness change, known climatic patterns and geologic events and the patterns of measured elevation change associated with geologically stable regions.

To compare our Time-SIFT results to Sisson et al.'s (2011) study of ice volume change spanning 1970 to 2007/2008, we created a DoD from the LiDAR DEM and our 1970 DEM, clipping it to the outlines of the 2007/2008 glacier extents and visualizing it with the same color scale as the previous study (Figures 1.6 and 1.7). Despite areas of missing data from the 1970 DEM in the higher elevations, general patterns of glacier advance/retreat and surface elevation changes are consistent between datasets. One notable pattern from this whole mountain perspective is that only two glaciers on the northeast slopes of Mount Rainier show cumulative advance from 1970 to 2007/2008, the Winthrop and Emmons glaciers (Figures 1.6 and 1.7).

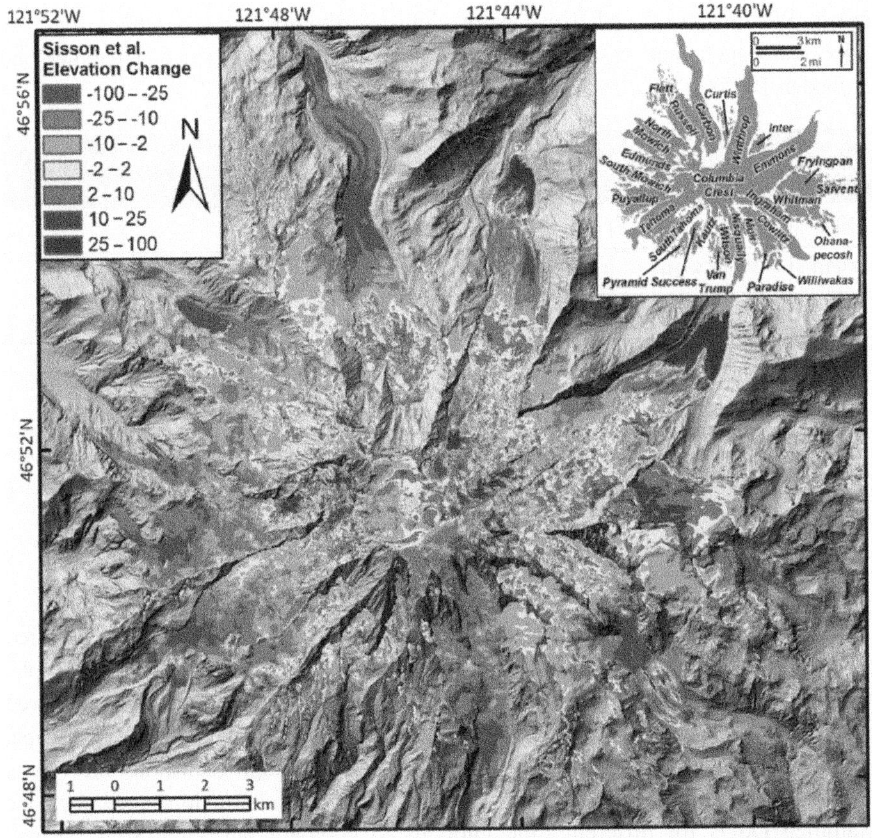

Figure 1.6. Comparison of ice thickness change for all glaciers on Mount Rainier between 1970 and 2007/2008, as measured in the previous study by Sisson et al. (2011), which differenced digital elevation models digitized from topographic maps derived from 1970 aerial images with airborne LiDAR collected in 2007 and 2008. For a color version of this figure, see www.iste.co.uk/mallet/multitemporal.zip

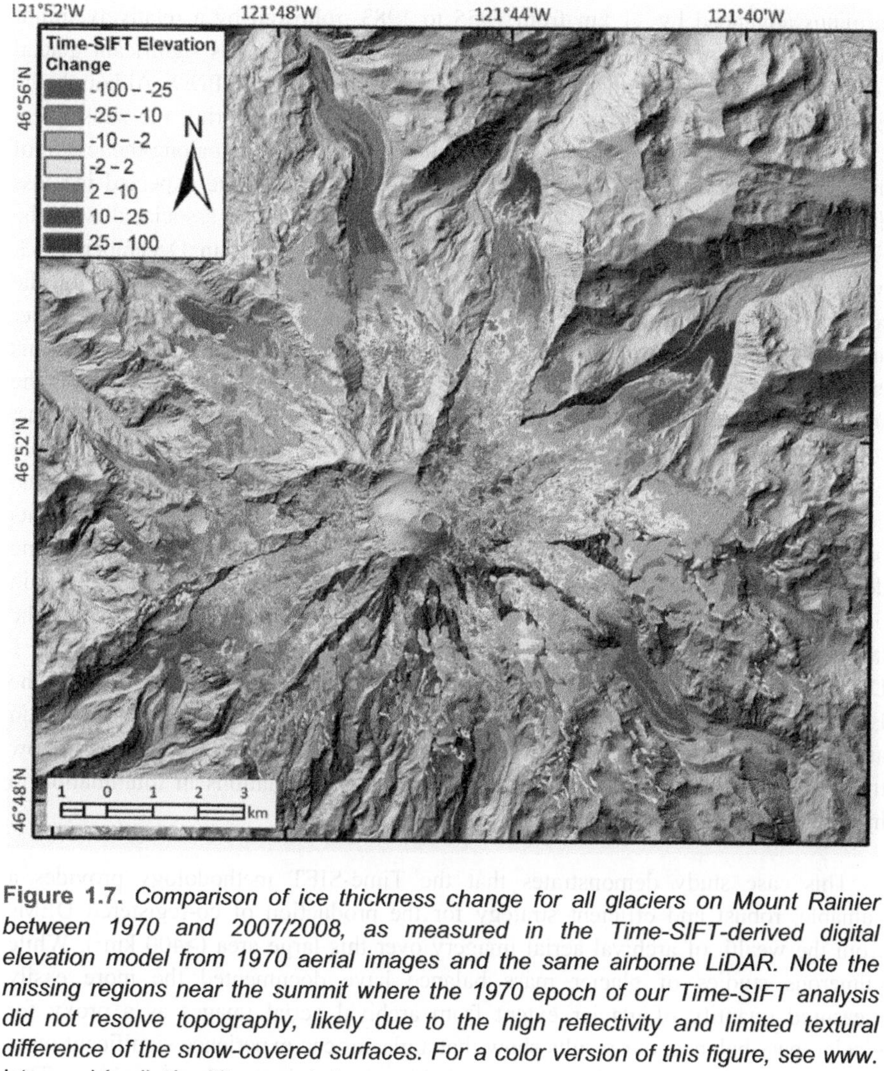

Figure 1.7. *Comparison of ice thickness change for all glaciers on Mount Rainier between 1970 and 2007/2008, as measured in the Time-SIFT-derived digital elevation model from 1970 aerial images and the same airborne LiDAR. Note the missing regions near the summit where the 1970 epoch of our Time-SIFT analysis did not resolve topography, likely due to the high reflectivity and limited textural difference of the snow-covered surfaces. For a color version of this figure, see www.iste.co.uk/mallet/multitemporal.zip*

For a finer-scale view of glacier change through time, we take a closer look at the Emmons Glacier (Figures 1.6–1.8). Examining the DoDs for our four time intervals, 1955–1970, 1970–1983, 1983–1984 and 1984–2007/2008, the glacier terminus advanced by ~1 km from 1955 to 1983, followed by a relatively minor retreat of ~0.1 km from 1984 to 2007/2008. The timing of this significant glacial advance is consistent with the 1947–1977 cool phase of the PDO. Although the Emmons Glacier does not show significant glacier retreat during the subsequent 1977–1998 warm PDO phase, it does exhibit significant thinning along the length of the glacier (Figures 1.8(C) and (D)) that is consistent with the expected ice loss during a PDO warm phase. Another known elevation change associated with the Emmons Glacier is the occurrence of a large rockfall event in December 1963, where an estimated 11 million cubic meters of rock collapsed from a peak on the south side of the Emmons Glacier, sending debris and mudflows across the glacier and down into the valley below (Crandell and Fahnestock 1963). The effects of this event are clearly visible in the 1955–1970 DoD, where a dark red zone on the southern edge (Figure 1.8C) shows a focused zone of elevation loss located just south of the glacier margin at a location consistent with the known rockfall origin. Further evidence of this rockfall is located downstream from the glacier (northeast corner of 1955-1970 DoD, Figure 1.8(C)), where the rockfall and mudflow deposits partially filled the main valley channel and partially eroded the young moraine deposits on the valley margins (Crandell and Fahnestock 1965). The stable terrain outside the glaciers, and in particular the regions above the treeline and below the late summer snowpack, provide an estimate of the "noise" present in the Time-SIFT DEMs of this case study. For example, a sparsely vegetated valley to the north of the Emmons Glacier exhibits topographic noise on the order of ~2 m. Steeper terrain and vegetated regions exhibited higher variability, likely due to differences in how trees are resolved or filtered in different epochs and variations in illumination in high-relief zones.

This case study demonstrates that the Time-SIFT methodology provides a valuable, robust and efficient strategy for the production of co-registered DEMs from the wealth of archival aerial imagery over this large area (>400 km^2). While numerous studies of glacier mass balance have documented the more easily measured quantity of glacier extent from archival aerial images as a proxy for glacier mass balance, our results show that a glacier can experience significant mass loss through thinning without a remarkable change in terminus position. This observation emphasizes the potential of Time-SIFT in assisting with the extraction of richer 3-D historical datasets, for quantifying historical geologic and geomorphic change.

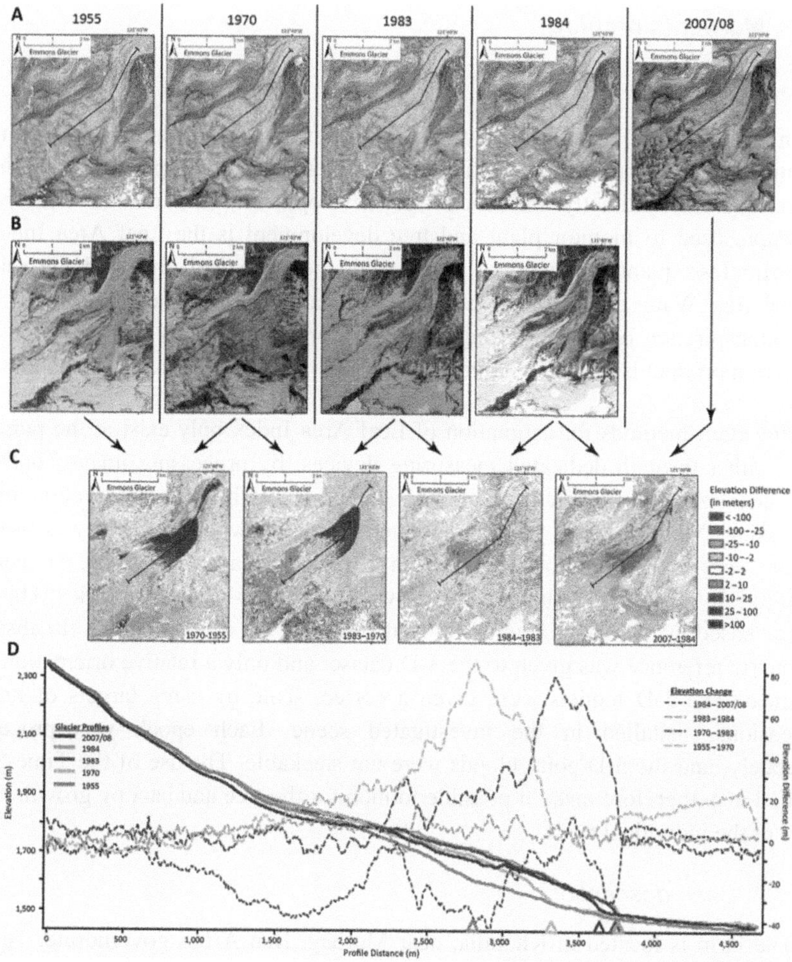

Figure 1.8. *Analysis of Time-SIFT derived digital elevation models of the Emmons Glacier on Mount Rainier. (A) Slope-maps of the lower Emmons Glacier for each Time-SIFT epoch and the 2007/2008 LIDAR data. (B) Orthoimages of the lower Emmons Glacier for each epoch. (C) Elevation changes that occurred in the time intervals between each Time-SIFT epoch and the LIDAR data. (D) Topographic profiles of the Emmons Glacier surface for each epoch (left axis) and the corresponding elevation difference between epochs (right axis). Triangles along the x-axis denote the glacier terminus positions and are color coded for the respective epoch. The elevation changes illustrate significant glacial advance and thickening below 1,900 m elevation from 1955 to 1983. From 1984 to 2007/2008, the terminus experienced only minor retreat, but saw significant thinning along much of its length, especially below 2,000 m. For a color version of this figure, see www.iste.co.uk/mallet/multitemporal.zip*

1.3.3. *Mornag's orchard*

1.3.3.1. *Objective*

In the technical management of orchards, farmers need to optimize both the quantities of water used for irrigation and the quantity of crop protection products. These quantities directly influence tree development. One of the widespread indicators used to monitor plant and tree development is the Leaf Area Index, a dimensionless quantity computed as the ratio between leaf area and the related ground area. Water demand is indeed directly related to the amount of foliage, with evapotranspiration occurring through the leaf stomata. Similarly, the need for crop protection product is directly related to the surface area of the foliage.

However, methods for estimation of Leaf Area Index only exist at the punctual scale, either through dedicated measuring devices, by processing images obtained with hemispherical cameras, or at the field scale, with earth observation image processing. Mabrouk et al. (2017) conducted a study where a UAV was used to acquire very high resolution 3-D data on a peach orchard in Mornag, Tunisia, in order to test the ability of such data to provide Leaf Area Index time series. This test case is based on the data gathered for that study. In the original study, no absolute geometric reference was given to the 3-D dataset and only a relative orientation was obtained. The 3-D models were given a correct scale by using targets of known dimensions, installed in the investigated scene. Each epoch was processed separately, and the 3-D point clouds were not stackable. The use of the Time-SIFT method may therefore make it possible to monitor the tree and canopy growth at the scale of the individual tree.

1.3.3.2. *Case description*

The farm is located in Khelidia, near Mornag, Ben Arous governorate, Tunisia (36°39'50"N 10°12'32"E). This region has a semi-arid climate with mild winters and hot summers. The field plot on which the experiment was conducted is a 0.77 ha field planted in 2010 with peach trees of the Sweet Cap variety.

In total, five image acquisitions at five different epochs of the first semester of the year 2015 were done. Flights plans were conducted at low altitude (between 10 and 30 m) with nadir and oblique images in order to favor the rendering of the tree crowns (see Table 1.7). The ancillary data used in Mabrouk et al. (2017), which consisted of targets of known dimensions installed at different places for each acquisition, were not used for this case study.

Epoch (date)	Images (#)	Flying height (m)	Focal length (mm)	Approximate scale	Ground sampling distance (cm)
2015/04/20	107	24.8	8.8	1:2818	0.61
2015/05/05	581	11.6	8.8	1:1318	0.32
2015/05/15	96	15.3	8.8	1:1738	0.31
2015/05/26	205	12.7	8.8	1:1443	0.33
2015/06/04	209	14.7	8.8	1:1670	0.33

Table 1.7. *Image dataset parameters for the "Mornag's orchard" study*

1.3.3.3. *Application of the Time-SIFT method*

For this test case, all images were acquired with the same UAV and the same camera. As a result, all images were imported in a single multi-temporal block. Unfortunately, considering both the grass growth between the trees and the tree growth (mainly leaf growth and increased leaf coverage), the first attempts resulted in missed orientations for some of the images or epochs within the multi-temporal block, depending on the tested parameters. Therefore, a trial-and-error approach was adopted until a functional set of values for the Photoscan parameters was found. It appears that increasing the number of desired tie points allowed for the success of the orientation of all images of the multi-temporal block, which is depicted in Table 1.8.

Processing step	Property	Value
Alignment	Accuracy	High
	Pair preselection	No
	Key point limit	40,000
	Tie point limit	40,000
	Adaptive camera model fitting	Yes
Camera calibration	Camera groups	Split by epoch
Dense cloud	Quality	High
	Depth filtering	Moderate
DEM	Pixel size (m)	0.01
	Interpolation	Enabled

Table 1.8. *Time-SIFT parameters in Mornag's orchard case study*

1.3.3.4. *Results*

Once the alignment of the whole block was achieved, alignment quality was found to be quite good, with a reprojection error of approximately 0.5 pixels (Table 1.9). These results were favored by the fact that oblique imagery was acquired at each epoch, ensuring a better estimation of image block geometry (James et al. 2019). It is also worth noting that if point cloud density is lower (and DEM resolution coarser) for the first acquisition, done at a higher altitude, the RMS reprojection error stays within the same range as other epochs, which relates well with the multi-scale robustness of the Time-SIFT method.

Epoch	Dense cloud (points)	Dense cloud (pts/m^2)	DEM (m/pixels)	RMS reprojection error (pixels)
2015/04/20	32.0×10^6	6,650	0.0120	0.53
2015/05/05	95.6×10^6	23,700	0.0065	0.50
2015/05/15	39.4×10^6	26,400	0.0065	0.43
2015/05/26	77.9×10^6	22,300	0.0067	0.47
2015/06/04	63.4×10^6	22,700	0.0064	0.51

Table 1.9. *Time-SIFT results for Mornag's orchard*

In terms of thematic results, having perfectly stackable point clouds, DEMs and orthophotos allow for a detailed 3-D monitoring at the tree scale. Indeed, when examining the DEMs at the tree level, we can clearly see the canopy development through the growing season, with branches growing and being covered by more leaves through time (Figure 1.9). It is worth noting that no ground information was used to obtain these results. The scale of the 3-D models was provided by the GPS position recorded onboard the UAV, and apart from this information, only image information was used in the processing. This example proves that the Time-SIFT method can be used to apply a very light protocol that allows us to monitor tree crops at the tree scale. In the previous studies based on the same dataset, Mabrouk et al. (2018) showed that this kind of information may be favorably used to pilot the irrigation and/or pesticide application scheme.

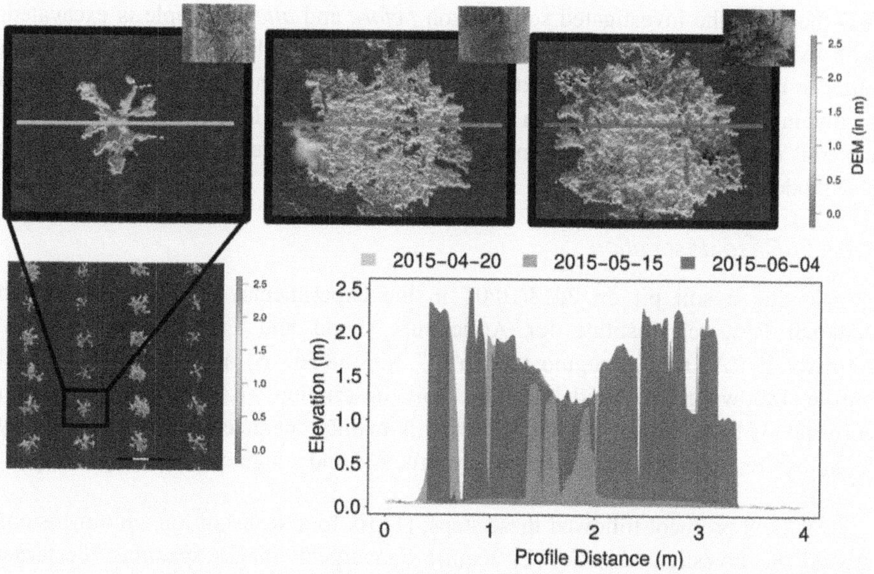

Figure 1.9. *Monitoring of a single tree crown development. Plan views of a digital elevation model (bottom left) with zoomed views at three dates (top) and extracted profiles (bottom right). Tree development within this time period mainly consists of foliage growth. For a color version of this figure, see www.iste.co.uk/mallet/multitemporal.zip*

1.3.4. *Soil bulk density*

1.3.4.1. *Objective*

In soil studies, especially for the monitoring of hydrological processes, soil bulk density is one of the major physical parameters. Among the existing methods, the core sampling method relies on the use of a metal cylinder with a known volume (most often 100 cm^3) that is hammered into the target soil horizon. The soil sample is then dried and weighed in the lab, with the bulk density calculated by dividing this mass by the sample volume. This is the most common method due to its efficiency; however, core sampling cannot be applied in all sediments, such as when the soil is too pebbly. In these cases, pedologists rely on the use of other methods which are based on the estimation of soil sample volumes. These methods share the same requirement: volume estimation must be done from the top of the soil horizon, which in turn requires the investigator to excavate the soil until the desired horizon. We investigate a new method for estimating sample volumes for bulk density measurements using photogrammetry. In simple terms, the protocol is to create a

3-D model of the investigated soil horizon *before* and *after* a sample is excavated, with the volume of the soil sample being the difference between both models. In order to minimize the need for external data, we implement the Time-SIFT method to obtain a direct co-registration of both 3-D models. Hence, the required field material is limited to a camera and targets of known dimensions used to scale the 3-D models.

1.3.4.2. *Case description*

We dug a soil pit on 2018/10/01 in the experimental fields of the French National Research Institute for Agriculture, Food and Environment (INRAE, formerly INRA) near Mauguio (43°36′30″ N 3°58′56″ E) and described the soil profile following the guidelines for soil description (FAO 2006). The soil corresponds to a fluvisol (WSR 2015) and is mainly characterized by a sandy clay loam texture, medium subangular blocky structure and a high discernible porosity.

Our field protocol followed these steps: (1) fix four scales of known dimensions around the investigated zone, (2) acquire convergent images spanning the target area, (3) excavate five separate samples from different locations on the target surface and (4) acquire a second set of convergent images (see Table 1.10 for the acquisition parameters). The four targets of known dimension provide the scale to the 3-D model and were not moved between image acquisitions. To validate our results, we also collected three core samples to allow us to compare this method with the traditional approach.

Epoch (time) 2018/10/01	Images (#)	Acquisition distance (m)	Focal length (mm)	Approximate scale	Ground sampling distance (cm)
13:27	34	1	18	1:55	0.02
13:55	43	1	18	1:55	0.02

Table 1.10. *Image dataset parameters for the "soil bulk density case" study*

1.3.4.3. *Application of the Time-SIFT method*

For this case, the Time-SIFT multi-temporal block contains only two epochs separated by 28 minutes. Agisoft Photoscan Pro automatically detected the targets and we input scale information using these targets after image alignment. We split the aligned block into separate before and after blocks for subsequent dense construction. The convergent acquisition favored accurate lens autocalibration. We retained the default processing parameters; these are reported in Table 1.11.

Processing step	Property	Value
Alignment	Accuracy	High
	Pair preselection	No
	Key point limit	40,000
	Tie point limit	4,000
	Adaptive camera model fitting	Yes
Camera calibration	Camera groups	Split by epoch
Optimization	Lens parameters	f, cx, cy, k1-k3, p1, p2
	Marker accuracy (pix)	n/a
	Marker accuracy (m)	n/a
	Tie point accuracy (pix)	1
Dense cloud	Quality	High
	Depth filtering	Mid
DEM	Pixel size (m)	0.01
	Interpolation	Enabled

Table 1.11. *Time-SIFT parameters in soil bulk density case study*

1.3.4.4. *Results*

The Time-SIFT method produced high-resolution accurate models for volume calculation without any need for cumbersome parameter tuning or additional processes to co-register 3-D models. We found that systematic image collection across the full target zone, with a convergent imaging geometry is critical to capture the full details of the surface and support the accurate estimation of lens parameters, all of which is required for high-quality SfM results. The success in lens auto-calibration is demonstrated by a reprojection error of 0.86 pixels (Table 1.12) and a Root Mean Square Error (RMSE) of 0.2 mm, which is the same scale as the ground sampling distance.

Epoch time (2018/10/01)	Dense cloud (points)	Dense cloud (pts/cm^2)	DEM (cm/pixels)	RMS reprojection error (pixels)	RMS reprojection error on scales (cm)
13:27	10.1×10^6	1065	0.214	0.86	0.021
13:55					

Table 1.12. *Time-SIFT results for soil bulk density*

The profile view illustrates the accurate registration of the stable surfaces and the negligible noise at this scale (Figure 1.10). We assessed the quality of volume estimation with the Time-SIFT method by comparing it with classical soil volume estimation methods. Volume measurements between 100 and 1,400 cm^3 made with the Time-SIFT method produced an RMSE of 15 cm^3 (full study details in Coulouma et al. (2021)). The Time-SIFT methods produced a mean soil bulk density of 1.53 with a standard deviation of 0.04 (n = 5). These measurements are consistent with the mean bulk density of 1.59 with a standard deviation of 0.03 (n = 15) obtained from the core sampling methods (Coulouma et al. 2021).

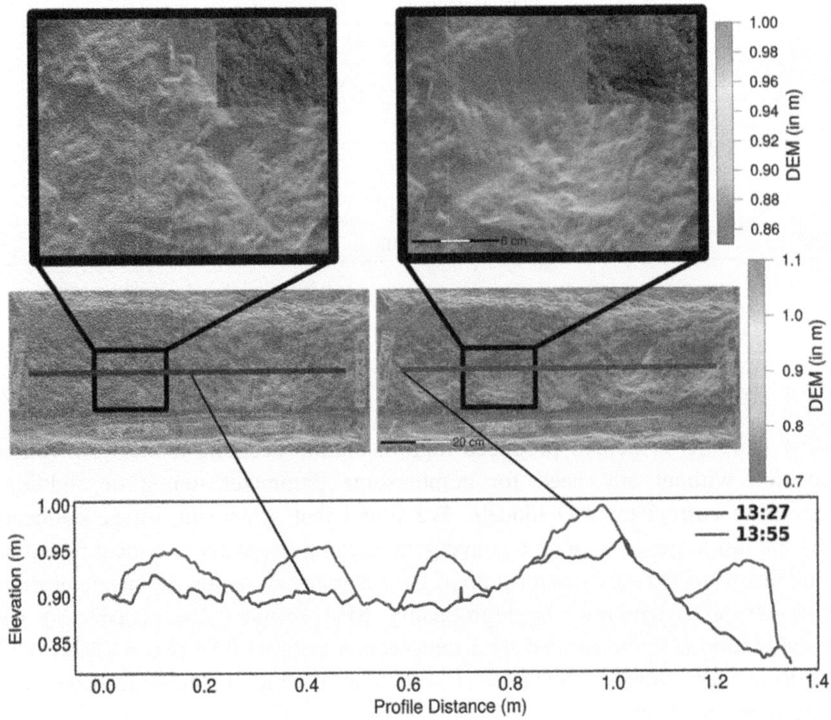

Figure 1.10. *Profiles of the five excavations and close-up on a single sampling site. Excavations were done in the vertical wall of the soil pit. At this scale, no difference between the two profiles can be detected on the stable zones. For a color version of this figure, see www.iste.co.uk/mallet/multitemporal.zip*

In the estimation of soil bulk density by photogrammetry, the Time-SIFT method allows the operator to follow a simplified protocol. Indeed, once the scaling target is set, the operator freely acquires a convergent set of images *before* and *after* the soil sampling without any constraint, but the full stereoscopic coverage of the investigated horizon. The Time-SIFT method allows for a co-registration of the two datasets, which can then be compared to estimate sample volumes only with image information and the use of scale bars.

1.3.5. *Benthic species in rocky shore*

1.3.5.1. *Objective*

In the last decade, with the improvement of photographic and video surveys, the evaluation of species habitats at fine resolution scale (i.e. cm) has become a reality (e.g. Laliberte and Rango 2011; Ventura et al. 2018). This facilitates better comprehension about species habitat selection and behavior at the individual level. In complex habitats, such as the rocky-shore intertidal zone in coastal ecosystems, the microtopographic features of the substrate potentially determine the population dynamics and small-scale distribution of benthic species by interacting with seasonal environmental changes (e.g. temperature, waves action and sand coverage) (Harley 2008; Gomes et al. 2018) and determining inter-specific interactions (Fletcher and Underwood 1987). Intertidal rocky-shore ecosystems are extremely dynamic. They are subject to many environmental factors across temporal scales from hours (e.g. tide) to years (e.g. climate-related events). To better understand spatiotemporal species distribution on complex habitats, it is thus imperative to represent the microhabitat structure by evaluating not only its static features such as the substrate roughness (Meager et al. 2011), but also its dynamics. Here, we aim to test the Time-SIFT methodology to evaluate its adequacy to study the population dynamics of the mussel *Mytilus sp.* in a rocky-shore intertidal area in Brittany based on a monthly survey using high-resolution aerial images taken by drone. We develop a time series of co-registered 3-D models to address the following question: How does the structure and dynamics of a small-scale habitat influence the distribution patterns and structure (e.g. individual size and density) of mussel population?

1.3.5.2. *Case description*

The monitored rocky-shore area is located in *Le Petit Minou* beach, close to the semi-enclosed coastal Bay of Brest (Western Brittany, France) (48°20′19″ N 4°37′03″ W). This location presents diverse features and is influenced by the dynamics of the beach environment, which changes through time as a result of waves and tidal cycles. We used a DJI® Phantom 4 pro platform to take images at low altitude every 1 s approximately (about 15- 30-m height, ground resolution: ~0.7 cm, covered surface: ~7,500 m^2) to obtain an overlap of about 80% between consecutive images. All images were collected at low tide to reduce the effect of water in the contrast and visualization of the rocky shore. We selected four months of drone-survey images for testing with the Time-SIFT methodology. We seek a perfect overlap of the geographic coordinates to identify potential topographic dynamics and consequences on mussels' distribution. Flight parameters are reported in Table 1.13.

Epoch (date)	Images (#)	Flying height (m)	Focal length (mm)	Approximate scale	Ground sampling distance (cm)
2019/02/26	90	29.3	8.8	1:3330	0.7
2019/03/22	103	28.3	8.8	1:3215	0.7
2019/05/06	156	24.5	8.8	1:2785	0.6
2019/06/05	287	15.6	8.8	1:1775	0.4

Table 1.13. *Flight parameters for the rocky-shore case study*

1.3.5.3. *Application of the Time-SIFT method*

As with the soil bulk density test case (section 1.2.4.), application of the Time-SIFT method was straightforward, successfully producing co-registered 3-D models utilizing the default parameters in Agisoft Photoscan Pro (Table 1.14). This was initially unexpected due to the seemingly high degree of change in portions of the study area, including new sand deposits and areas inundated by the sea.

Processing step	Property	Value
Alignment	Accuracy	High
	Pair preselection	No
	Key point limit	40,000
	Tie point limit	4,000
	Adaptive camera model fitting	Yes
Camera calibration	Camera groups	Split by epoch
Dense cloud	Quality	High
	Depth filtering	Moderate
DEM	Pixel size (m)	0.01
	Interpolation	Enabled

Table 1.14. *Time-SIFT parameters in the rocky-shore case study*

1.3.5.4. *Results*

The accurate co-registration of individual epochs is illustrated by the topographic profiles in Figure 1.11. This is possible because all epoch images were acquired using the same planning and purpose, and because each image covers both stable (rock) and changing areas (sand, mussels and sea-water). Success in lens auto-calibration is supported by the corresponding small reprojection error (Table 1.15).

Epoch	Dense cloud (points)	Dense cloud (pts/m^2)	DEM (m/pixels)	RMS reprojection error (pixels)
2019/02/26	32.0×10^6	6,650	0.012	0.53
2019/03/22	95.6×10^6	23,700	0.0065	0.50
2019/05/06	39.4×10^6	26,400	0.0065	0.43
2019/06/05	77.9×10^6	22,300	0.0067	0.47

Table 1.15. *Time-SIFT results for rocky shore*

The results show excellent performance of the Time-SIFT method to compare temporal variations, from centimeter to meter scales (Figure 1.11). Significant variations occurred in the accumulation of sand between the rocks, with accumulation during spring time, reducing the surface area of rocky habitats (Figure 1.11, right and bottom). Because the sand burial or a heavy load of suspended sand can stress mussels and cause mortality (Zardi et al. 2006), such sand inundation may have significant consequences on the studied mussel population. The geographic overlap may permit a comparison of mussel's distribution at the scale of centimeters to evaluate the dynamic of mussel patches coverage and identify and quantify the consequences of such sand mobilization on the mussel beds (orthoimages in Figure 1.11, left). The Time-SIFT method allows the study area to be sub-sampled and thus will provide the opportunity to compare contrasting habitats to assess whether there is a direct relationship between the evolution of topography related to sand mobilization and mussel beds distribution and evolution.

Although drone images bring an easy and cheap way to study local scale habitat features (cm) at median spatial scale (>100 m length), typically, the number and distribution of GCPs determine the accuracy of the topography reconstruction of coastal linear landforms using SfM photogrammetry (Jaud et al. 2018). However, in coastal areas, recording the GCPs is time consuming and frequently restricted by tides and coastal topography. As a result, this setting is an important case study for the Time-SIFT method. Under these conditions and as observed in this case study, the Time-SIFT approach facilitates high-quality temporal comparison of the orthophotos and 3-D reconstructions where ground control opportunities are limited. This opens up new perspectives in using archival aerial imagery and/or collecting new drone imagery to monitor change. Quantifying the dynamics of habitat features at high resolution is essential to better integrate small-scale processes into species distribution modeling approaches, particularly in the context of climate change and its effects on coastal ecosystems.

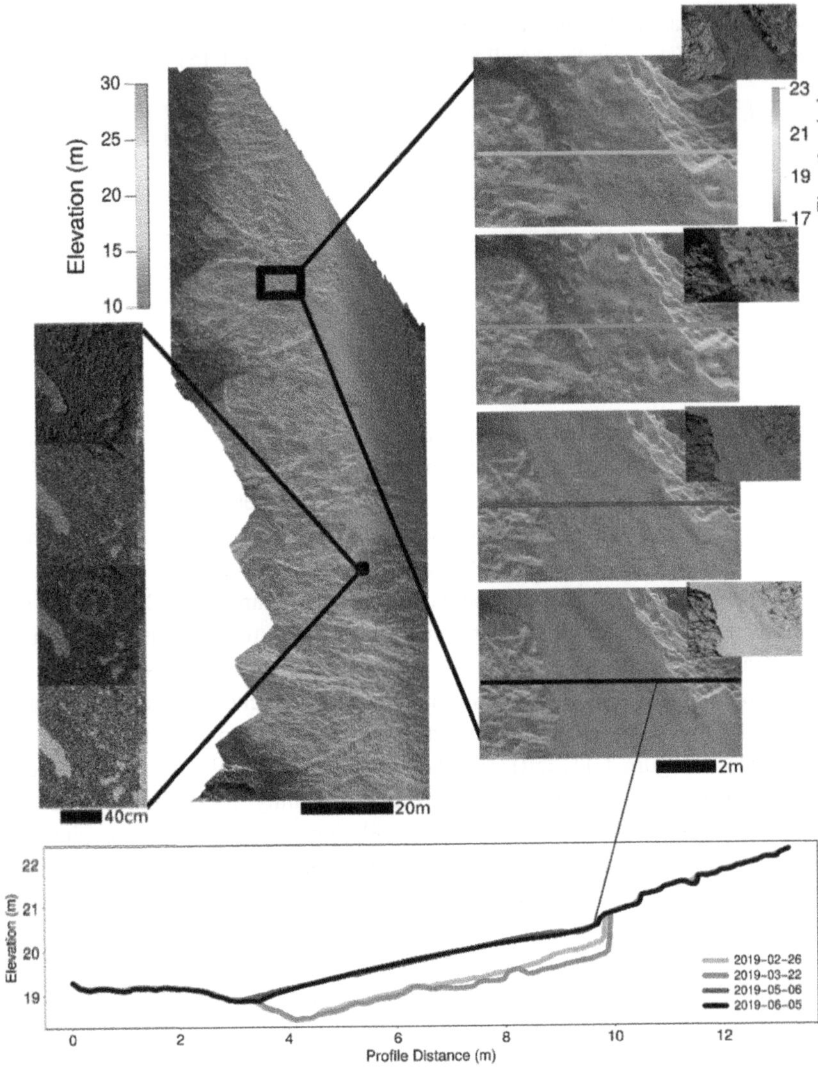

Figure 1.11. *Multi-temporal digital elevation models and orthophotos showing evolution of the mussel beds distribution (left) and geomorphologic changes (right and bottom). Red circles on the orthoimages indicate the disappearance of mussels in the selected area. For a color version of this figure, see www.iste.co.uk/mallet/multitemporal.zip*

1.3.6. *Test cases synthesis*

With a successful application of the Time-SIFT method on five contrasting test cases, we derive several conclusions about the potential of this method.

Time-SIFT is able to handle a wide range of different spatiotemporal scales

For our five varied spatiotemporal cases, Time-SIFT succeeded in obtaining complete multi-temporal blocks. Considering that the Time-SIFT method is almost based on image information only and that it takes advantage of the invariance properties of the SIFT-like algorithms, this finding was expected. Moreover, the spatiotemporal sampling was adapted to the spatiotemporal variability of the investigated phenomena, which strongly favored the success of the Time-SIFT method. In the first case, the high level of noise, still lower than the intensity of the studied phenomena, should be reconciled to the quality of the archival images. However, for the last four cases and despite the good quality of the images, conditions of success were far from optimal considering the relative amount of change in the image time series, whether the change affected large parts of glaciers, a cultivated orchard, a soil pit wall with excavations or even a tidal shore with significant sand deposits. Perhaps surprisingly, adjusting the processing parameters through trial-and-error was only necessary for the Mornag's orchard test site. The reason may be that, for this test case, the change concerned not only the growing trees, but also the ground coverage between the trees, where grass grew during the period of analysis. Perhaps the most remarkable result is the one obtained for the rocky-shore test case. Images taken near the sea constitute a "worst case" in photogrammetry, due to changes within epochs due to waves and the fact that no GCPs can be set on the ocean side of the images.

Time-SIFT can mitigate limitations imposed by the lack of ground control points

The test cases for Mornag's orchard (section 1.2.3), soil bulk density (section 1.2.4) and the rocky shore (section 1.2.5) employed a processing routine where no GCPs were used. These studies defined the scaling for their 3-D model through the use of either the UAV GPS data or an object of known dimensions in the images (soil bulk density test case). This may be one of the most powerful features of the Time-SIFT method, considering that GCPs may be difficult to obtain due to diverse factors such as the use of archival imagery, for which ground information was lost or never collected, or due to a limited accessibility to the ground. Finally, as in the "rocky-shore" test case, GCPs may not be available on some parts of the covered domain simply because there is no stable ground within the target area (foreshore and areas permanently covered by the sea).

The Time-SIFT method therefore provides an unprecedented potential for unlocking the processing of stereoscopic imagery, given that absolute reference and/or scale can be given to the multi-temporal block only after obtaining the relative orientation of the whole image dataset.

Time-SIFT produces robust co-registration of multiple image epochs

For all the test cases, reprojection error falls within the pixel size, or is at worst, the order of magnitude of the pixel. It means that intrinsic co-registration of the image block can be done with a precision that is given by the pixel size. This result is consistent with the known quality of the SIFT feature positioning – approximately 0.3 pixels – depending on the feature detection and matching algorithm (Appolonio et al. 2014). In some cases, this quality may not be sufficient, such as when the target phenomena have a magnitude which is only slightly bigger than the pixel size. In these cases, the approach by Truong et al. (2018) using image correlation to refine the positions of the SIFT points may provide favorable results.

Examining 2-D profiles across where change has not occurred provides an extrinsic assessment of the quality of the results and an estimate of the order of magnitude of the noise in the modeled DEMs. For the Peyne test cases, this finding is supported by the fine DEM and coarse DEM error values quantified in Feurer and Vinatier (2018a, 2018b).

1.4. Conclusion

In this chapter, we proposed five test cases to demonstrate the ability of the Time-SIFT method to provide coherent multi-temporal 3-D and/or 2-D datasets with limited or no external registration data. Our results show that the Time-SIFT method consistently succeeds in the simultaneous orientation of all images of different epochs within a significant range of spatiotemporal scales. This work demonstrates that the Time-SIFT method can be successfully applied for a wide variety of cases and illustrates the range of opportunities for application of time series analysis of photogrammetrically derived digital elevation models. Additional work could further characterize optimal conditions and primary sources of error for application of the Time-SIFT method. In particular, sensitivity analyses are required to better determine a priori the required set of parameters that would enable the success of the method and to determine the impact of the dataset characteristics such as the number of images by epoch, the number of epochs, or the image characteristics of each epoch.

We expect that the Time-SIFT method will help to unlock the existing multi-temporal stereoscopic image archive for a broad range of change detection studies. Candidate studies for testing the Time-SIFT method include tracking large landslide displacements (Dewitte et al. 1998; Lucieer et al. 2014) and seasonal evolution of narrow landscape elements such as ditches, hedgerows, channels (Vinatier et al. 2018), and documentation of structured objects such as houses or buildings. Finally, to support access to the Time-SIFT method, we have opened it up as a module to the Agisoft Metashape software (Vinatier and Feurer 2023).

1.5. References

Apollonio, F.I., Ballabeni, A., Gaiani, M., Remondino, F. (2014). Evaluation of feature-based methods for automated network orientation. *International Archives of the Photogrammetry, Remote Sensing & Spatial Information Sciences*, XL-5, 47–54. doi: 10.5194/isprsarchives-XL-5-47-2014.

Bakker, M. and Lane, S.N. (2017). Archival photogrammetric analysis of river-floodplain systems using Structure from Motion (SfM) methods. *Earth Surface Processes and Landforms*, 42(8), 1274–1286.

Bemis, S.P., Micklethwaite, S., Turner, D., James, M.R., Akciz, S., Thiele, S.T., Bangash, H.A. (2014). Ground-based and UAV-based photogrammetry: A multi-scale, high-resolution mapping tool for structural geology and paleoseismology. *Journal of Structural Geology*, 69(Part A), 163–178.

Brodu, N. and Lague, D. (2012). 3D terrestrial lidar data classification of complex natural scenes using a multi-scale dimensionality criterion: Applications in geomorphology. *ISPRS Journal of Photogrammetry and Remote Sensing*, 68, 121–134.

Carrivick, J.L., Smith, M.W., Quincey, D.J. (2016). *Structure from Motion in the Geosciences*. John Wiley & Sons, New York.

Chandler, J.H. and Brunsden, D. (1995). Steady state behaviour of the Black Ven mudslide: The application of archival analytical photogrammetry to studies of landform change. *Earth Surface Processes and Landforms*, 20(3), 255–275.

Chanut, M.-A., Kasperski, J., Dubois, L., Dauphin, S., Duranthon, J.-P. (2017). Quantification des déplacements 3D par la méthode PLaS – Application au glissement du Chambon (Isère). *Revue française de géotechnique*, 150, 4.

Cook, K.L. and Dietze, M. (2019). A simple workflow for robust low-cost UAV-derived change detection without ground control points. *Earth Surface Dynamics*, 7(4), 1009–1017.

Coulouma, G., Feurer, D., Vinatier, F., Huttel, O. (2021). Assessing new sensor-based volume measurement methods for high-throughput bulk density estimation in the field under various soil conditions. *European Journal of Soil Science*, 72, 2049–2061.

Cowley, D.C. and Stichelbaut, B.B. (2012). Historic aerial photographic archives for European archaeology. *European Journal of Archaeology*, 15(2), 217–236.

Crandell, D.R. and Fahnestock, R.K. (1965). Rockfalls and avalanches from little Tahoma Peak on Mount Rainier, Washington. Technical report, 1221-A, U.S. Government Printing Office.

Dewitte, O., Jasselette, J.-C., Cornet, Y., Van Den Eeckhaut, M., Collignon, A., Poesen, J., Demoulin, A. (2008). Tracking landslide displacements by multi-temporal DTMs: A combined aerial stereophotogrammetric and LIDAR approach in western Belgium. *Engineering Geology*, 99(1–2), 11–22.

Eltner, A., Kaiser, A., Castillo, C., Rock, G., Neugirg, F., Abellan, A. (2016). Image-based surface reconstruction in geomorphometry – Merits, limits and developments. *Earth Surface Dynamics*, 4(2), 359–389.

FAO (2006). *Guidelines for Soil Description*, 4th edition. FAO, Rome.

Feurer, D. and Vinatier, F. (2018a). The Time-SIFT method: Detecting 3-D changes from archival photogrammetric analysis with almost exclusively image information. *arXiv preprint. arXiv:1807.09700.*

Feurer, D. and Vinatier, F. (2018b). Joining multi-epoch archival aerial images in a single SfM block allows 3-D change detection with almost exclusively image information. *ISPRS Journal of Photogrammetry and Remote Sensing*, 146, 495–506.

Filhol, S., Perret, A., Girod, L., Sutter, G., Schuler, T.V., Burkhart, J.F. (2019). Time-lapse photogrammetry of distributed snow depth during snowmelt. *Water Resources Research*, 55, 7916–7926. doi: 10.1029/2018WR024530.

Fischer, L., Eisenbeiss, H., Kaab, A., Huggel, C., Haeberli, W. (2011). Monitoring topographic changes in a periglacial high-mountain face using high-resolution DTMs, Monte Rosa East Face, Italian Alps. *Permafrost and Periglacial Processes*, 22(2), 140–152.

Fischler, M.A. and Bolles, R.C. (1981). Random sample consensus: A paradigm for model fitting with applications to image analysis and automated cartography. *Communications of the ACM*, 24(6), 381–395.

Fletcher, W.J. and Underwood, A.J. (1987). Interspecific competition among subtidal limpets: Effect of substratum heterogeneity. *Ecology*, 68(2), 387–400.

Fonstad, M.A., Dietrich, J.T., Courville, B.C., Jensen, J.L., Carbonneau, P.E. (2013). Topographic structure from motion: A new development in photogrammetric measurement. *Earth Surface Processes and Landforms*, 38(4), 421–430.

Giordano, S., Le Bris, A., Mallet, C. (2018). Toward automatic georeferencing of archival aerial photogrammetric surveys. *ISPRS Annals of Photogrammetry, Remote Sensing and Spatial Information Sciences*, IV(2), 105–112.

Gomes, I., Peteiro, L., Bueno-Pardo, J., Albuquerque, R., Perez-Jorge, S., Oliveira, E.R., Alves, F.L., Queiroga, H. (2018). What's a picture really worth? On the use of drone aerial imagery to estimate intertidal rocky shore mussel demographic parameters. *Estuarine, Coastal and Shelf Science*, 213, 185–198.

Haas, F., Hilger, L., Neugirg, F., Umstadter, K., Breitung, C., Fischer, P., Hilger, P., Heckmann, T., Dusik, J., Kaiser, A. et al. (2016). Quantification and analysis of geomorphic processes on a recultivated iron ore mine on the Italian island of Elba using long-term ground-based lidar and photogrammetric SfM data by a UAV. *Natural Hazards and Earth System Sciences*, 16(5), 1269–1288.

Harley, C.D.G. (2008). Tidal dynamics, topographic orientation, and temperature-mediated mass mortalities on rocky shores. *Marine Ecology Progress Series*, 371, 37–46.

James, M.R., Chandler, J.H., Eltner, A., Fraser, C., Miller, P.E., Mills, J.P., Noble, T., Robson, S., Lane, S.N. (2019). Guidelines on the use of structure-from-motion photogrammetry in geomorphic research. *Earth Surface Processes and Landforms*, 44, 2081–2084. doi: 10.1002/esp.4637.

Jaud, M., Passot, S., Allemand, P., Le Dantec, N., Grandjean, P., Delacourt, C. (2019). Suggestions to limit geometric distortions in the reconstruction of linear coastal landforms by SfM photogrammetry with PhotoScan® and MicMac® for UAV surveys with restricted GCPs pattern. *Drones*, 3(1), 2. doi: 10.3390/drones3010002.

Josberger, E.G., Bidlake, W.R., March, R.S., Kennedy, B.W. (2007). Glacier mass-balance fluctuations in the Pacific Northwest and Alaska, USA. *Annals of Glaciology*, 46, 291–296.

Laliberte, A.S. and Rango, A. (2011). Image processing and classification procedures for analysis of sub-decimeter imagery acquired with an unmanned aircraft over arid rangelands. *GIScience & Remote Sensing*, 48(1), 4–23.

Lowe, D.G. (2004). Distinctive image features from scale-invariant keypoints. *International Journal of Computer Vision*, 60(2), 91–110.

Lucieer, A., Jong, S.M.D., Turner, D. (2014). Mapping landslide displacements using Structure from Motion (SfM) and image correlation of multi-temporal UAV photography. *Progress in Physical Geography*, 38(1), 97–116.

Mabrouk, H., Feurer, D., Massuel, S., Ben Jemaa, F. (2017). Un exemple de l'utilisation des drones dans la gestion technique d'un verger de pêcher en Tunisie. *Watch Letter*, 38, 73–79.

Mantua, N.J. and Hare, S.R. (2002). The Pacific Decadal Oscillation. *Journal of Oceanography*, 58(1), 35–44.

Marcus, W.A. and Fonstad, M.A. (2008). Optical remote mapping of rivers at sub-meter resolutions and watershed extents. *Earth Surface Processes and Landforms: The Journal of the British Geomorphological Research Group*, 33(1), 4–24.

Meager, J.J., Schlacher, T.A., Green, M. (2011). Topographic complexity and landscape temperature patterns create a dynamic habitat structure on a rocky intertidal shore. *Marine Ecology Progress Series*, 428, 1–12.

Nagarajan, S. and Schenk, T. (2016). Feature-based registration of historical aerial images by area minimization. *ISPRS Journal of Photogrammetry and Remote Sensing*, 116(Supplement C), 15–23.

Newman, M., Alexander, M.A., Ault, T.R., Cobb, K.M., Deser, C., Di Lorenzo, E., Mantua, N.J., Miller, A.J., Minobe, S., Nakamura, H. et al. (2016). The Pacific Decadal Oscillation, revisited. *Journal of Climate*, 29(12), 4399–4427.

Parente, L., Chandler, J.H., Dixon, N. (2021). Automated registration of SfM-MVS multitemporal datasets using terrestrial and oblique aerial images. *The Photogrammetric Record*, 36(173), 12–35.

Rupnik, E., Daakir, M., Pierrot Deseilligny, M. (2017). MicMac – A free, open-source solution for photogrammetry. *Open Geospatial Data, Software and Standards*, 2, 14. doi: 10.1186/s40965-017-0027-2.

Salach, A. (2017). SAPC – Application for adapting scanned analogue photographs to use them in structure from motion technology. *International Archives of the Photogrammetry, Remote Sensing & Spatial Information Sciences*, 42, 197–204. doi: 10.5194/isprs-archives-XLII-1-W1-197-2017.

Sevara, C., Verhoeven, G., Doneus, M., Draganits, E. (2017). Surfaces from the visual past: Recovering high-resolution terrain data from historic aerial imagery for multitemporal landscape analysis. *Journal of Archaeological Method and Theory*, 25, 611–642. doi: 10.1007/s10816-017-9348-9.

Sisson, T.W., Robinson, J.E., Swinney, D.D. (2011). Whole-edifice ice volume change A.D. 1970 to 2007/2008 at Mount Rainier, Washington, based on LiDAR surveying. *Geology*, 39(7), 639–642.

Smith, M.W., Carrivick, J.L., Quincey, D.J. (2016). Structure from motion photogrammetry in physical geography. *Progress in Physical Geography*, 40(2), 247–275.

Truong Giang, N., Muller, J.-M., Rupnik, E., Thom, C., Pierrot-Deseilligny, M. (2018). Second iteration of photogrammetric processing to refine image orientation with improved tie-points. *Sensors*, 18(7), 2150.

Vargo, L.J., Anderson, B.M., Horgan, H.J., Mackintosh, A.N., Lorrey, A.M., Thornton, M. (2017). Using structure from motion photogrammetry to measure past glacier changes from historic aerial photographs. *Journal of Glaciology*, 63(242), 1105–1118.

Ventura, D., Bonifazi, A., Gravina, M., Belluscio, A., Ardizzone, G. (2018). Mapping and classification of ecologically sensitive marine habitats using unmanned aerial vehicle (UAV) imagery and object-based image analysis (OBIA). *Remote Sensing*, 10(9), 1331.

Verhoeven, G. (2011). Taking computer vision aloft – Archaeological three-dimensional reconstructions from aerial photographs with photoscan. *Archaeological Prospection*, 18(1), 67–73.

Vinatier, F. and Arnaiz, A.G. (2018). Using high-resolution multitemporal imagery to highlight severe land management changes in Mediterranean vineyards. *Applied Geography*, 90, 115–122.

Vinatier, F. and Feurer, D. (2023). Time-SIFT module for Agisoft Metashape software. Module, Zenodo. doi: 10.5281/zenodo.8359983.

Vinatier, F., Dollinger, J., Rudi, G., Feurer, D., Belaud, G., Bailly, J.-S. (2018). The use of photogrammetry to construct time series of vegetation permeability to water and seed transport in agricultural waterways. *Remote Sensing*, 10(12), 2050.

Westoby, M.J., Brasington, J., Glasser, N.F., Hambrey, M.J., Reynolds, J.M. (2012). Structure-from-motion photogrammetry: A low-cost, effective tool for geoscience applications. *Geomorphology*, 179, 300–314.

WSR (2015). World reference base for soil resources 2014, update 2015. World Soil Resources report no. 106. FAO, Rome.

Zardi, G.I., Nicastro, K.R., Porri, F., McQuaid, C.D. (2006). Sand stress as a non-determinant of habitat segregation of indigenous (*Perna perna*) and invasive (*Mytilus galloprovincialis*) mussels in South Africa. *Marine Biology*, 148(5), 1031–1038.

2

Hierarchical Crop Mapping from Satellite Image Sequences with Recurrent Neural Networks

Mehmet OZGUR TURKOGLU, Stefano D'ARONCO, Konrad SCHINDLER and Jan Dirk WEGNER

EcoVision Lab, ETH Zurich and University of Zurich, Switzerland

2.1. Introduction

Monitoring agricultural land use is of high importance for food production, biodiversity and forestry (Gómez et al. 2016). An increasing world population, climate change and changes in food consumption habits put yet uncultivated areas under pressure, while leading to intensification in existing agricultural areas (Laurance et al. 2014). Cropland expansion and intensive use of agricultural areas are often connected with negative ecological impacts like deforestation and biodiversity loss, but also degradation of ecosystem services like ground and surface water quality (Herzog et al. 2008; Dise et al. 2011). Detailed and up-to-date knowledge about agricultural areas is essential to inform policies on reducing the environmental impacts of intensive agriculture (Gómez et al. 2016). Examples of policy-driven incentives (through subsidies) are extensively used in grassland portions per farm to promote biodiversity, or a certain crop mix in the rotation (Finger and Lehmann 2012), in order to prevent soil degradation. In order to guide policy and to monitor progress toward more sustainable land use, more accurate and frequently-updated country-scale monitoring of agricultural activities is essential. Traditionally, this

Multitemporal Earth Observation Image Analysis,
coordinated by Clément MALLET and Nesrine CHEHATA.

monitoring is based on farmer self-reporting and spot checking by the authorities in the field, which is laborious, costly and prone to errors.

Modern machine learning methods in combination with publicly available satellite imagery provide new possibilities for accurate, spatially dense monitoring of agricultural sites at high temporal resolution and low costs. One particularly promising recent sensor is Sentinel-2, due to its ground sampling distance of 10 m and a revisit rate of 3–5 days (see Figure 2.1). In general, the spectral signal of the vegetation as captured by the satellite has specific characteristics as a function of (i) soil structure and composition (e.g. soil brightness, soil water content, soil type, etc.), (ii) vegetation structure (e.g. canopy cover, leaf area index (LAI), plant height, leaf angle, etc.) and (iii) leaf biochemistry (e.g. chlorophyll, water content, nitrogen content, etc.) (Thenkabail et al. 2013). Not only does each plant species have its own spectral signature, but it also changes in a species-specific pattern over time as a function of its phenological stage (Walter et al. 2015; Anderegg et al. 2020). Instead of merely analyzing images at a single point in time, sequences (time series) of satellite images therefore provide valuable additional evidence to identify crop species.

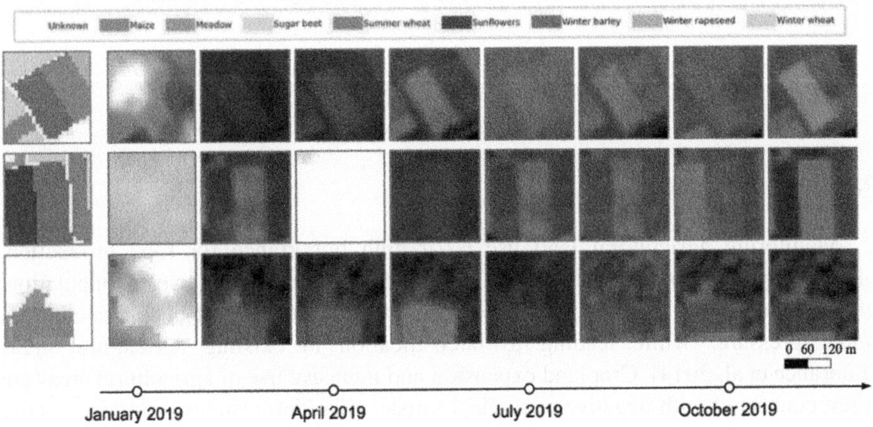

Figure 2.1. *Example of Sentinel-2 satellite images of the* ZueriCrop *dataset. Each row shows randomly sampled images (false color composite: NIR-Red-Green) from a satellite image time series. The first column shows the ground-truth where different colors correspond to different crop types. For a color version of this figure, see www.iste.co.uk/mallet/multitemporal.zip*

A challenging aspect from the perspective of a supervised classification approach is, however, that the distribution of agricultural land use categories is highly skewed. While few categories make up most of the cultivated area, most categories occur only rarely and cover small portions of land. In reality, this leads to very long tailed class distributions with a large number of different crop types. As a result, supervised classifiers that are trained on such an imbalanced dataset are prone to perform poorly on the rare classes because the dominating categories make up most of the training data. Often, the class distribution is not only imbalanced, but the underrepresented classes also come with very few reference data (i.e. labels), making the training procedure of any data-driven method more challenging. A viable way to reduce these effects are hierarchical classification strategies (Srivastava and Salakhutdinov 2013; Zhu and Bain 2017; Wehrmann et al. 2018; Koo et al. 2018; Roy et al. 2020) that can be further sub-divided into methods based on ensembles of classifiers, re-weighting schemes, etc., which we shall explore further in this chapter.

The basic idea of using a label hierarchy is that although many classes are rare, they may belong to the same super-class at a coarser level and share common features. This observation can be used in order to regularize the training of a model and improve the generalization error, especially for the rare classes. Imposing prior knowledge about the class structure through hierarchical attributes adds (soft) constraints to the model and encourages it to pool shared information from related classes. Having such "coarse-granularity features" earlier in the hierarchy means that it is easier to learn due to the larger training set, while having a fine-grained classification later in the hierarchy can focus on discriminating fewer sub-classes, thus using the rare training examples more efficiently. Designing a label hierarchy for crop classification based on visual similarity and agricultural expertise is a promising direction to explore. For instance, apple orchards, pear orchards and chestnut orchards all occur only rarely individually (in Switzerland) but belong to the same, coarser class of *orchards* and share many features. We point out another advantage of a hierarchical scheme: we can use the class scores to determine how reliable the predictions at different hierarchy levels are and still use the coarse outputs if the fine-grained ones are too unreliable. For example, if the decision between *apple* and *pear* is uncertain, the coarser label *orchard* may still have high confidence. For many applications in practice like summary statistics or the calculation of subsidies by public administration, this coarser level of granularity is often good enough. In case fine-grained annotations are still needed, trustworthy annotations at the coarser level can significantly accelerate the manual labeling process. Looking again at the orchard example, human annotators would only have to assign orchard pixels to the classes apple, pear and chestnut, instead of looking at the entire country and all possible categories.

In this chapter, which is based on papers by Turkoglu et al. (2021a, 2021b), we review different deep learning methods for image sequence analysis and provide an introduction to recurrent network architectures. We further describe a recurrent network approach for mapping agricultural areas that is *hierarchical*. It exploits a tree-structured label hierarchy built by domain experts and encodes image evidence via convolutional layers. The proposed architecture has multiple levels of representation that consist of stacked, convolutional recurrent neural networks (RNNs). The different network levels predict successively finer label resolutions in the hierarchical tree. In order to extract meaningful evidence from the satellite image sequence, the hierarchical, multi-label architecture stacks several RNN cells to form a deep architecture. However, this gain of network capacity comes at a cost: deeper RNN architectures are brittle to set up and train due to vanishing or exploding gradients for many RNN cell types as we show later in this chapter. To reduce this problem, we design a new RNN cell type called STAR cell (Turkoglu et al. 2021b) that facilitates a stable training process when dealing with deep architectures. We analyze the benefits of both the label hierarchy and the STAR cell type experimentally for the classification of agricultural areas. We demonstrate that the approach compares favorably to the state of the art on a real-world dataset and show that the proposed multi-label RNN is more effective than widely used techniques for coping with imbalanced data distributions, such as data augmentation or class-balanced loss functions.

In the following, we will first review existing approaches in literature for crop mapping from satellite data and look at techniques that counter imbalanced data distributions. We shall also provide a brief introduction to deep learning methods for image sequence analysis and explain in-depth how to exploit a crop label hierarchy for improving model performance, especially for rare classes. In section 2.5, we will evaluate the proposed method both quantitatively and qualitatively.

2.2. Literature

Deep learning for sequence modeling has gained much attention in recent years. The main advantage is that (i) deep learning no longer relies on hand-engineered features to encode temporal (and/or spatial) patterns and (ii) the large model capacity allows us to learn very complex, highly nonlinear relationships given sufficient (labeled) training data and computational resources. Recurrent neural networks, such as long short-term memory (LSTM) (Hochreiter and Schmidhuber 1997), have been shown to be one of the most effective solutions: with their ability to learn a hidden representation that *memorizes* important patterns, they are able to model long-term dependencies and deal with sequences of arbitrary length while preserving scalability. A large amount of different recurrent cells have been proposed with the aim of improving performance, reducing the number of parameters, and/or facilitating the training procedure (see, for instance, Chung et al. (2014); Li et al. (2018a); Chang et al. (2019)). When dealing with image-like input sequences, it is important to take into account spatial patterns in the data. This can be effectively done by using convolutional layers. In Sainte Fare Garnot et al. (2019) and Rustowicz et al. (2019), convolutions and recurrence are used separately: spatial features are extracted with a convolutional neural network (CNN) and then gated recurrent units (Chung et al. 2014) are used on the resulting feature vectors to model the temporal evolution. Interdonato et al. (2019) instead extract features using both CNN and RNN and concatenate them to produce the final feature vector that is then fed to a fully connected neural network. Rußwurm and Körner (2018a) and Turkoglu et al. (2021b) employ a convolutional variant of RNNs that model spatial and temporal dependencies simultaneously.

Recurrent cells, though widely used, are not the only deep learning architectures that can be employed for sequential data. Another type of neural architecture for processing sequential data are temporal convolutional networks (TCNs) that have been successfully applied to a wide range of applications like audio synthesis (Oord et al. 2016), language modeling (Bai et al. 2018) or video action segmentation (Farha and Gall 2019). TCNs are 1D fully convolutional network architectures, where the convolution operation is performed along the time dimension. Two key properties are typical for TCNs: (i) they use causal convolutions where an output at time t only depends on input elements from time t and earlier in the previous layer and (ii) TCNs can use dilated convolutions to increase the receptive field of the network, which is beneficial to model long-term patterns over time. TCN has multiple advantages compared to recurrent neural networks like being amenable to parallelization in the temporal domain, unlike RNNs where predictions for later time-steps must wait for their predecessors to complete. Therefore, both training and inference of TCNs can be much faster compared to RNNs. Another significant advantage are more stable gradients during training, which means that TCNs are usually easier to train than RNN architectures. On the other hand, RNNs seem to model long-term patterns better as we will see when comparing their performance to TCNs later in this chapter. Another viable option for sequence modeling is the *Transformer* method (Vaswani et al. 2017), originally proposed for machine translation tasks and strongly related to the self-attention mechanism (Bahdanau et al. 2015). Bahdanau et al. (2015) originally introduced the self-attention mechanism for RNNs. The main idea is that instead of using only the last hidden representation, a context vector is employed to obtain a weighted sum of all hidden states, also called attention scores. This contextual evidence helps us to focus on the relevant parts of the input sequence for the current (output) prediction. Weights are adaptive and determined by the feed-forward neural network, which is jointly trained with all of the other components without using any extra supervision signal. Inspired by this original idea, Vaswani et al. (2017) introduced the fully attention-based *Transformer* model. We evaluate and compare the different properties of RNN, TCN and Transformer approaches for crop classification.

Crop classification with multi-temporal satellite data has been widely studied in remote sensing. Traditional machine learning approaches with handcrafted features (Inglada et al. 2015; Wardlow and Egbert 2008; Vuolo et al. 2018) predominantly rely on vegetation indices like the Normalized Difference Vegetation Index (NDVI) (Foerster et al. 2012; Ustuner et al. 2014; Peña-Barragán et al. 2011; Conrad et al. 2010). Different strategies have been explored to better model the temporal evolution as further evidence for classification, such as temporal windows (Conrad et al. 2014), hidden Markov models and dynamic time warping (Siachalou et al. 2015; Belgiu and Csillik 2018), and conditional random fields (Giordano et al. 2020). For instance, Wardlow and Egbert (2008) extract features, which are time series of NDVI, from MODIS data collected over the growing season of crops, and they perform classification with a decision tree. Similarly, Conrad et al. (2014) investigate the optimum number of acquisition dates and most suitable temporal windows for the discrimination of crops from RapidEye satellite time series data. These traditional methods share the fact that their performances are constrained by the limited discriminativeness and robustness of the hand-crafted features, as well as by the limited expressive power of conventional classifiers.

More recently, deep learning methods have shown their ability to effectively solve many pattern recognition task. Their main advantage is twofold: (i) they no longer rely on hand-engineered features to encode spectral, spatial or temporal patterns; (i) their large capacity makes them able to learn very complex, highly nonlinear relationships, if given sufficient labeled training data and computational resources. Rußwurm and Körner (2017) use a recurrent neural network with long short-term memory (LSTM) to encode temporal dependencies in the data, while Rußwurm and Körner (2018a) improve the result on the same dataset by encoding both temporal *and* spatial dependencies via convolutional LSTM and gated recurrent units (GRUs). In Rustowicz et al. (2019) and Sainte Fare Garnot et al. (2019), satellite images are first processed individually with a CNN to obtain per-image features; then temporal dependencies between these features are modeled with a separate RNNs. Further options are temporal CNNs that combine features also across time with convolutions (Pelletier et al. 2019), or models that use the attention principle (Vaswani et al. 2017) to aggregate information across time (Rußwurm et al. 2019; Rußwurm and Körner 2020). Sainte Fare Garnot et al. (2020) and Sainte Fare Garnot and Landrieu (2020b) combine pixel-set encoder and Transformer (Vaswani et al. 2017) and show improved performance over RNN-based approaches. Finally, in our previous work (Turkoglu et al. 2021b), we build a deep RNN with a new cell structure termed STAR that trains better than LSTM- and GRU-type models while being more parameter efficient. This makes it possible to train deeper models, which translates into improved performance across a range of sequence modeling tasks, including crop classification.

Handling imbalanced datasets is generally a problem for supervised classification tasks. It is common for real-world applications that some classes occur more often than others (e.g. animal species and crop types). Another reason for imbalanced datasets in practice is that some labels may simply be easier to collect and thus occur more often in the dataset. This leads to long-tailed class distributions being the norm, rather than the exception, for large, real-world datasets (Xiao et al. 2010). Under standard training regimes, machine learning models tend to ignore rare, underrepresented classes and focus on the dominant classes to maximize cumulative performance across the entire dataset (Wang et al. 2017; Ren et al. 2018; Dong et al. 2018). While those shortcomings are easily overlooked when evaluating with global performance metrics like overall accuracy, they become obvious with class-balanced metrics like average class precision or F1-score. In fact, for many practical applications a class-balanced evaluation is essential, as rare classes have the same (or even higher) importance as frequent ones. This is also true for our agricultural mapping problem where crops that have high financial or ecological value (for instance, orchards and vegetables) are rare compared to pastoral grasslands or wheat fields. Two major strategies have been developed to solve this problem: (i) algorithm-level approaches and (ii) data-level approaches. A typical algorithm-level approach is cost-sensitive learning where the loss function is re-weighted by a factor inversely proportional to the class frequencies (Ling and Sheng 2008; Huang et al. 2016; Khan et al. 2017, 2019), that is, training samples of rare classes receive higher weight. An inherent consequence of resampling or reweighting training samples according to rarity is an unspecific model bias toward the rare classes. Data-level approaches, which are often implemented through ensembles of classifiers, try to balance the dataset either by oversampling minority classes (Chawla et al. 2002; Douzas and Bacao 2018; Cui et al. 2019) or by undersampling the majority classes (He and Garcia 2009). Undersampling dominant classes runs the risk of missing large parts of the data distribution, thus hurting model performance. On the other hand, oversampling rare classes reaches its limit if the number of available samples in a class is too low. Thus in the extreme case where few samples of the rare classes are available, such approaches reach their limit, and other sorts of regularization methods need to be used. As we show in this chapter, leveraging the hierarchical structure of the labels is a viable solution since, for the coarser label granularity, multiple classes are combined together augmenting the number of per class samples available. We propose to leverage the hierarchical structure of the labels to counter class imbalance and to improve performance for rare classes.

Hierarchical classification in remote sensing has been investigated before. Melgani and Bruzzone (2004) develop hierarchical tree-based classification strategies using binary classifiers like support vector machines (SVM) for hyperspectral remote sensing data. Chen et al. (2009) propose a rule-based method that hierarchically classifies land cover and land use from LIDAR and WorldView-2 data. Similarly, Wu et al. (2016) apply a rule-based classification of LIDAR data for *building* classification followed by a classification of *road*, *vegetation* and *bare soil* with SVMs by additionally incorporating WorldView-2. Another rule-based, hierarchical method is proposed in (Heupel et al. 2018) for crop classification from four different satellite sensors: Landsat-7 and Landsat-8, Sentinel-2A and Rapid-Eye. The first-level classifier decides whether it is a *winter crop* or a *summer crop* and the second-level classifiers predict and organize them into eight fine-grained classes, for example, *potato* and *corn*. Their method exhaustively relies on hand-crafted features and expert knowledge. Jiao et al. (2019) apply rule-based decision-making to classify land covers of coastal wetlands into four coarse classes, which are subdivided into more fine-grained classes with SVMs. Goel et al. (2018) study hierarchical metric learning for classification of remote sensing data. They use iterative max-margin clustering to organize the classes in a hierarchical fashion and subsequently learn different distance metric transformations for the classes present at the non-leaf nodes of the tree. Another idea is using individual random forests at different hierarchy levels for land cover and land use classification (Sulla-Menashe et al. 2011, 2019) from MODIS data. More recently, Demirkan et al. (2020) investigated the benefit of hierarchical classification with SVMs and random forests for land cover and land use classification from Sentinel-2 data. Similarly, (Gbodjo et al. 2020) use individual RNNs at different hierarchy levels and exploit multiple complementary sources of information, radar (Sentinel-1) and optical data (Sentinel-2) in order to achieve better performance. Although hierarchical classification in remote sensing has been studied before; all methods use traditional, hand-crafted features and decision trees designed for specific scenes and datasets by domain experts. Complex workflows that employ multiple independent classifiers at different stages are costly to compute, need a lot of manual tuning for each new dataset and erroneous decisions at early stages can hardly be compensated for later on in the pipeline. Additionally, none of the existing workflows dealt with a hierarchical approach for a realistic, large-scale, imbalanced dataset with a large number of classes. Finally, it is worth mentioning that Sainte Fare Garnot and Landrieu (2020a) simultaneously developed a method that leverages label hierarchy for a metric-guided prototype learning and show improved performance on S2-Agri crop dataset (Sainte Fare Garnot et al. 2020) in a prototype learning setting.

Hierarchical classification has been studied thoroughly for many years and a large body of literature exists. For example, Amit and Geman (1997); Sahbi and Geman (2006); and Gangaputra and Geman (2006) proposed methods relying on tree-like hierarchies. Since the comeback of neural networks, various works have adapted these earlier ideas to deep learning. Srivastava and Salakhutdinov (2013) propose a tree-like hierarchy in CNNs for image classification. Their method learns to organize the classes into a tree hierarchy, and this tree structure imposes a prior over the classifier parameters, improving results for the minority classes. Yan et al. (2015) embed deep CNNs into a two-level category hierarchy. Easily distinguishable classes are separated with a coarser classifier, while a dedicated fine classifier separates the other in more difficult cases. Xiao et al. (2014); and Roy et al. (2020) study hierarchical networks composed of deep CNNs in the context of incremental learning. Chen et al. (2019) propose a training strategy that leverages the information from a label hierarchy. It maximizes the probability of the ground truth class, and at the same time, neutralizes the probabilities of the other classes in a hierarchical fashion, making the model take advantage of the label hierarchy explicitly. Another recent work (in the field of hierarchical text classification) is of Mao et al. (2019), where the hierarchy is explored at both training and inference time with a Markov decision process and deep reinforcement learning. Zhu and Bain (2017) and Wehrmann et al. (2018) propose multi-stage deep CNNs. Like ours, their models have multiple outputs of different granularity as well as multiple objectives. Koo et al. (2018) further improve on this idea by combining a CNN that extracts a hierarchical image representation with an RNN to capture the hierarchical tree of labels.

To the best of our knowledge, our method is one of the first methods that explicitly encodes the inherent (and for domain experts well-known) label hierarchy for crop classification in satellite image sequences in the deep learning setting. Our method differs from all existing literature in that it is based on an integrated, convolutional and recurrent model (convRNN) that is able to capture all relevant spatiotemporal correlations.

2.3. Background: sequence modeling with recurrent neural networks

In this section, we will describe how sequential data can be encoded with recurrent neural networks. In general, we are interested in a function $f_W : \mathcal{X}^T \rightarrow \mathcal{Y}^T$ that maps the input sequence of length T: $(x_0, x_1, ..., x_T)$ where $x_t \in \mathbb{R}^{c_x}$ (c_x is the input dimension, e.g., number of bands for satellite data) to a target representation $\hat{y}_t$ using the learnable parameters W. Depending on whether the problem at hand is a regression or a classification task, $\hat{y}_t$ can take values on a discrete or continuous set. We are often interested in predicting only a single label for the entire sequence; we are only interested in $\hat{y}_T$, so the subscript T is usually omitted. Although not equivalent, it is usually straightforward to adapt a model to suit both regression and classification. Since crop prediction is a classification problem, a standard cross-entropy (CE) loss function is used:

$$L = -\sum_{c=1}^{C} y_c \log(\hat{y}_c), \qquad [2.1]$$

where C is the number of classes and y_c is the ground truth label. The training procedure consists in tuning the network parameters W, usually via gradient descent, in order to minimize the average of the loss computed over a training set.

The main idea behind recurrent networks is to find a latent representation h_t, which can be recurrently updated at each time step t and that includes all of the necessary information from the past in order to predict future labels $\hat{y}_t, ..., \hat{y}_T$. T is the overall length of the input sequence, which can vary. The basic building block of an RNN is a computational *unit* (or *cell*) that combines two inputs: the data of the current time step in the sequence and the unit's own output from the previous time step (see Figure 2.2). Formally, an RNN cell is a nonlinear transformation that maps the input signal x_t at time t and the hidden state of the previous time step $t-1$ to the current hidden state h_t:

$$h_t = f(x_t, h_{t-1}, W), \qquad [2.2]$$

with W the trainable parameters of the cell. In order to map the hidden state h_t to the final label prediction $\hat{y}_t$, a multi-layer perceptron (MLP) can be used. As mentioned earlier, it depends on the task whether the final state h_T, the complete sequence of states $\{h_t\}$, or a single sequence label (typically defined as the average $\frac{1}{T}\sum_t h_t$), is the desired target for which loss $\mathcal{L}$ is computed. The simplest form of an RNN, usually called vanilla RNN (vRNN), can be defined with the following expression:

$$h_t = \tanh(W_x x_t + W_h h_{t-1} + b). \quad [2.3]$$

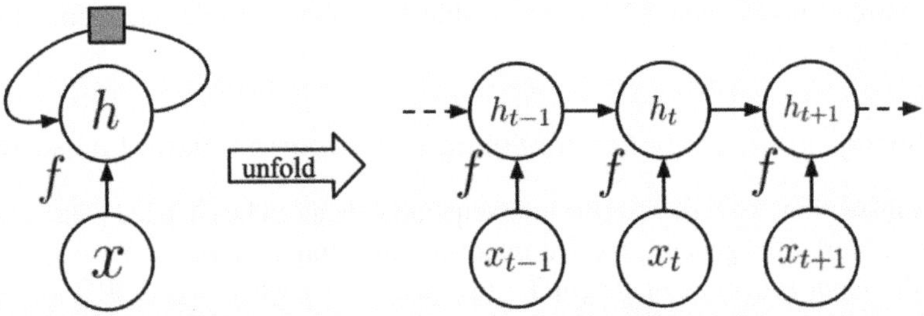

Figure 2.2. *Computation structure of recurrent neural networks (RNNs). It processes information from the input x by incorporating it into the state h that is passed forward through time. The square indicates a delay of a single time step. On the right, the same network is represented as an unfolded computational graph, where each node is now associated with one particular time instance*

While RNNs can in principle handle sequences of arbitrary and varying length, they are (in their basic form) challenged by long-term dependencies, since learning those would require the propagation of gradients over many time steps. To mitigate this limitation, *gated* architectures have been proposed. The most prominent examples are LSTM (Hochreiter and Schmidhuber 1997) and GRU (Chung et al. 2014) (see Figure 2.3). Both methods use gating mechanisms to store and propagate information over longer time intervals, thus reducing the vanishing gradient problem. Although gated architectures mitigate the vanishing gradient, they may still suffer from exploding gradients. When the gradient back propagates over many time steps, its magnitude can become very high leading to the exploding gradient problem, which makes learning deeper models hard. To solve this problem, the simple yet effective technique is *gradient clipping* (Pascanu et al. 2013).

2.3.1. *Convolutional recurrent neural networks*

Images as input for sequence modeling can be represented as three-dimensional tensors: $X \in \mathbb{R}^{w \times h \times c_X}$ where w and h are width and height of the input images and

c_X is the number channels. If the model output is an image too, $\hat{y}_t$ is a three-dimensional tensor as well. Convolutional recurrent neural networks (convRNN) are the convolutional version of RNNs, designed to represent spatiotemporal data. ConvRNNs have been used for different spatiotemporal modeling tasks like weather forecasting (Xingjian et al. 2015), video action recognition (Li et al. 2018b), video forecasting (Su et al. 2020) and the prediction of heat diffusion (Saha et al. 2020). ConvRNNs differ from standard RNNs in that the matrix multiplications are replaced with the convolution operator, denoted with $*$:

$$Z(i,j,k) = (X * F)(i,j,k) = \sum_{u,v,l} F(u,v,l,k) \times X_{i-u,j-v,l}, \quad [2.4]$$

where F represents the filter parameters, which is a four-dimensional tensor, and X is the input tensor, for example, the input image (Figure 2.4). It is often possible to convert a recurrent cell type into its convolutional version (Xingjian et al. 2015; Siam et al. 2017; Turkoglu et al. 2021b) by replacing the linear operation, which is a matrix multiplication in a fully connected layer, with a convolution layer.

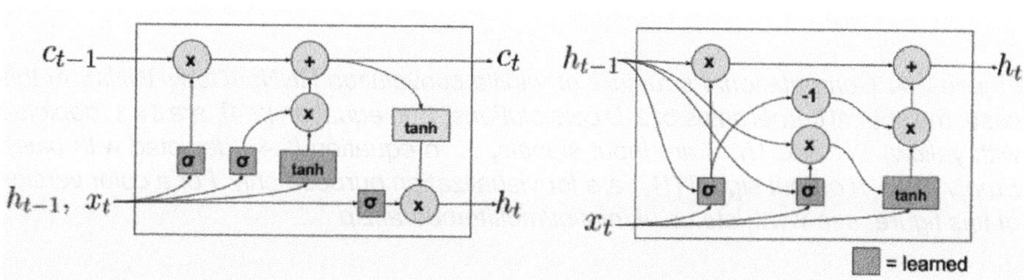

$$i_t = \sigma(W_{xi}x_t + W_{hi}h_{t-1} + b_i)$$
$$f_t = \sigma(W_{xf}x_t + W_{hf}h_{t-1} + b_f)$$
$$o_t = \sigma(W_{xo}x_t + W_{ho}h_{t-1} + b_o)$$
$$z_t = \tanh(W_{xz}x_t + W_{hz}h_{t-1} + b_z)$$
$$c_t = f_t \circ c_{t-1} + i_t \circ z_t$$
$$h_t = o_t \circ \tanh(c_t)$$

(a) LSTM

$$f_t = \sigma(W_{xf}x_t + W_{hf}h_{t-1} + b_f)$$
$$r_t = \sigma(W_{xr}x_t + W_{hr}h_{t-1} + b_r)$$
$$z_t = \tanh(W_{xz}x_t + W_{hz}(r_t \circ h_{t-1}) + b_z)$$
$$h_t = f_t \circ h_{t-1} + (1 - f_t) \circ z_t$$

(b) GRU

Figure 2.3. *RNN cell structures and dynamics of LSTM and GRU. σ is the Sigmoid function, $\circ$ denotes the Hadamard (element-wise) product and i_t, f_t, o_t are, respectively, the input gate, forget gate and output gate activations. z_t is the update. LSTM additionally has a separate cell state c_t, whereas GRU has a reset gate r_t instead of the input and output gates*

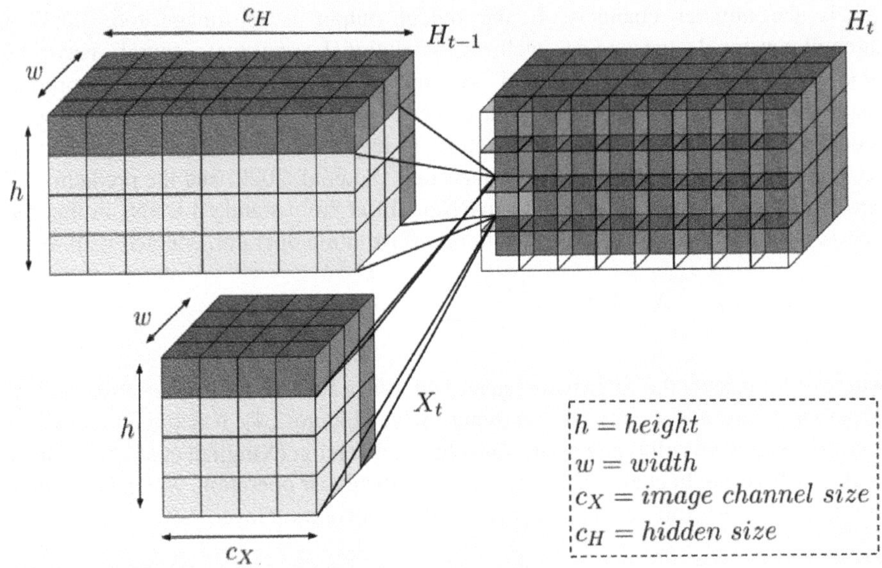

Figure 2.4. *Computational structure of vanilla convolution RNNs (ConvRNNs). In this case, the spatial kernel sizes of 2-D convolutions, F in equation [2.4], are 3×3 (depicted with yellow). X_t and H_{t-1} are input signals, X in equation [2.4] (depicted with blue). Empty grids in output signal (H_t) are for visualization purpose only. For a color version of this figure, see www.iste.co.uk/mallet/multitemporal.zip*

2.3.2. *Deep multi-layer RNNs*

Abstract, discriminative features are often better represented by deeper architectures (Bengio 2009). In the same way that multiple hidden layers can be stacked in traditional feed-forward networks, multiple recurrent cells can also be stacked on top of each other. The output (or the hidden state) of the lower cell is connected to the input of the next, higher cell. Several works (Chung et al. 2015; Zilly et al. 2017; Shahroudy et al. 2016) have shown the ability of deeper recurrent architectures to extract more complex, expressive features from the input and make better predictions. However, these architectures are usually composed of only two or three layers because training deeper recurrent architectures is still very challenging. More specifically, deep RNNs suffer from two main shortcomings: (i) they are difficult to train because of gradient instability, that is, the gradient either explodes or vanishes during training; and (ii) the large number of parameters contained in each single cell makes deep architectures extremely resource-intensive. Both properties restrict the practical use of deep RNNs and particularly their usage for image-like input data, which generally requires multiple, stacked convolutional layers to extract discriminative, abstract representations. When stacking multiple RNN cells on top of each other, the hidden state of the lower level $l - 1$ is passed on as input to the next-higher level l (Figure 2.5). In mathematical terms, this corresponds to the recurrent relation

$$h_t^l = f(h_t^{l-1}, h_{t-1}^l, W)\,. \qquad [2.5]$$

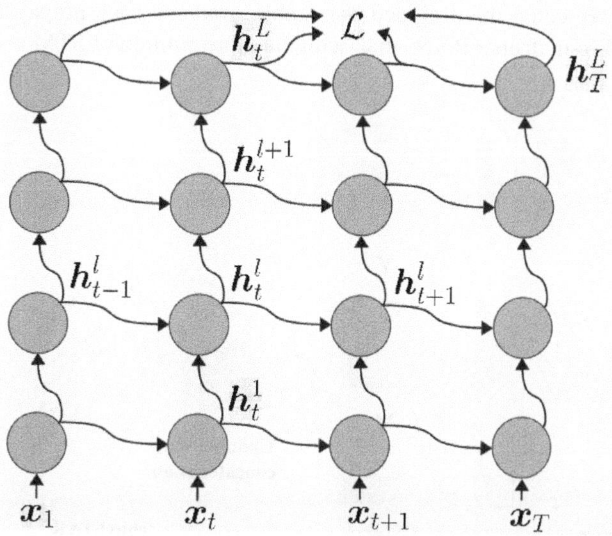

Figure 2.5. *Computation structure of deep multi-layer RNNs*

Temporal unfolding leads to a two-dimensional lattice with depth L and length T (Figure 2.5). The forward pass runs from left to right and from bottom to top. Gradients flow in the opposite direction: at each cell the gradient of the loss w.r.t. the output of the cell is used to compute the gradient w.r.t. (i) the weights, (ii) the input and (iii) the previous hidden state. The latter two gradients are then propagated through the respective gates to the preceding cells in time and depth. This two-dimensional lattice poses more challenges then a single RNN layer. In Turkoglu et al. (2021b), the magnitude of these gradients changes across the lattice is investigated. The analysis shows that common RNN cells are biased toward attenuating or amplifying the gradients and thus prone to destabilizing the training of deep recurrent networks. We introduce a new type of RNN cell called STAR: Stackable Recurrent Cell. This recurrent cell has been designed by conducting a mathematical analysis on the gradient magnitude as it propagates through the network during training, and it has been shown to have a more stable gradient propagation over the different layers when compared to other recurrent architectures.

2.4. Hierarchical multi-stage convolutional recurrent network

In order to achieve a hierarchical representation similar to multi-stage CNN classification networks (Wehrmann et al. 2018; Zhu and Bain 2017), but for image sequences instead of single images, we propose a multi-stage classification network consisting of multiple convolutional RNN layers. The network has N stages, each made of a two-layer convSTAR (Turkoglu et al. 2021b) architecture (Figure 2.6). Because STAR cells are designed for stable gradient back-propagation, we can construct and train deeper RNNs than with the more common LSTM or GRU.

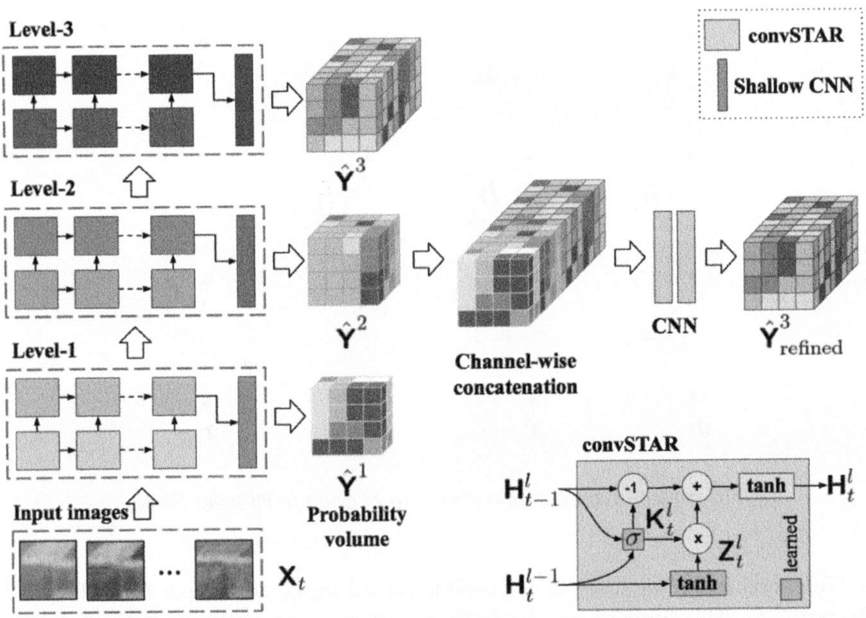

Figure 2.6. *Hierarchical, multi-stage convolutional STAR network (ms-convSTAR). For a color version of this figure, see www.iste.co.uk/mallet/multitemporal.zip*

More formally, a convSTAR cell at convolution layer l and at time t takes as input the hidden state tensor H_t^{l-1} of the previous layer $(l-1)$, where the input H_t^0 corresponds to the channels of the multi-spectral input image X_t. That tensor is first nonlinearly transformed and then linearly combined with the previous hidden state H_{t-1}^l. The weights of the linear combination are modulated by a gating variable K_t^l that depends on both inputs and controls the information flow:

$$K_t^l = \sigma(W_x * H_t^{l-1} + W_h * H_{t-1}^l + B_K) \quad [2.6]$$

$$Z_t^l = \tanh(W_z * H_t^{l-1} + B_z) \quad [2.7]$$

$$H_t^l = \tanh(H_{t-1}^l + K_t^l \circ (Z_t^l - H_{t-1}^l)), \quad [2.8]$$

where σ is the sigmoid nonlinearity, $\circ$ is the Hadamard (element-wise) product, and W and B are the trainable weight and bias tensors. The output tensors (hidden states) at stage n are fed as inputs to the next stage $(n+1)$. The final hidden state $(H_T^2)^n$ of each hierarchical level is fed to a conventional CNN classifier to obtain label probabilities $\hat{Y}^n$. All convolutional kernels are of size 3×3 and each convSTAR layer has 64 filters. A shallow 1-layer CNN is used to convert the final hidden state to a label. The network is trained with the cross-entropy (CE) loss, leading to the objective function

$$L = \sum_{n=1}^{N} \lambda_n CE(Y^n, \hat{Y}^n) = -\sum_{n=1}^{N} \sum_{c=1}^{C} \lambda_n Y_c^n \log(\hat{Y}_c^n), \quad [2.9]$$

where λ_n represents hyper-parameters that determine the relative influence of different hierarchy levels on the tree, with $\sum \lambda_n = 1$. By taking into account the loss at all levels of granularity, the network imposes the label hierarchy as a (soft) prior to guide the feature encoding. For example, the features of *apple orchards* and *pear orchards* should support an assignment to the coarser *orchard* label, too.

2.4.1. *Label refinement*

The model described so far embeds the label hierarchy in the network. However, by itself this does not guarantee consistency between the predictions at different stages. As an extreme example, the network could learn to simply ignore the input from the coarser hierarchy level and build independent classifiers for different granularities. The predicted labels at test time can then violate the parent-child relations of the hierarchy, e.g., label a pixel as *sunflower* and at the same time a *orchard*. To imprint the preference for coherent labels across the hierarchy levels, we add a *label refinement network* which is a simple CNN which refines the output of the multi-level ms-convSTAR network. The refinement stage takes all of the predictions $\{\hat{Y}^1 \ldots \hat{Y}^N\}$ from the individual network stages and produces a final prediction for the finest label granularity, $\hat{Y}^N_{\text{refined}}$, while modeling their interactions. More specifically, 3-dimensional probability volumes from three stages are concatenated in the channel dimension and are fed to the CNN (see Figure 2.6). Formally, it is defined as

$$\hat{Y}^N_{\text{refined}} = \hat{Y}^N + F(concat[\hat{Y}^1, \hat{Y}^2, ..., \hat{Y}^N]), \quad [2.10]$$

where F is a CNN depicted in Figure 2.7. The final loss function of the network is the weighted sum of the ms-convSTAR losses and the refinement loss, with a hyper-parameter γ for the relative influence of the refinement loss:

$$L = \sum_{n=1}^{N} \lambda_n CE(Y^n, \hat{Y}^n) + \gamma CE(Y^N, \hat{Y}^N_{\text{refined}}), \qquad [2.11]$$

where $\hat{Y}^N_{\text{refined}}$ represents the output of the refinement label layer and has exactly the same dimension as $\hat{Y}^N$ (see Figure 2.6).

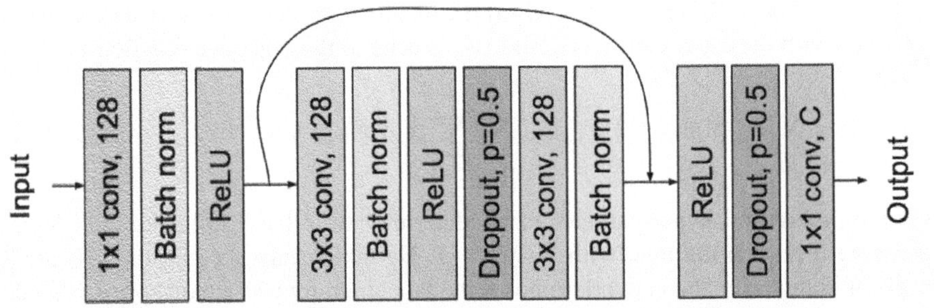

Figure 2.7. *Label refinement CNN architecture. For a color version of this figure, see www.iste.co.uk/mallet/multitemporal.zip*

Note that, empirically, we observed that refining coarser predictions does not affect the final model performance significantly, so in the experiment we run the refinement only for the finest label granularity.

2.4.2. *Implementation details*

We have implemented our network architecture in PyTorch. For all experiments, we use Adam (Kingma and Ba 2014) as optimizer with batch size 4 and run the training for 30 epochs. The learning rate is set to 0.001 at the start of the training and divided by 10 every 10 epochs. The model is regularized with weight decay of 0.0001. Gradient magnitudes are clipped to 5 to prevent the exploding gradients. The hyper-parameters of the loss function (equation [2.9]) are empirically determined and set to $\lambda_1 = 0.1$, $\lambda_2 = 0.3$ and $\lambda_3 = \gamma = 0.6$. Input image patches are flipped randomly with 66% chance during training for data augmentation. All source codes, trained models and the dataset are available online at: https://github.com/0zgur0/ms-convSTAR.

2.5. Experiment

2.5.1. *Dataset*

We create a new, hierarchical crop classification dataset named the *ZueriCrop* dataset. The dataset covers a 50 km × 48 km area in the Swiss cantons of Zurich and Thurgau between January 2019 and December 2019. It contains 28,000 Sentinel-2A image patches of size 24 pixels × 24 pixels, each observed 71 times over a period of 52 weeks (approximately uniformly distributed). In the experiment, we use only the four spectral channels available at 10 m resolution (red, green, blue, and near-infrared) since we observed that adding other channels does not significantly improve the final performance. We do not use any cloud detection method to discard patches with a high cloud cover because RNN architectures are robust to uninformative inputs (see section 2.5.7). The dataset contains 48 agricultural land cover classes, which can be seen as organized in a hierarchical structure as shown in Figure 2.9 and 116,000 individual agricultural fields. The average farm and field sizes in Switzerland are 21 hectares and 1.5 hectares, respectively, with approximately 70% of all Swiss agricultural land being grassland (Bundesamt für Statistik 2020), of which 12% are temporary grasslands used in rotation with other crops (Stumpf et al. 2020). This situation leads to a diverse set of crop classes with a highly skewed, imbalanced class distribution (see Figure 2.8 and Turkoglu et al. 2021a for more information about the dataset).

2.5.2. *Crop class hierarchy*

We organize the 48 crop classes of the *ZueriCrop* dataset into a hierarchy based on expert knowledge of the Swiss agricultural system (Figure 2.9). While our specific hierarchy here is hand-crafted and tailored for this specific dataset in Switzerland, it could be replaced with crop class hierarchies of other countries or scenarios without changing the method itself. Moreover, it could also possibly be learned from training data, which would be rather appealing from a technical point of view.

The first level of the class hierarchy was chosen with two goals in mind: (i) separate the main categories found in the Swiss agricultural landscape and (ii) group crop types according to their visual appearance in satellite images. For example, *field crops* are grown in crop rows on fields, a feature that can be picked up by remote sensing. Another example is *grassland*, the largest class in the dataset, which is a very heterogeneous class containing many different grassland types, mixtures and other land use scenarios. Their main feature is that they are permanently green with high biomass and generally low plant height. Other classes on the first level are as follows: *orchards*, *special crops* and *forest*. Orchards can usually be identified by their specific planting patterns. Special crops are a pool of marginally grown specialist crops (e.g. berries and herbs) and have a very diverse set of sub-classes.

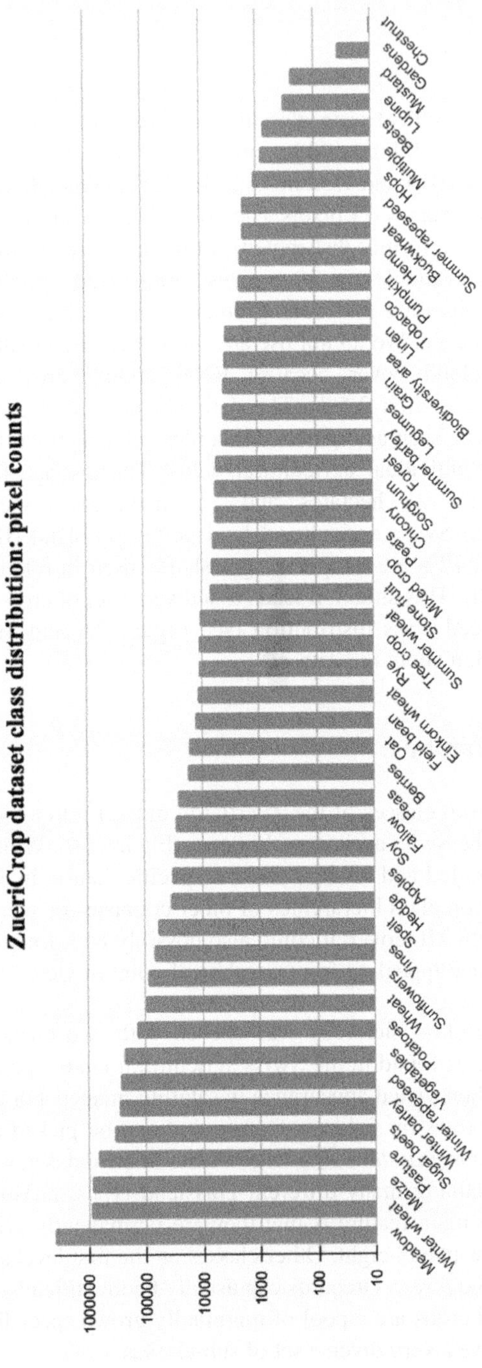

Figure 2.8. *Class distribution in the ZueriCrop dataset: Histogram of pixels. Note the logarithmic scale of the y-axis. For a color version of this figure, see www.iste.co.uk/mallet/multitemporal.zip*

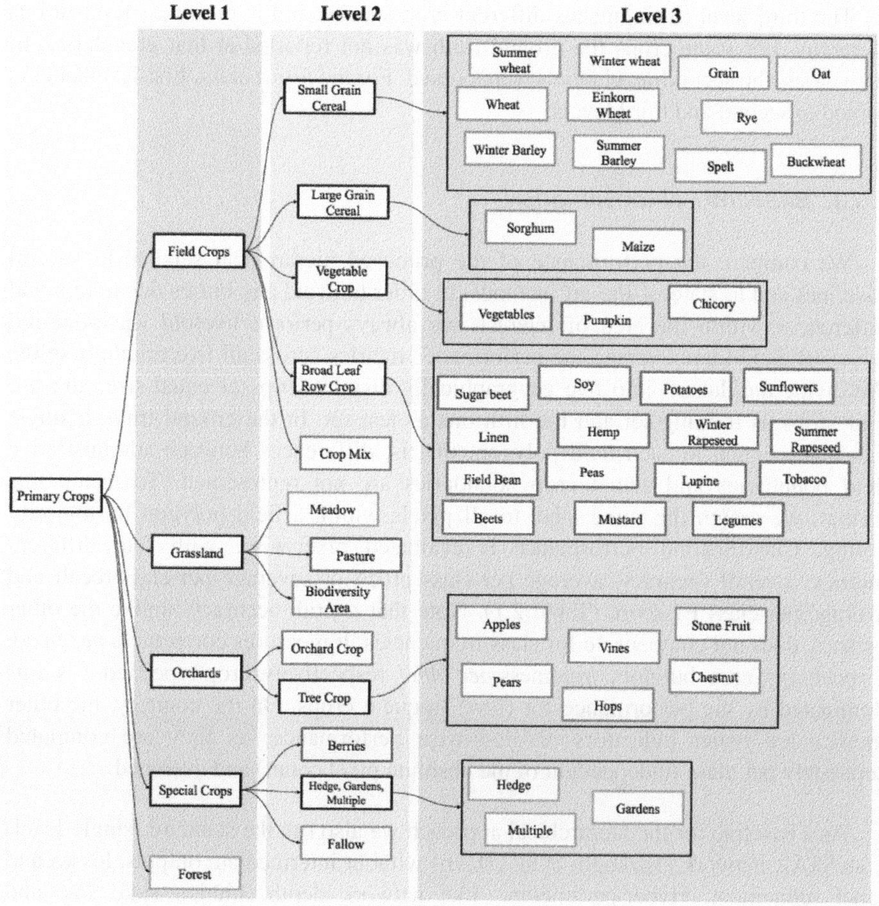

Figure 2.9. *The hierarchy of all crop classes of the* ZueriCrop *dataset. Black box indicates intermediate label levels, while color boxes indicate the finest granularity. For a color version of this figure, see www.iste.co.uk/mallet/multitemporal.zip*

The second level of the class hierarchy contains more refined versions of the preceding classes. Classes in the second level were selected to represent the plant family and agronomic use and practices in cultivation. All crops of the *small grain cereal* class are cultivated in a very similar manner in rows with little row spacing, similar plant seed density per square meter and similar erectile canopy structure with ears appearing at the top of the plant habitus (small grain as fruits). Similarly, the *broad leaf row crop* class contains dicotyledonous plant species that are cultivated in rows and possess, in contrast to grain crops, a horizontal leaf surface pattern.

The third level distinguishes different crop species and is the finest level in our hierarchy. For some crops the ground truth was not reported at that granularity. In such cases, the second-level labels were copied. For the *forest* class, first-level label is copied to second and third levels.

2.5.3. *Baselines and evaluation*

We compare the performance of the proposed hierarchical scheme to several baselines and to state-of-the-art methods. In order to avoid any biases due to regional differences within the area of interest, we always perform fivefold leave-one-out cross-validation and average the performance metrics across all five train/test splits. We divide the dataset into five geographically disjoint strips of equal size, and use four strips as training set and the fifth one as test set. In the ground truth, a single label is assigned to each field polygon, that is, differences between administrative field boundaries and actual crop boundaries are not represented. To match this format, we assign the same label to all pixels within a field polygon by majority voting. Classification performance is evaluated pixel-wise, with four different metrics: overall accuracy, average per-class precision, average per-class recall and average per-class F1-score (Table 2.1). Note that overall accuracy, unlike the other metrics, does not compensate for class frequencies. It measures correctness *per pixel*, respectively, area, but not correctness *per label*, respectively, crop type, and it is thus dominated by the performance for (few) frequent crops. On the contrary, the other metrics are better indicators of class-wise performance, as they are computed separately per class (independent of the absolute pixel count) and averaged.

As a baseline for the hierarchical approach we also run the standard (single-level) convSTAR network (Turkoglu et al. 2021b) without intermediate outputs, losses and label refinement. Hyper-parameters, like network depth, hidden state size and learning rate schedule are all set the same as for the hierarchical method. As further baseline experiments, we add data augmentation and class-balanced losses to the standard convSTAR network, which are alternative, widely used techniques to improve the performance on imbalanced datasets. When balancing the loss, the contribution of a training example is weighted inversely proportional to the class frequency, such that every class (in principle) contributes an equal share of the total loss. For data augmentation, training patches are sampled inversely proportional to the class frequency to achieve the same effect. The class frequencies are found by counting pixels over the entire training set.

The use of a class-balanced loss function causes a subtle difference, as it affects the gradient magnitude reducing the overall learning rate, which can potentially harm the training. Therefore, we also test another variant, denoted as Class-balance loss-2, where, after including the weight of the loss function for each class, we compute the median effective learning rate and set it to match the default learning rate. Moreover, we also evaluate the recently proposed, more robust re-weighting scheme of Cui et al. (2019). This state-of-the-art technique uses the effective number of samples for each class to rebalance the loss. As the method has an additional, empirically chosen hyper-parameter β, we test different settings $\beta \in \{0.99, 0.999, 0.9999\}$.

The baselines described so far serve to isolate the impact of the proposed hierarchical labeling scheme, and to that end use the same convSTAR backbone. As an additional external reference, we also compare against random forest and other state-of-the-art deep learning methods that have been developed for multi-temporal crop classification. Rußwurm and Körner (2017) use LSTM, Pelletier et al. (2019) propose a temporal convolutional network and Rußwurm et al. (2019); and Rußwurm and Körner (2020) design a Transformer network. All four methods process pixels individually, unlike our approach that aggregates context information from image patches with convolutions. Convolutions and recurrence are used separately by Rustowicz et al. (2019), who extract features from the images with a convolutional architecture in the style of U-Net (Ronneberger et al. 2015) and apply convolutional LSTM to the resulting feature vectors to represent their temporal evolution. Also, U-Net (Ronneberger et al. 2015) itself is compared since it has become a standard architecture for image segmentation and has been used in remote sensing (e.g. Stoian et al. 2019; Flood et al. 2019). Early temporal fusion, spectral channels of different timestamps which are concatenated in the channel direction, is applied to deal with multi-temporal data. An integrated, convolutional variant of GRU has already been used by Rußwurm and Körner (2018a). This work is, from a technical point of view, similar to our convSTAR baseline, except that it uses a different form of recurrence and employs a bi-directional approach.

2.5.4. *Performance comparison*

We find that among the baselines, simple data augmentation performs best, but all baselines are clearly outperformed by the proposed, hierarchical ms-convSTAR, on all performance metrics (see Table 2.1). Most significantly, there are significant improvements on those metrics that compensate for class frequencies and average measure per-class performance. Our proposed ms-convSTAR increases mean class precision by $>$11 percentage points, and mean class recall by $>$10 percentage points. Accordingly, the F1-score (their harmonic mean) increases by $>$11 percentage points. These results indicate that our method improves in particular the classification of less frequent classes, which was the initial motivation for using the crop label hierarchy and developing ms-convSTAR. Data augmentation, where we oversample the minority classes during training (i.e. patches with rare classes are sampled inversely proportional to their frequencies), improves the F1-score by 2.9 percentage points compared to the baseline convSTAR. However, it degrades the overall accuracy significantly by 2.3 percentage points, because oversampling rare classes induces a global bias toward those classes and degrades performance for the dominant classes. Adding a standard class-balanced loss to the baseline convSTAR approach significantly decreases performance across all measures, whereas the state-of-the-art class-balanced loss technique (Cui et al. 2019) can improve the F1-score a little (by 2.2 percentage points with $\beta = 0.9999$) but slightly reduces the overall accuracy (by 0.2 percentage point). In summary, even though a class-balanced loss or data augmentation brings a mild improvement for minority classes, the gains are not very big, and they tend to reduce the overall performance in return. In contrast, our proposed ms-convSTAR greatly boosts performance for mean precision, mean recall and F1-score, while improving overall accuracy, too.

Method	Precision (%)	Recall (%)	F1 (%)	Accuracy (%)
convSTAR	40.2	37.3	37.2	87.3
+ Data augmentation	48.3	39.3	41.1	85.0
+ Class-balanced loss	26.9	32.7	28.2	75.6
+ Class-balanced loss-2	26.5	31.7	27.3	74.0
+ Cui 2019, $\beta = 0.99$	42.2	37.5	36.5	87.3
+ Cui 2019, $\beta = 0.999$	39.4	35.7	35.6	87.4
+ Cui 2019, $\beta = 0.9999$	43.3	39.1	39.4	87.1
ms-convSTAR w/o LR	59.3	47.1	49.8	87.6
ms-convSTAR	**60.1**	**49.8**	**52.4**	**88.0**

Table 2.1. *Performance comparison between ms-convSTAR (bottom row) and nonhierarchical baseline methods. We compare against standard (1-level) convSTAR (top row) as well as further baselines that extend convSTAR with different techniques intended to compensate class imbalance. Precision, recall and F1-score are mean values over all classes. All numbers are averaged over 5 cross-validation folds. The best score for each metric is shown in* ***bold***

Figure 2.10 illustrates the difference between the confusion matrix of results achieved with proposed ms-convSTAR (Figure 2.15) and with its baseline counterpart convSTAR. ms-convSTAR improves the performance for many less frequent classes such as *stone fruit*, *legumes* and *tobacco*, while preserving the same performance level for frequent classes like *meadow* or *maize*. We note that for exceedingly rare classes, the performance does not improve. These classes have too few pixels (typically $< 1,000$) to be represented well. Moreover, they are often only present in some of the five cross-validation stripes, and as a result may not appear in either the training set or the test set. For completeness, we nevertheless leave those classes in the dataset and report performance numbers.

In Table 2.2, we compare the proposed ms-convSTAR (bottom row) to a number of state-of-the-art methods for crop classification from image time series. ms-convSTAR significantly improves performance across all measures. Again the gains are mostly due to better classification of rare classes, as indicated by an improvement of $>$ 9.9 percentage points in F1-score. Model implementations are taken from official codebases, and similar hyperparameters with the proposed method are used as the hidden size, the initial learning rate, the number of epochs, etc. For other model-specific hyperparameters (e.g. number of layers in TCN (Pelletier et al. 2019), number of heads in Transformer (Rußwurm et al. 2019)), values from original papers are used (see Table 2.7 for more details). Random forest parameter settings follow Rußwurm and Körner (2020). To achieve better classification performance for the random forest, we augmented the raw input reflectance with the NDVI. The NDVI is computed for each time step and concatenated to the corresponding input reflectance as an additional channel. The training dataset is randomly undersampled in such a way that the number of per class samples is limited to 10K ($\approx$ the median of the number of per class samples) as done in Rußwurm and Körner (2020).

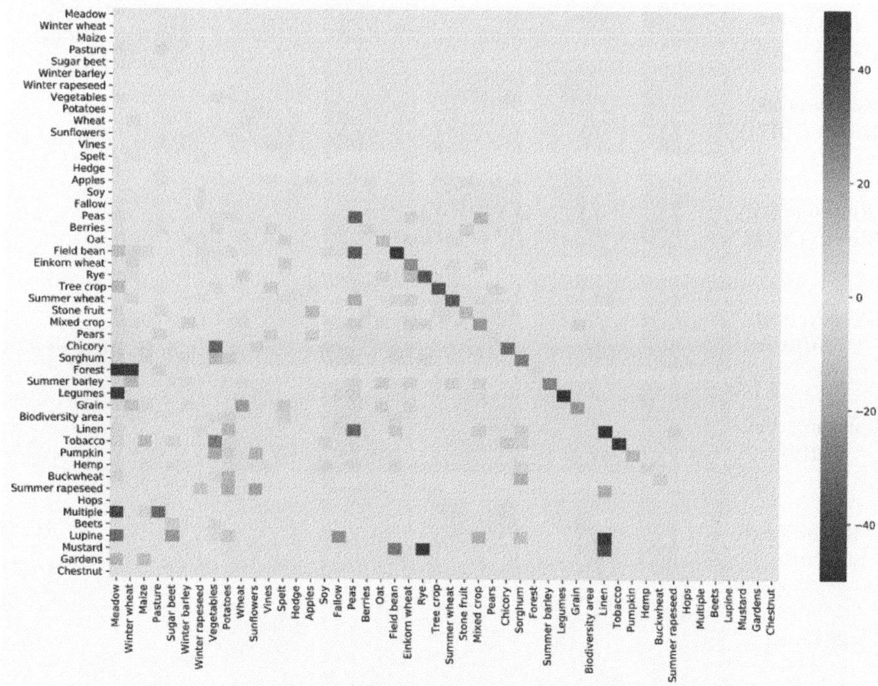

Figure 2.10. *Benefit of ms-convSTAR: difference between normalized confusion matrices. Averaged over five cross-validation folds. Green color denotes margins in favor of ms-convSTAR (higher correctness on the diagonal, respectively, lower confusion off the diagonal), and brown color denotes margins in favor of the baseline counterpart (convSTAR). For a color version of this figure, see www.iste.co.uk/mallet/multitemporal.zip*

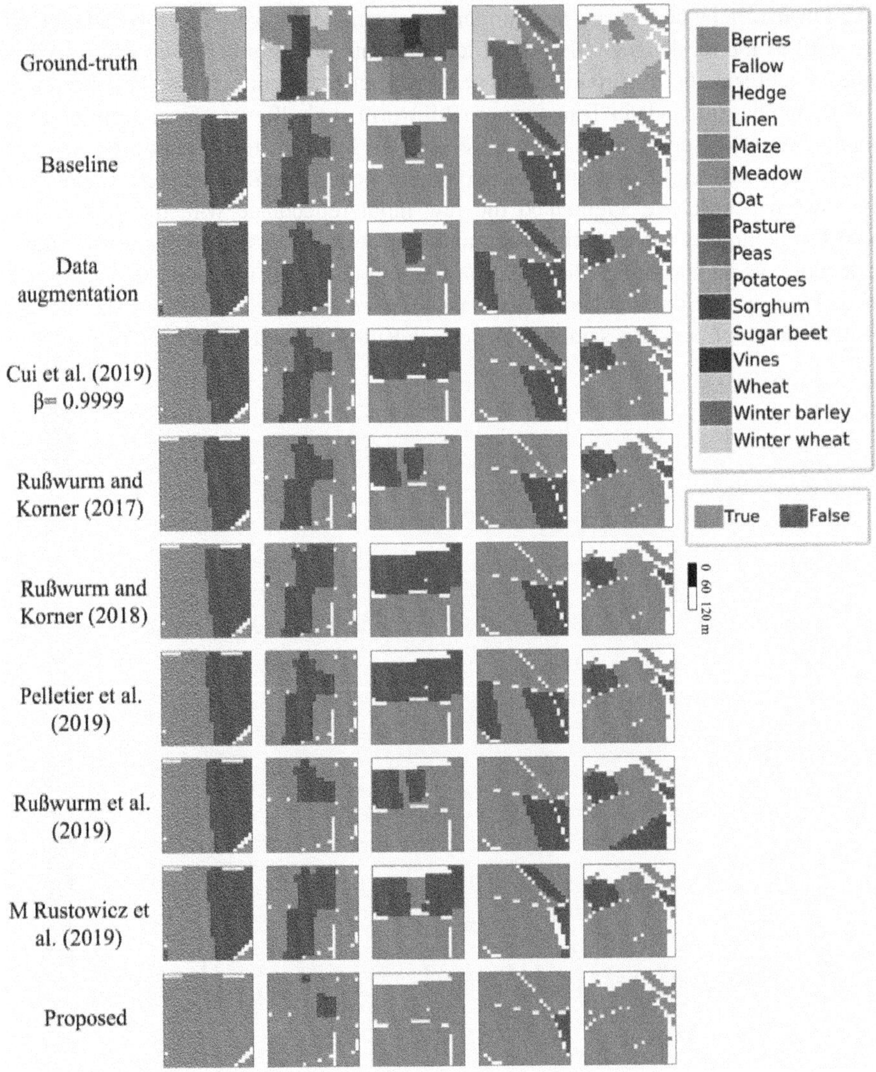

Figure 2.11. *Visual results for five different samples (train-test fold 2). Green color indicates correct classification, red color indicates misclassified pixels (respectively, fields). For a color version of this figure, see www.iste.co.uk/mallet/multitemporal.zip*

We show qualitative comparisons for several output samples in Figure 2.11. Our new method does what it is designed for: it correctly predicts rare classes such as *linen* or *sorghum*, where all other methods fail. Another qualitative comparison to the closest competitor (Rußwurm and Körner 2018a) is shown in Figure 2.16 for a larger region. We point out that both approaches, ms-convSTAR and (Rußwurm and Körner 2018a), typically misclassify the same fields. But the number of mistakes is significantly smaller with ms-convSTAR.

Method	Precision(%)	Recall(%)	F1(%)	Accuracy(%)
Random Forest*	46.4	<u>40.7</u>	38.9	78.8
LSTM (Rußwurm and Körner 2017)	37.7	27.9	29.2	84.1
TCN (Pelletier et al. 2019)	39.2	27.7	29.3	83.5
Transformer (Rußwurm et al. 2019)	<u>56.8</u>	38.4	42.3	85.4
2D-CNN (U-Net)	34.6	25.7	26.7	82.2
U-Net+convLSTM (Rustowicz et al. 2019)	47.7	32.8	35.2	85.0
Bi-convGRU (Rußwurm and Körner 2018a)	55.0	39.6	<u>42.5</u>	<u>86.4</u>
ms-convSTAR	**60.1**	**49.8**	**52.4**	**88.0**

Table 2.2. *Performance comparison of ms-convSTAR (bottom row) with state-of-the-art methods. Precision, recall and F1-score are mean values over all classes. All numbers are averaged over 5 cross-validation folds. The best score for each metric is shown in* ***bold*** *and the second best is <u>underlined</u>. *Random Forest is trained with a balanced dataset by undersampling the majority classes which leads to improved class-wise performance*

2.5.5. *Ablation study*

As an ablation study, we first evaluate the performance gain due to the label refinement module by also running our hierarchical ms-convSTAR model without the label refinement module (denoted as ms-convSTAR w/o LR). Results in Table 2.1 show that the CNN-based label refinement consistently improves all performance metrics, particularly per-class performance.

We then investigate the importance of the RNN cell type used in the network (Table 2.3). We construct the hierarchical multi-stage network (ms-convX) and its nonhierarchical baseline counterparts (convX) using the most popular RNN cells: LSTM and GRU. We experimentally evaluate three-layer and six-layer convRNN versions and corresponding ms-convRNN versions (Table 2.3). While convLSTM and convGRU produce acceptable results for the three-layer version without hierarchy, they perform very poorly for the deeper, six-layer versions. As has been shown in Turkoglu et al. (2021b), both cell types LSTM and GRU suffer from gradient vanishing problems, which get more severe with deeper architectures. Using our proposed hierarchical approach fixes these gradient problems and leads to substantial improvements with respect to their baseline versions without hierarchy. Unlike convLSTM and convGRU, convSTAR does not suffer from gradient issues at any point and performs well for all settings, outperforming all existing approaches in overall accuracy (Table 2.2). Combining convSTAR with the proposed hierarchical approach leads to superior performance with a clear margin above all other methods on the *ZueriCrop* dataset.

	Method	Precision (%)	Recall (%)	F1 (%)	Accuracy (%)
3 layers	convLSTM	31.5	27.9	28.1	84.6
	ms-convLSTM	37.7	30.1	31.1	83.9
	convGRU	43.1	38.2	38.6	87.2
	ms-convGRU	49.8	38.6	40.8	85.8
	convSTAR	48.3	42.6	43.5	87.8
	ms-convSTAR	54.6	42.7	45.3	86.8
6 layers	convLSTM	1.1	2.3	1.4	47.2
	ms-convLSTM	40.5	33.6	34.7	85.2
	convGRU	15.8	15.7	15.0	71.1
	ms-convGRU	52.2	42.9	44.7	86.9
	convSTAR	40.2	37.3	37.2	87.3
	ms-convSTAR	**60.1**	**49.8**	**52.4**	**88.0**

Table 2.3. *Performance comparison of multi-stage convRNNs (ms-convRNNs). All numbers are averaged over 5 cross-validation folds. The best score for each metric is shown in* **bold** *and the second best is underlined*

2.5.6. *Simultaneous multi-level classification*

An attractive property of ms-convSTAR is that it simultaneously predicts per-pixel labels at three different levels in the class hierarchy. Performance of proposed ms-convSTAR at different levels of granularity is given in Table 2.4. As we expected, the accuracy of the model improves while going to the coarser levels.

Level	Precision (%)	Recall (%)	F1 (%)	Accuracy (%)
1st	80.3	52.9	57.0	96.3
2nd	73.0	51.0	54.5	89.2
3rd	60.1	49.8	52.4	88.0

Table 2.4. *Performance of proposed ms-convSTAR at different levels of granularity. All numbers are averaged over 5 cross-validation folds*

Such a feature also makes it possible to chose the granularity of the labels according to the output (confidence) scores. In this way, we can produce reliable maps where most pixels are assigned a label and all labels have sufficiently high confidence – at the cost of only assigning a coarse-grained label to some pixels were the fine-grained answer is too uncertain. For instance, if the model is uncertain about deciding for either *apple* or *pear* orchard, the coarser label *orchard* can be assigned with much higher confidence. In a number of applications, this coarser level of annotation is good enough. An example is the summary statistics computed by the Swiss federal administration, where a large degree of coverage with coarse labels is critical. Moreover, coarse but correct and (nearly) complete answers are a lot more useful for downstream GIS processing: for instance, mapping orchards is much easier if the user knows which polygons have not yet received a fine-grained label and must be checked.

To quantify this effect, we measure the area that is classified above a given confidence score and compare ms-convSTAR models with different numbers of hierarchy levels. We set a prediction confidence value p, and, for each pixel switch to the next-coarser label if the confidence level p is not achieved. Figure 2.12 shows the overall accuracy and the coverage rate (the proportion of pixels with a confident label at any granularity) for two confidence levels, $p = 0.6$ and $p = 0.9$. As can be seen if we wish to retain only confident predictions and set $p = 0.9$, the three-level hierarchy brings a large benefit. While accuracy (computed over covered areas) stays the same, coverage improves by $\approx 0.20 - 0.25$, although, obviously, some areas are only predicted at a coarse label resolution.

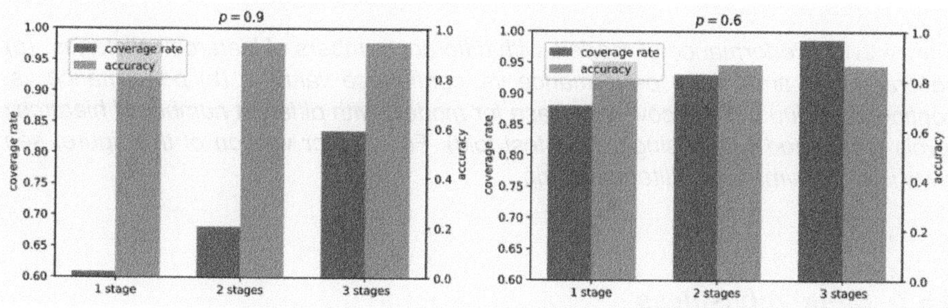

$p = 0.9$ $p = 0.6$

Figure 2.12. *Performance of models with different numbers of hierarchical stages at fixed confidence values: (a) $p = 0.9$, (b) $p = 0.6$. The fraction of classified pixels (coverage rate, blue) grows with an increasing number N of hierarchy levels (computed with a single train-test fold). For a color version of this figure, see www.iste.co.uk/mallet/multitemporal.zip*

Figure 2.13(a) shows how the coverage rate and the model performance change as functions of the confidence value p. We can see that, as expected, when decreasing p, the coverage increases, whereas the accuracy decreases. Figure 2.13(b) shows how the model performance improves with increasing p in the full coverage case (every pixel is classified, regardless of the p value, at the coarsest available level). The full, three-stage model boosts performance significantly with increasing p; in contrast, a basic one-stage model does not show this behavior. This analysis suggests that the imposed label hierarchy is actually meaningful for the problem, in the sense that the coarser classes are indeed easier to discriminate than their finer sub-classes. Moreover, the results once more support our hierarchical multi-level scheme: additional hierarchy levels consistently help to predict more pixels correctly and confidently, for any value of p. An interesting future research question in this context could be how to add new crop classes incrementally and how to update the model hierarchy accordingly.

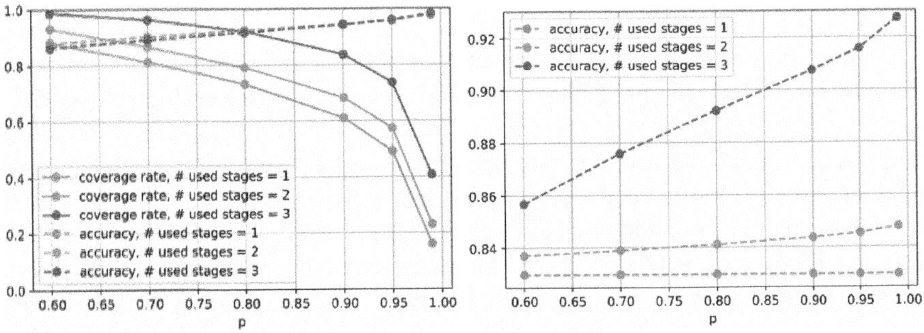

Figure 2.13. *Performance of models with different numbers of hierarchical stages: (a) coverage rate and model performance vs. confidence value p; (b) performance vs. confidence value p in full coverage case for models with different number of hierarchy levels (computed with a single train-test fold). For a color version of this figure, see www.iste.co.uk/mallet/multitemporal.zip*

2.5.7. *Cloud robustness*

RNNs normally exhibit good robustness against missing data due to clouds in the context of crop mapping (see Rußwurm and Körner (2018b), for instance). We have conducted two experiments to further analyze robustness against clouds. In Table 2.5, we summarize the results of an experiment where we have trained the proposed model with the cloud mask as an additional input channel, so as to explicitly give the model the information which pixels are obscured by clouds. The performance is practically the same, suggesting that the model learns to detect the presence of clouds, such that the additional input brings no benefit. In Figure 2.14, we look at the issue from another angle: we apply a threshold to the Sen2Cor cloud score to determine how many images are cloudy in the time series for every pixel and quantify how the varying number of cloud-free observations influences the performance. To that end, we rank the classified pixels according to the number of cloudy observations in the time series. As can be seen, the accuracy remains almost constant as pixels are increasingly affected by cloud cover.

Method	Precision (%)	Recall (%)	F1 (%)	Accuracy (%)
ms-convSTAR	59.8	**49.7**	52.1	**88.0**
ms-convSTAR w/ CM	**60.2**	48.8	**52.2**	87.9

Table 2.5. *Performance comparison: the proposed model vs. the same model with the cloud masks (CM) as additional input. Numbers are averaged over 5 cross-validation folds. The best score for each metric is shown in* ***bold***

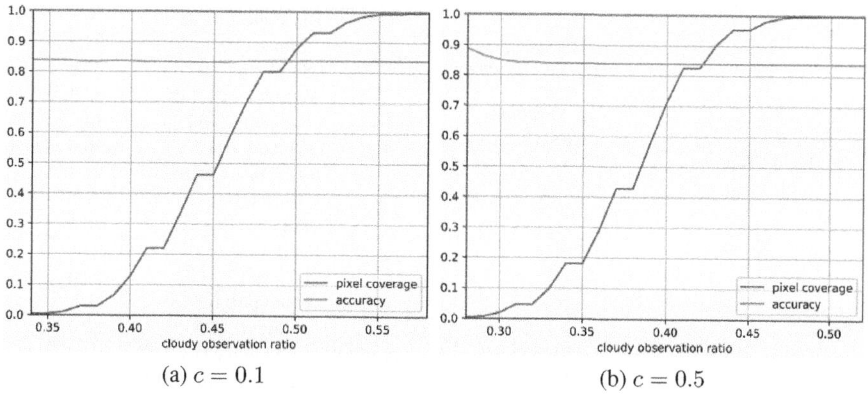

Figure 2.14. *Robustness against clouds. We plot the classification accuracy against the proportion of cloud-free observations. Along the x-axis, pixels are ordered according to how often they are obscured by clouds. The blue curves depict the cumulative number of pixels with at most a certain fraction of cloudy views. The orange curves show the accuracy of the model evaluated over those pixels. The test was repeated with two different thresholds c for the Sen2Cor cloud scores. For a color version of this figure, see www.iste.co.uk/mallet/multitemporal.zip*

2.6. Summary and future outlook

In this chapter, we have given an overview of different deep learning concepts for sequence data analysis. We presented a hierarchical classification approach for multi-temporal crop classification from satellite images and demonstrated how the design of the STAR cell can help training deeper RNN models. Our ms-convSTAR method is a multi-stage convolutional recurrent neural network that leverages an explicit, hierarchical tree structure of the labels. Besides classifying the input data simultaneously at multiple interdependent hierarchical levels, the method also features a CNN-based label-refinement component to merge information across the hierarchy. In the experiments on *ZueriCrop*, ms-convSTAR has shown improved per-class performance and outperforms competing state-of-the-art methods. In particular, it greatly improves the classification of many rare classes.

Level-1	Acc (%)	Level-2	Acc (%)	Level-3	Acc (%)
Field Crops (41.39%)	98.2	Small Grain Cereal (16.89%)	97.7	Summer Wheat (0.13%)	38.0
				Winter Wheat (10.77%)	95.7
				Wheat (1.20%)	39.4
				Einkorn Wheat (0.14%)	49.8
				Summer Barley (0.05%)	26.1
				Winter Barley (3.49%)	96.4
				Grain (0.05%)	21.7
				Rye (0.14%)	68.1
				Spelt (0.70%)	84.0
				Oat (0.21%)	76.4
				Buckwheat (0.02%)	7.0
		Large Grain Cereal (10.37%)	94.3	Maize (10.30%)	94.7
				Sorghum (0.07%)	27.0
		Vegetable Crop (2.85%)	83.4	Vegetables (2.76%)	82.3
				Pumpkin (0.03%)	14.2
				Chicory (0.07%)	30.8
		Broad Leaf Row Crop (11.20%)	95.1	Sugar Beet (4.26%)	97.2
				Beets (0.01%)	0.0
				Potatoes (1.65%)	86.6
				Sunflowers (1.05%)	92.8
				Linen (0.04%)	54.4
				Hemp (0.02%)	3.1
				Soy (0.37%)	87.4
				Winter Rapeseed (3.21%)	97.6
				Summer Rapeseed (0.02%)	0.0
				Field Bean (0.15%)	88.3
				Peas (0.32%)	82.4
				Lupine (< 0.01%)	0.0
				Tobacco (0.03%)	48.5
				Mustard (< 0.01%)	0.0
				Legumes (0.05%)	78.0
		Crop Mix (0.08%)	16.7	Crop Mix (0.08%)	39.7
Grassland (56.00%)	97.9	Meadow (47.76%)	95.3	Meadow (47.76%)	94.9
		Pasture (8.20%)	37.4	Pasture (8.20%)	42.4
		Biodiversity A. (0.04%)	0.0	Biodiversity A. (0.04%)	0.0
Orchards (1.54%)	49.6	Orchard Crop (1.41%)	76.3	Apples (0.40%)	69.3
				Pears (0.07%)	2.3
				Vines (0.83%)	82.2
				Hops (0.01%)	0.0
				Stone Fruit (0.09%)	24.1
				Chestnut (< 0.01%)	0.0
		Tree Crop (0.13%)	20.9	Tree Crop (0.13%)	37.2
Special Crops	8.3	Hedge, Gardens, Multiple (0.43%)	1.0	Hedge (0.42%)	1.5
				Gardens (< 0.01%)	0.0
				Multiple (0.01%)	0.0
		Berries (0.22%)	18.8	Berries (0.22%)	25.8
		Fallow (0.36%)	65.8	Fallow (0.36%)	70.3
Forest (0.07%)	0.0	Forest (0.07%)	0.0	Forest (0.07%)	1.6

Table 2.6. *Class-wise performance of proposed ms-convSTAR at different levels of granularity. All numbers are averaged over 5 cross-validation folds. Class frequencies are given in parenthesis*

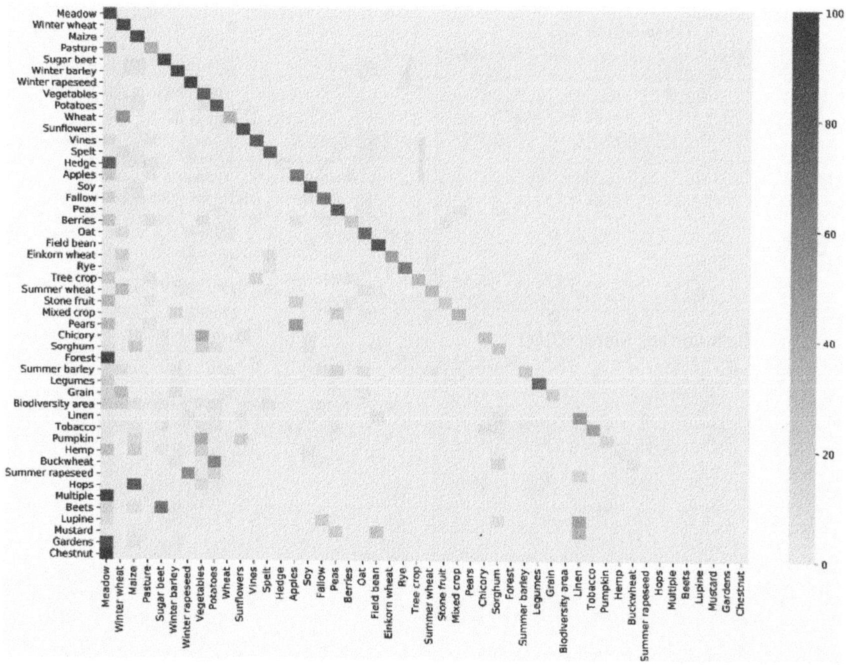

Figure 2.15. *Confusion matrix for the proposed method. Averaged over five cross-validation folds. Rows show true labels and columns show predicted labels. The sum of each row is normalized to* 1. *For a color version of this figure, see: http://www.iste.co.uk/mallet/multitemporal.zip*

Method	Hidden	Layer	Kernel	Batch	Note
Random Forest					max_depth:60 max_features:auto min_sample_leaf:1 min_sample_split:3 n_estimators:1000
LSTM	128	1		576	
TCN	64	3	5	576	dropout: 0.5
Transformer	64	4		576	head: 4
Bi-convGRU	64	1	3x3	4	
U-Net		12	3x3	4	filters: 64, 64, 128, 128, 256, 256
U-Net+convLSTM	256	12+1	3x3	4	filters: 64, 64, 128, 128, 256, 256

Table 2.7. *Parameters of baseline methods*

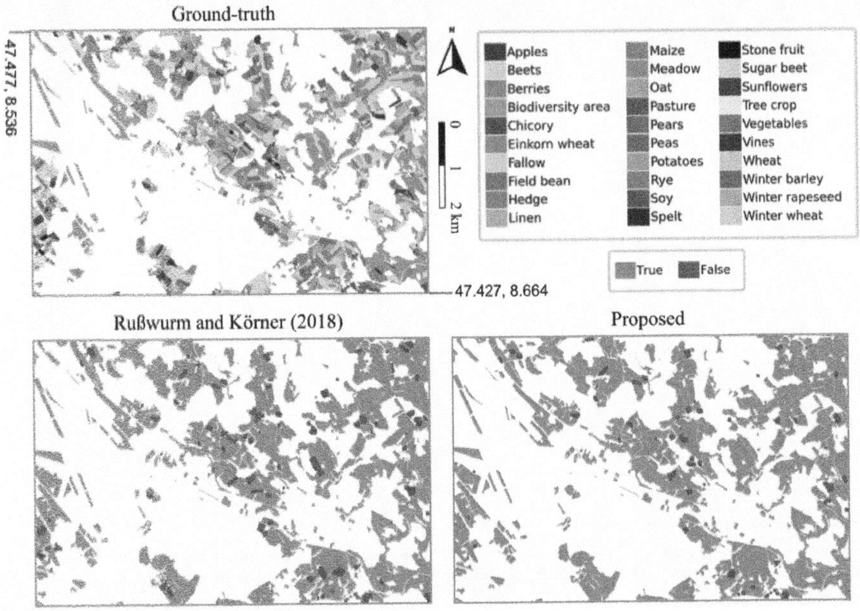

Figure 2.16. *Qualitative comparison: ms-convSTAR versus the best-performing alternative on the* ZueriCrop *data. For a color version of this figure, see: http://www.iste.co.uk/mallet/multitemporal.zip*

The label hierarchy in this work is specific to Switzerland, but different hierarchies can be defined for other geographic regions. In that context, it should be noted that hard-coding a particular label hierarchy into the network architecture is an inherent part of the current design, but lacks flexibility when moving to other environments. In future work, it may be interesting to explore the possibility to learn the label hierarchy in a data-driven fashion, for instance, with unsupervised learning techniques or in conjunction with the network training.

2.7. References

Amit, Y. and Geman, D. (1997). Shape quantization and recognition with randomized trees. *Neural Computation*, 9(7), 1545–1588.

Anderegg, J., Yu, K., Aasen, H., Walter, A., Liebisch, F., Hund, A. (2020). Spectral vegetation indices to track senescence dynamics in diverse wheat germplasm. *Frontiers in Plant Science*, 10, 1749.

Bahdanau, D., Cho, K., Bengio, Y. (2015). Neural machine translation by jointly learning to align and translate. In *ICLR, International Conference on Learning Representations*, San Diego, California.

Bai, S., Kolter, J.Z., Koltun, V. (2018). An empirical evaluation of generic convolutional and recurrent networks for sequence modeling. *arXiv preprint arXiv:1803.01271.*

Belgiu, M. and Csillik, O. (2018). Sentinel-2 cropland mapping using pixel-based and object-based time-weighted dynamic time warping analysis. *Remote Sensing of Environment*, 204, 509–523.

Bengio, Y. (2009). Learning deep architectures for AI. *Foundations and Trends® in Machine Learning*, 2(1), 1–127.

Bundesamt für Statistik (2020). *Landwirtschaft und Ernährung – Taschenstatistik 2020.* Bundesamt für Statistik (BFS), Bern.

Chang, B., Chen, M., Haber, E., Chi, E.H. (2019). AntisymmetricRNN: A dynamical system view on recurrent neural networks. *arXiv preprint arXiv:1902.09689.*

Chawla, N.V., Bowyer, K.W., Hall, L.O., Kegelmeyer, W.P. (2002). SMOTE: Synthetic minority over-sampling technique. *Journal of Artificial Intelligence Research*, 16, 321–357.

Chen, Y., Su, W., Li, J., Sun, Z. (2009). Hierarchical object oriented classification using very high resolution imagery and LIDAR data over urban areas. *Advances in Space Research*, 43(7), 1101–1110.

Chen, H.-Y., Tsai, L.-H., Chang, S.-C., Pan, J.-Y., Chen, Y.-T., Wei, W., Juan, D.-C. (2019). Learning with hierarchical complement objective. *arXiv preprint arXiv:1911.07257.*

Chung, J., Gulcehre, C., Cho, K., Bengio, Y. (2014). Empirical evaluation of gated recurrent neural networks on sequence modeling. In *NIPS Workshop*, Montreal.

Chung, J., Gulcehre, C., Cho, K., Bengio, Y. (2015). Gated feedback recurrent neural networks. In *International Conference of Machine Learning*, Lille.

Conrad, C., Fritsch, S., Zeidler, J., Rücker, G., Dech, S. (2010). Per-field irrigated crop classification in arid central Asia using spot and ASTER data. *Remote Sensing*, 2(4), 1035–1056.

Conrad, C., Dech, S., Dubovyk, O., Fritsch, S., Klein, D., Löw, F., Schorcht, G., Zeidler, J. (2014). Derivation of temporal windows for accurate crop discrimination in heterogeneous croplands of Uzbekistan using multitemporal rapideye images. *Computers and Electronics in Agriculture*, 103, 63–74.

Cui, Y., Jia, M., Lin, T.-Y., Song, Y., Belongie, S. (2019). Class-balanced loss based on effective number of samples. In *Proceedings of the IEEE Conference on Computer Vision and Pattern Recognition*, Long Beach, California.

Demirkan, D.Ç., Koz, A., Düzgün, H.S. (2020). Hierarchical classification of sentinel 2-a images for land use and land cover mapping and its use for the CORINE system. *Journal of Applied Remote Sensing*, 14(2), 026524.

Dise, N.B., Ashmore, M., Belyazid, S., Bleeker, A., Bobbink, R., Vries, W.D., Erisman, J.W., Spranger, T., Stevens, C.J., Berg, L.V.D. (2011). *Nitrogen as a Threat to European Terrestrial Biodiversity*. Cambridge University Press, Cambridge.

Dong, Q., Gong, S., Zhu, X. (2018). Imbalanced deep learning by minority class incremental rectification. *IEEE Transactions on Pattern Analysis and Machine Intelligence*, 41(6), 1367–1381.

Douzas, G. and Bacao, F. (2018). Effective data generation for imbalanced learning using conditional generative adversarial networks. *Expert Systems with Applications*, 91, 464–471.

Farha, Y.A. and Gall, J. (2019). MS-TCN: Multi-stage temporal convolutional network for action segmentation. In *Proceedings of the IEEE Conference on Computer Vision and Pattern Recognition*, Institute for Electrical and Electronics Engineers (IEEE), New York.

Finger, R. and Lehmann, N. (2012). Policy reforms to promote efficient and sustainable water use in Swiss agriculture. *Water Policy*, 14(5), 887–901.

Flood, N., Watson, F., Collett, L. (2019). Using a U-net convolutional neural network to map woody vegetation extent from high resolution satellite imagery across Queensland, Australia. *International Journal of Applied Earth Observation and Geoinformation*, 82, 101897.

Foerster, S., Kaden, K., Foerster, M., Itzerott, S. (2012). Crop type mapping using spectral–temporal profiles and phenological information. *Computers and Electronics in Agriculture*, 89, 30–40.

Gangaputra, S. and Geman, D. (2006). A design principle for coarse-to-fine classification. In *2006 IEEE Computer Society Conference on Computer Vision and Pattern Recognition (CVPR'06)*, Institute for Electrical and Electronics Engineers (IEEE), New York.

Gbodjo, Y.J.E., Ienco, D., Leroux, L., Interdonato, R., Gaetano, R., Ndao, B. (2020). Object-based multi-temporal and multi-source land cover mapping leveraging hierarchical class relationships. *Remote Sensing*, 12(17), 2814.

Giordano, S., Bailly, S., Landrieu, L., Chehata, N. (2020). Improved crop classification with rotation knowledge using sentinel-1 and -2 time series. *Photogrammetric Engineering & Remote Sensing*, 86(7), 431–441.

Goel, A., Banerjee, B., Pižurica, A. (2018). Hierarchical metric learning for optical remote sensing scene categorization. *IEEE Geoscience and Remote Sensing Letters*, 16(6), 952–956.

Gómez, C., White, J.C., Wulder, M.A. (2016). Optical remotely sensed time series data for land cover classification: A review. *ISPRS Journal of Photogrammetry and Remote Sensing*, 116, 55–72.

He, H. and Garcia, E.A. (2009). Learning from imbalanced data. *IEEE Transactions on Knowledge and Data Engineering*, 21(9), 1263–1284.

Herzog, F., Prasuhn, V., Spiess, E., Richner, W. (2008). Environmental cross-compliance mitigates nitrogen and phosphorus pollution from Swiss agriculture. *Environmental Science and Policy*, 11(7), 655–668.

Heupel, K., Spengler, D., Itzerott, S. (2018). A progressive crop-type classification using multitemporal remote sensing data and phenological information. *PFG – Journal of Photogrammetry, Remote Sensing and Geoinformation Science*, 86(2), 53–69.

Hochreiter, S. and Schmidhuber, J. (1997). Long short-term memory. *Neural Computation*, 9(8), 1735–1780.

Huang, C., Li, Y., Change Loy, C., Tang, X. (2016). Learning deep representation for imbalanced classification. In *Proceedings of the IEEE Conference on Computer Vision and Pattern Recognition*, Institute for Electrical and Electronics Engineers (IEEE), New York.

Inglada, J., Arias, M., Tardy, B., Hagolle, O., Valero, S., Morin, D., Dedieu, G., Sepulcre, G., Bontemps, S., Defourny, P. et al. (2015). Assessment of an operational system for crop type map production using high temporal and spatial resolution satellite optical imagery. *Remote Sensing*, 7(9), 12356–12379.

Interdonato, R., Ienco, D., Gaetano, R., Ose, K. (2019). DuPLO: A dual view point deep learning architecture for time series classification. *ISPRS Journal of Photogrammetry and Remote Sensing*, 149, 91–104.

Jiao, L., Sun, W., Yang, G., Ren, G., Liu, Y. (2019). A hierarchical classification framework of satellite multispectral/hyperspectral images for mapping coastal wetlands. *Remote Sensing*, 11(19), 2238.

Khan, S.H., Hayat, M., Bennamoun, M., Sohel, F.A., Togneri, R. (2017). Cost-sensitive learning of deep feature representations from imbalanced data. *IEEE Transactions on Neural Networks and Learning Systems*, 29(8), 3573–3587.

Khan, S.H., Hayat, M., Zamir, S.W., Shen, J., Shao, L. (2019). Striking the right balance with uncertainty. In *Proceedings of the IEEE Conference on Computer Vision and Pattern Recognition*, Institute for Electrical and Electronics Engineers (IEEE), New York.

Kingma, D.P. and Ba, J. (2014). Adam: A method for stochastic optimization. In *ICLR, International Conference on Learning Representations*, Appleton, Wisconsin.

Koo, J., Klabjan, D., Utke, J. (2018). Combined convolutional and recurrent neural networks for hierarchical classification of images. *arXiv preprint arXiv:1809.09574*.

Laurance, W.F., Sayer, J., Cassman, K.G. (2014). Agricultural expansion and its impacts on tropical nature. *Trends in Ecology and Evolution*, 29(2), 107–116.

Li, S., Li, W., Cook, C., Zhu, C., Gao, Y. (2018a). Independently recurrent neural network (INDRNN): Building a longer and deeper RNN. In *Proceedings of the IEEE Conference on Computer Vision and Pattern Recognition*, Institute for Electrical and Electronics Engineers (IEEE), New York.

Li, Z., Gavrilyuk, K., Gavves, E., Jain, M., Snoek, C.G. (2018b). VideoLSTM convolves, attends and flows for action recognition. *Computer Vision and Image Understanding*, 166, 41–50.

Ling, C.X. and Sheng, V.S. (2008). Cost-sensitive learning and the class imbalance problem. *Encyclopedia of Machine Learning*, 2011, 231–235.

Mao, Y., Tian, J., Han, J., Ren, X. (2019). Hierarchical text classification with reinforced label assignment. *arXiv preprint arXiv:1908.10419*.

Melgani, F. and Bruzzone, L. (2004). Classification of hyperspectral remote sensing images with support vector machines. *IEEE Transactions on Geoscience and Remote Sensing*, 42(8), 1778–1790.

Oord, A.V.D., Dieleman, S., Zen, H., Simonyan, K., Vinyals, O., Graves, A., Kalchbrenner, N., Senior, A., Kavukcuoglu, K. (2016). Wavenet: A generative model for raw audio. *arXiv preprint arXiv:1609.03499*.

Pascanu, R., Mikolov, T., Bengio, Y. (2013). On the difficulty of training recurrent neural networks. In *ICML'13: Proceedings of the 30th International Conference on Machine Learning*, 28, Appleton, Wisconsin.

Pelletier, C., Webb, G.I., Petitjean, F. (2019). Temporal convolutional neural network for the classification of satellite image time series. *Remote Sensing*, 11(5), 523.

Peña-Barragán, J.M., Ngugi, M.K., Plant, R.E., Six, J. (2011). Object-based crop identification using multiple vegetation indices, textural features and crop phenology. *Remote Sensing of Environment*, 115(6), 1301–1316.

Ren, M., Zeng, W., Yang, B., Urtasun, R. (2018). Learning to reweight examples for robust deep learning. In *Proceedings of the 35th International Conference on Machine Learning*, PMLR, 80, Stockholm.

Ronneberger, O., Fischer, P., Brox, T. (2015). U-net: Convolutional networks for biomedical image segmentation. In *International Conference on Medical Image Computing and Computer-assisted Intervention*, Springer, Berlin, Heidelberg.

Roy, D., Panda, P., Roy, K. (2020). Tree-CNN: A hierarchical deep convolutional neural network for incremental learning. *Neural Networks*, 121, 148–160.

Rußwurm, M. and Körner, M. (2017). Temporal vegetation modelling using long short-term memory networks for crop identification from medium-resolution multi-spectral satellite images. In *Proceedings of the IEEE Conference on Computer Vision and Pattern Recognition Workshops*, Institute for Electrical and Electronics Engineers (IEEE), New York.

Rußwurm, M. and Körner, M. (2018a). Multi-temporal land cover classification with sequential recurrent encoders. *ISPRS International Journal of Geo-Information*, 7(4), 129.

Rußwurm, M. and Körner, M. (2018b). Convolutional LSTMs for cloud-robust segmentation of remote sensing imagery. In *NIPS Workshop*, Montreal.

Rußwurm, M. and Körner, M. (2020). Self-attention for raw optical satellite time series classification. *ISPRS Journal of Photogrammetry and Remote Sensing*, 169, 421–435.

Rußwurm, M., Lefèvre, S., Körner, M. (2019). Breizhcrops: A satellite time series dataset for crop type identification. *Proceedings of the 36th International Conference on Machine Learning*, PMLR 97, Long Beach, California.

Rustowicz, R., Cheong, R., Wang, L., Ermon, S., Burke, M., Lobell, D. (2019). Semantic segmentation of crop type in Africa: A novel dataset and analysis of deep learning methods. In *Proceedings of the IEEE Conference on Computer Vision and Pattern Recognition Workshops*, Institute for Electrical and Electronics Engineers (IEEE), New York.

Saha, P., Dash, S., Mukhopadhyay, S. (2020). PhICNet: Physics-incorporated convolutional recurrent neural networks for modeling dynamical systems. *arXiv preprint arXiv:2004.06243*.

Sahbi, H. and Geman, D. (2006). A hierarchy of support vector machines for pattern detection. *Journal of Machine Learning Research*, 7(75), 2087–2123.

Sainte Fare Garnot, V. and Landrieu, L. (2020a). Leveraging class hierarchies with metric-guided prototype learning. *arXiv preprint arXiv:2007.03047*.

Sainte Fare Garnot, V. and Landrieu, L. (2020b). Lightweight temporal self-attention for classifying satellite images time series. In *International Workshop on Advanced Analytics and Learning on Temporal Data*, Springer, Berlin, Heidelberg.

Sainte Fare Garnot, V., Landrieu, L., Giordano, S., Chehata, N. (2019). Time-space tradeoff in deep learning models for crop classification on satellite multi-spectral image time series. In *IGARSS 2019-2019 IEEE International Geoscience and Remote Sensing Symposium*, 6247–6250, Yokohama.

Sainte Fare Garnot, V., Landrieu, L., Giordano, S., Chehata, N. (2020). Satellite image time series classification with pixel-set encoders and temporal self-attention. In *Proceedings of the IEEE/CVF Conference on Computer Vision and Pattern Recognition*, Institute for Electrical and Electronics Engineers (IEEE), New York.

Shahroudy, A., Liu, J., Ng, T.-T., Wang, G. (2016). NTu Rgb+ D: A large scale dataset for 3D human activity analysis. In *CVPR*, Institute for Electrical and Electronics Engineers (IEEE), New York.

Siachalou, S., Mallinis, G., Tsakiri-Strati, M. (2015). A hidden Markov models approach for crop classification: Linking crop phenology to time series of multi-sensor remote sensing data. *Remote Sensing*, 7(4), 3633–3650.

Siam, M., Valipour, S., Jagersand, M., Ray, N. (2017). Convolutional gated recurrent networks for video segmentation. In *2017 IEEE International Conference on Image Processing (ICIP)*, Institute for Electrical and Electronics Engineers (IEEE), New York.

Srivastava, N. and Salakhutdinov, R.R. (2013). Discriminative transfer learning with tree-based priors. In *Advances in Neural Information Processing Systems, Conference on Neural Information Processing Systems*, Lake Tahoe.

Stoian, A., Poulain, V., Inglada, J., Poughon, V., Derksen, D. (2019). Land cover maps production with high resolution satellite image time series and convolutional neural networks: Adaptations and limits for operational systems. *Remote Sensing*, 11(17), 1986.

Stumpf, F., Schneider, M.K., Keller, A., Mayr, A., Rentschler, T., Meuli, R.G., Schaepman, M., Liebisch, F. (2020). Spatial monitoring of grassland management using multi-temporal satellite imagery. *Ecological Indicators*, 113, 106201.

Su, J., Byeon, W., Huang, F., Kautz, J., Anandkumar, A. (2020). Convolutional tensor-train LSTM for spatio-temporal learning. *arXiv preprint arXiv:2002.09131*.

Sulla-Menashe, D., Friedl, M.A., Krankina, O.N., Baccini, A., Woodcock, C.E., Sibley, A., Sun, G., Kharuk, V., Elsakov, V. (2011). Hierarchical mapping of Northern Eurasian land cover using MODIS data. *Remote Sensing of Environment*, 115(2), 392–403.

Sulla-Menashe, D., Gray, J.M., Abercrombie, S.P., Friedl, M.A. (2019). Hierarchical mapping of annual global land cover 2001 to present: The MODIS collection 6 land cover product. *Remote Sensing of Environment*, 222, 183–194.

Thenkabail, P.S., Mariotto, I., Gumma, M.K., Middleton, E.M., Landis, D.R., Huemmrich, K.F. (2013). Selection of hyperspectral narrowbands (HNBs) and composition of hyperspectral twoband vegetation indices (HVIS) for biophysical characterization and discrimination of crop types using field reflectance and hyperion/EO-1 data. *IEEE Journal of Selected Topics in Applied Earth Observations and Remote Sensing*, 6(2), 427–439.

Turkoglu, M.O., D'Aronco, S., Perich, G., Liebisch, F., Streit, C., Schindler, K., Wegner, J.D. (2021a). Crop mapping from image time series: Deep learning with multi-scale label hierarchies. *Remote Sensing of Environment*, 264, 1–19.

Turkoglu, M.O., D'Aronco, S., Wegner, J., Schindler, K. (2021b). Gating revisited: Deep multi-layer RNNs that can be trained. *IEEE Transactions on Pattern Analysis and Machine Intelligence*, 44(8), 4081–4092.

Ustuner, M., Sanli, F.B., Abdikan, S., Esetlili, M., Kurucu, Y. (2014). Crop type classification using vegetation indices of rapideye imagery. *The International Archives of Photogrammetry, Remote Sensing and Spatial Information Sciences*, 40(7), 195.

Vaswani, A., Shazeer, N., Parmar, N., Uszkoreit, J., Jones, L., Gomez, A.N., Kaiser, Ł., Polosukhin, I. (2017). Attention is all you need. In *Advances in Neural Information Processing Systems, Conference on Neural Information Processing Systems*, Long Beach, California.

Vuolo, F., Neuwirth, M., Immitzer, M., Atzberger, C., Ng, W.-T. (2018). How much does multi-temporal sentinel-2 data improve crop type classification? *International Journal of Applied Earth Observation and Geoinformation*, 72, 122–130.

Walter, A., Liebisch, F., Hund, A. (2015). Plant phenotyping: From bean weighing to image analysis. *Plant Methods*, 11(1), 1–11.

Wang, Y.-X., Ramanan, D., Hebert, M. (2017). Learning to model the tail. In *Advances in Neural Information Processing Systems, Conference on Neural Information Processing Systems*, Long Beach, California.

Wardlow, B.D. and Egbert, S.L. (2008). Large-area crop mapping using time-series MODIS 250 m NDVI data: An assessment for the US central great plains. *Remote Sensing of Environment*, 112(3), 1096–1116.

Wehrmann, J., Cerri, R., Barros, R. (2018). Hierarchical multi-label classification networks. In *International Conference on Machine Learning, International Conference on Learning Representations*, Appleton, Wisconsin.

Wu, M., Sun, Z., Yang, B., Yu, S. (2016). A hierarchical object-oriented urban land cover classification using worldview-2 imagery and airborne LiDAR data. In *IOP Conference Series: Earth and Environmental Science*, 46. IOP Publishing, Bristol.

Xiao, J., Hays, J., Ehinger, K.A., Oliva, A., Torralba, A. (2010). Sun database: Large-scale scene recognition from abbey to zoo. In *2010 IEEE Computer Society Conference on Computer Vision and Pattern Recognition*, Institute for Electrical and Electronics Engineers (IEEE), New York.

Xiao, T., Zhang, J., Yang, K., Peng, Y., Zhang, Z. (2014). Error-driven incremental learning in deep convolutional neural network for large-scale image classification. In *Proceedings of the 22nd ACM International Conference on Multimedia*, Association for Computing Machinery, New York.

Xingjian, S., Chen, Z., Wang, H., Yeung, D.-Y., Wong, W.-K., Woo, W.-C. (2015). Convolutional LSTM network: A machine learning approach for precipitation nowcasting. In *Advances in Neural Information Processing Systems, Conference on Neural Information Processing Systems*, Montreal.

Yan, Z., Zhang, H., Piramuthu, R., Jagadeesh, V., DeCoste, D., Di, W., Yu, Y. (2015). HD-CNN: Hierarchical deep convolutional neural networks for large scale visual recognition. In *Proceedings of the IEEE International Conference On Computer Vision*, Institute for Electrical and Electronics Engineers (IEEE), New York.

Zhu, X. and Bain, M. (2017). B-CNN: Branch convolutional neural network for hierarchical classification. *arXiv preprint arXiv:1709.09890.*

Zilly, J.G., Srivastava, R.K., Koutník, J., Schmidhuber, J. (2017). Recurrent highway networks. In *International Conference of Machine Learning*, Sydney.

3

Exploiting Multitemporal Multispectral High-resolution Satellite Data toward Annual Land Cover and Crop Type Mapping: A Case Study in Greece

Christina KARAKIZI, Konstantinos KARANTZALOS and Zacharias KANDYLAKIS

Remote Sensing Laboratory, National Technical University of Athens, Greece

3.1. Introduction

Accurate and regularly updated land cover and crop type maps are considered essential for several scientific communities, as well as for public and regional authorities in terms of supporting decision making, planning, sustainable development goals and natural resources management. Concurrently, free and open access to high spatial, temporal and spectral resolution satellite optical data has brought a revolution in research and operational applications exploiting satellite data. The disengagement of data availability and quality from high required cost has increased the amount of users and allowed the expansion of remote sensing applications to various earth-observation-related domains. At the same time, no cost limitations instigated the growth of remote sensing monitoring and mapping

Multitemporal Earth Observation Image Analysis,
coordinated by Clément MALLET and Nesrine CHEHATA.

applications, temporally as well as spatially, for example, at the country, continental and even global scale.

During the last five years, Landsat 8 (L8) and Sentinel-2 (S2) have represented the two main sensors offering openly high spatial resolution multispectral imagery, and thus a significant number of land cover mapping applications have been presented in the literature based on their multitemporal data (Chaves et al. 2020; Phiri et al. 2020). Comparing the S2 mission to L8 expanded the possibilities for remote sensing applications offering higher spatial resolution for most spectral bands, a higher revisit time of five days, but also three additional bands in the Red-edge region. At the same time, the combined use and integration of both sensors' data has become a trend (Xiong et al. 2017; Claverie et al. 2018; Frantz 2019; Griffiths et al. 2019; Nguyen and Henebry 2019) in order to develop dense time series for studying dynamic phenomena or for tackling the high frequency of cloud coverage in some parts of the world. However, differences in observation geometry, spectral bands and geolocation exist and pose many challenges for an integration that allows seamless exploitation (Griffiths et al. 2019).

Toward the seamless exploitation of multispectral-image time series, a number of pre-processing steps play a crucial role in their efficient analysis. Even for a single sensor, the acquisition conditions vary for every single image taken. Overall, data from the S2 and L8 sensors are affected by atmospheric, topographic, cloud and shadows effects that have the potential to reduce mapping accuracy and final product reliability (Phiri et al. 2018, 2020). Geolocation errors and bidirectional effects should also be considered, especially when integrating data from different sensors (Griffiths et al. 2019) or different swaths and orbits. Nonetheless, all the corrective preparation steps require additional computational cost and efforts toward the main application, and thus the availability of already processed data or analysis ready datasets is critical for users of such multitemporal imagery.

Once multitemporal data with consistent geometry and radiometry have been formed, the main analysis for geospatial mapping tasks is performed predominantly through data classification procedures. A recent review article, on the topic of S2 and L8 applications for land use and land cover mapping, has documented that supervised pixel-based classification was the principal classification approach adopted in the reviewed papers, in most cases employed through machine learning techniques, deep learning and artificial neural networks (NNs) (Chaves et al. 2020). Regardless of the machine learning architecture, supervised techniques and their outcomes are strongly dependent on the availability of reference data that meet the specific mapping target requirements (Pelletier et al. 2017). The lack of sufficient reference data has always constrained supervised learning, especially for annually dynamic categories, like crops (Waldner et al. 2017), and still remains one of the greatest challenges for mapping studies. The choice of classification features, for example, spectral bands, indices and ancillary data, is another fundamental aspect of the classification framework affecting classifier's performance and mapping accuracy, but also the complexity regarding handling multitemporal data of high dimensionality.

Aiming to address the high dimensionality issues and corresponding computational challenges that come when accessing, processing and sharing Earth Observation (EO) big data, a number of cloud service architectures and web technologies have been developed and offered to users freely or upon payment (Sudmanns et al. 2020). A flagship example constitutes the tremendous increase in interest and usage of the free, planetary-scale platform of Google Earth Engine (GEE) for mapping applications, especially over large geographical areas. Nonetheless, there are a few concerns associated with the use of GEE, including the reproducibility of pipelines, as the algorithms used are not open-source, and may be subject to change. Additionally, the large-scale use of such standardized frameworks results in limited variation in this field of research, while making users reliant on a platform, which is presently free but not guaranteed to remain so.

In this chapter, we discuss the main aspects toward the effective exploitation of multitemporal L8 and S2 data for land cover and crop type mapping at high spatial resolution. We detail and elaborate on the main, data pre-processing steps and then analyze the basic aspects of the classification framework based on the most recent advances, focusing primarily on the usage of S2 data. Further discussion is provided on open and challenging issues toward systematic land cover and crop type mapping at an annual basis, supported additionally by indicative experimental results and validation. In particular, experiments assessing scalability and geographic generalization are presented, including a benchmarking with different classification features and spatial stratification schemes, toward detailed, that is, with more than 30 thematic classes, land cover and crop type mapping in Greece.

3.2. From raw data to analysis ready datasets

In this section, the required pre-processing steps toward constructing geometrically and radiometrically consistent data, that is, analysis ready data (ARD), are detailed. First, we provide an overview for already available ARD products and processing frameworks for L8 and S2 data. Next, we delve into the procedures of atmospheric, radiometric and geometric corrections, as well as cloud and shadow detection, analyzing recent advances for every step and mainly focusing on S2 data. Additionally, we provide a discussion on different approaches to deal with the issue of temporal irregularity and gaps in image time series. In each section, we also elaborate on the selection of pre-processing methods on our part, based on the performed analysis, our related published work and experimental results. Figure 3.1 presents a global framework, showing all methodology steps analyzed throughout this chapter, including the ARD preparation described in section 3.2, and the learning and map production-validation steps discussed in section 3.3.

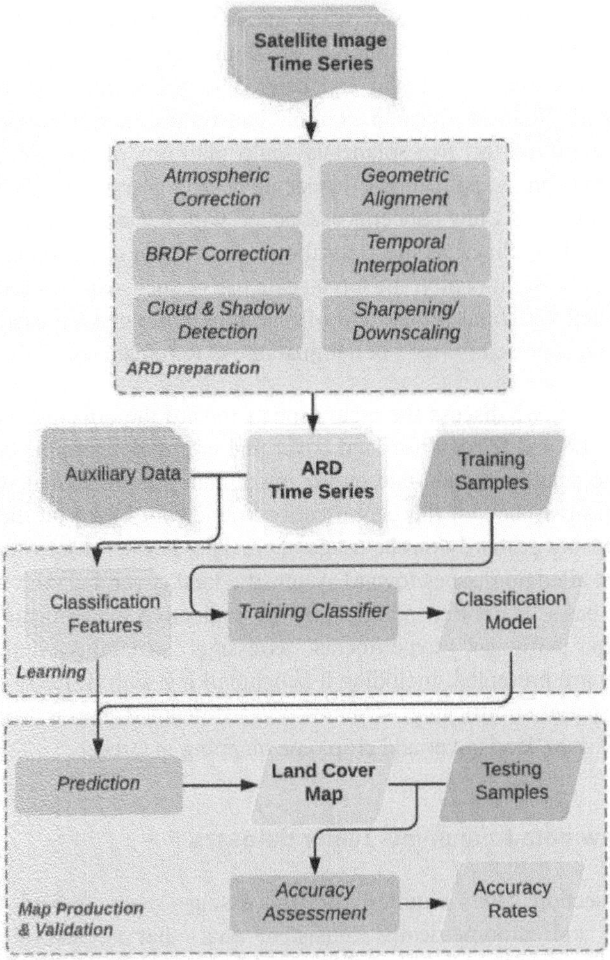

Figure 3.1. *Flow chart of the developed methodology for land cover and crop type mapping. For a color version of this figure, see www.iste.co.uk/mallet/multitemporal.zip*

3.2.1. *Availability of analysis ready data*

Pre-processing steps are crucial for efficient and coherent data analysis and exploitation; nonetheless, their implementation is highly demanding, with regard to computational and data-storage costs, for the production and handling of raw, intermediate and final products. Thus, the availability of ARD data with consistent geometry and radiometry is critical for users within the geospatial community.

Currently, providers of both S2 and L8 data offer higher level products, incorporating radiometric, geometric and atmospheric corrections. The European Space Agency (ESA) provides users with top-of-atmosphere Level-1C orthoimages and atmospherically corrected bottom-of-atmosphere Level-2A products (European Space Agency 2015). Similarly, the United States Geological Survey (USGS) offers radiometrically, geometrically and terrain-corrected Level-1 L8 products, as well as Level-2 surface reflectance L8 imagery. ARD L8 products are also available over the conterminous United States, Alaska and Hawaii (Wulder et al. 2019). Many recent works have exploited surface reflectance time series of S2 (Malinowski et al. 2020; Verde et al. 2020) or/and L8 data (de Oliveira Santos et al. 2019; Pflugmacher et al. 2019; Phalke et al. 2020; Tassi and Vizzari 2020) toward land cover and crop type mapping.

Other research initiatives are also offering the opportunity for users to freely exploit pre-processed data, prepared for further analysis and/or ARD and software frameworks toward ARD. Processed data from Theia Data and Services have been applied in numerous land cover and crop mapping published works (Pelletier et al. 2016; Inglada et al. 2017; Ienco et al. 2019; Stoian et al. 2019). For France and few other zones in the world, Theia offers S2 Level-2A data corrected for atmospheric effects and cloud masks based on MAJA software (Hagolle et al. 2017) acquired from 2016 onward, and surface reflectance of L8 data, acquired from March 2017 onward. Level 3A products for S2, that is, monthly, cloudless and surface reflectance syntheses, have been produced and provided by Theia on every 15th of the month, since July 2017 for France, August 2019 for Spain, September 2019 for Italy and October 2019 for the Maghreb and Western Sahel. Similarly, the EO Center of the German Aerospace Center (DLR) has offered free access to monthly cloud-free S2 mosaics (Level 3A products) of Germany since 2019 (Hagolle et al. 2021).

Regarding S2 and L8 combined use, one of the most prominent ARD datasets comes from the NASA initiative Harmonized L8 and S2 (HLS) project (Claverie et al. 2018). HLS ARD (https://hls.gsfc.nasa.gov/) consists of seamless products for S2 and L8 sensors, acquired through numerous processing steps, that is, atmospheric correction, cloud and cloud-shadow masking, spatial co-registration and common gridding, bidirectional reflectance distribution function (BRDF) normalization and spectral bandpass adjustment (Claverie et al. 2018). The HLS products have been successfully employed in recent research efforts for crop and Land Use/Land Cover (LULC) mapping (Griffiths et al. 2019; Nguyen and Henebry 2019). Other mapping projects (Rufin et al. 2019; Borges et al. 2020; Rosentreter et al. 2020; Schug et al. 2020) have used FORCE (Framework for Operational Radiometric Correction for Environmental monitoring) software, an "all-in-one" solution for the mass-processing and analysis of L8 and S2 image archives (Frantz 2019). FORCE (https://davidfrantz.github.io/code/force/) is open-source software, composed of state-of-the-art correction procedures, that is, cloud masking, radiometric, atmospheric, topographic and BRDF corrections. Furthermore, it incorporates data cubing, that is, spatial reorganization of the data into a non-overlapping grid system for enhanced efficiency and simplicity of ARD usage (Frantz 2019).

Already processed data and ARD are a very useful and cost-effective solution, regarding computational costs and labour time. However, different applications have different requirements and pose various challenges. Thus, the limitations inherent with such fixed solutions should be factored in when considering the use of ARD. To this end, the following sections provide more detailed insights on the critical pre-processing steps toward ARD. With regard to our experimental goals, we focus on the higher spatial and temporal resolution S2 data, since Greece, as a Mediterranean country, does not suffer from frequent cloud coverage, that is, combined use with L8 is not necessary.

3.2.2. *Atmospheric correction*

Atmospheric correction is the procedure of accounting for the parameters of the atmosphere that intervene between the sensor and the target (Earth's surface) in order to convert top-of-atmosphere reflectance values to bottom-of-atmosphere, that is, surface reflectance. The atmospheric correction of satellite imagery is considered a fundamental pre-processing step toward a reliable application of monitoring and/or mapping the surface of the Earth (Doxani et al. 2018). In this sense, currently the vast majority of relevant applications apply an atmospheric correction procedure before the main methodology framework or exploit already corrected surface reflectance products. Though not recommended, a few recent works concerning LULC and/or crop mapping exploited top-of-atmosphere reflectance data for their analysis (Huang et al. 2017; Xiong et al. 2017; Amani et al. 2020; Li et al. 2020).

Concerning already available surface reflectance products, ESA applies the Sen2Cor processor (Richter et al. 2012) based on algorithms proposed in the Atmospheric/Topographic Correction for Satellite Imagery (Richter and Schlapfer 2011; Drusch et al. 2012) to provide Level-2A products. This method performs atmospheric correction based on the libRadtran radiative transfer model (Mayer and Kylling 2005). The Sen2Cor processor (http://step.esa.int/main/snap-supported-plugins/sen2cor/) has been employed by the users to create surface reflectance products for similar studies (Preidl et al. 2020), including a recent publication from our laboratory (Karakizi et al. 2020).

Theia data, as described in section 3.2.1, have been/are produced using the MACCS/MAJA processor (https://logiciels.cnes.fr/en/content/maja). Various applications concerning LULC and crop mapping (Inglada et al. 2015, 2016, 2017; Matton et al. 2015; Defourny et al. 2019; Stoian et al. 2019) have employed this atmospheric correction framework, which is free and open. The MACCS processing chain (Hagolle et al. 2015) was jointly developed by CESBIO and CNES, while more recently CNES, CESBIO and DLR joined their efforts to develop a joint processor named MAJA, for the MACCS-ATCOR Joint Algorithm (Hagolle et al. 2017). MACCS/MAJA's distinct feature is that it was specifically designed to process time series of optical images at high resolution, acquired under quasi constant viewing angles. MAJA utilizes multitemporal information in order to estimate the aerosol optical thickness and correct the atmospheric effects, while also taking into account the adjacency effect and the illumination variations due to topography.

Apart from the above-mentioned frameworks, various other atmospheric correction methodologies for S2 data have been developed and implemented in several applications. Doxani et al. (2018) recently performed the Atmospheric Correction Inter-comparison eXercise (ACIX), as an international initiative aiming to analyze the surface reflectance products of various state-of-the-art atmospheric correction processors, for S2 and L8 time series over 19 selected sites. According to ACIX, eight processors, that is, CorA, FORCE, iCOR, LaSRC, MACCS, S2-AC2020, GFZ-AC and Sen2Cor, were benchmarked for S2 data. The overall assessment indicated that LaSRC, FORCE and MACCS provided accurate and robust surface reflectance estimates for all the study cases.

In the experiments presented in this chapter, we have employed the MAJA cloud screening and atmospheric correction framework, which has recently been released as fully open source software (version 4.2). Although, in previous studies, we have successfully utilized surface reflectance data derived from the providers' algorithms, that is, Sen2Cor for S2 and LaSRC for L8 (Karakizi et al. 2018a, 2018c, 2020, 2021), we have now integrated MAJA in our processing pipeline in order to exploit the entire time series information into the correction procedures. In particular, we utilize the flat surface reflectance files (FRE) produced by the MAJA pipeline, which are also corrected for slope effects.

3.2.3. *Directional effects and BRDF correction*

Most terrestrial surfaces are not Lambertian and thus directional reflectance effects are present in satellite reflectance retrievals due to variable solar-surface-sensor geometry (Roy et al. 2016). The accurate retrieval of albedo is typically achieved using the BRDF by taking into account the directional nature of the reflected radiation in order to minimize the anisotropic effects (Sütterlin et al. 2015). Data from L8 and S2 optical sensors do not provide adequate multi-angle observations to accurately capture the true effects of surface anisotropy, but this angular information can be derived from coarser resolution, wide swath, multi-angle imagers such as MODIS (Wulder et al. 2019).

One important characteristic of the S2 sensor is that images over a given site are acquired with constant observation angles at a constant local hour, and thus the directional effects are minimized. The variation of sun angles is also quite low because of the use of a sun-synchronous orbit (Hagolle et al. 2017). In this sense, higher level data provided by ESA (S2 Level-2A) do not incorporate a BRDF adjustment in their processing production chain. Similarly, most recent studies exploiting S2 time series toward land cover and crop type mapping do not apply extra processing steps toward BRDF correction.

Nonetheless, Hagolle et al. (2017), who aimed to merge S2 time series coming from different orbits with different viewing angles, have incorporated a directional correction in MAJA, v.2.0 onward, using a constant directional model following the idea from Claverie et al. (2015), generalized by Roy et al. (2016). Normalizing the BRDF effects is also critical for certain datasets and projects that integrate S2 and L8 data (Claverie et al. 2018; Frantz 2019; Nguyen and Henebry 2019), given the differing solar and viewing angles of the two sensors. Both HLS dataset and FORCE software utilize the retrieval of Nadir BRDF-adjusted reflectance through a global set of MODIS-derived BRDF kernel parameters based on the c-factor technique (Roy et al. 2016, 2017).

3.2.4. *Cloud and shadow detection*

Clouds are a pervasive, dynamic and unavoidable issue in satellite-borne optical imagery (Foga et al. 2017). It is commonly accepted that the presence of clouds and cloud shadows is considered as a source of noise for the reflectance time series. Thus, the majority of land cover and crop type mapping works that utilize optical satellite data apply either a cloud masking methodology or a threshold over which images are excluded for further processing. However, how and to what extent the presence of clouds affects the accuracy and effectiveness of mapping applications, that is, where to set the threshold, has not been much studied in the literature.

In a recently published work, we assessed the influence of cloud presence on the resulting classification accuracy for detailed (27 classes) LULC mapping, experimenting with gradually increasing cloud coverage thresholds for annual L8 time series (Karakizi et al. 2018c). Our results over a L8 scene in northern Greece indicated that the highest accuracy rates (over 80%) were achieved for the multitemporal datasets with up to 20% and 30% cloud coverage. Regarding the influence of cloud presence on the training and testing data, our experiments indicated that a cloud cover of over 25% per class resulted in lower F-score rates. In this sense, setting cloud thresholds comes as an easy choice that does not require additional processes and thus extra computational cost. Nonetheless, we acknowledge the fact that the threshold option can be limiting for a number of reasons, in various cases. LULC experiments from this particular study were conducted in a Mediterranean country of relatively low cloud cover, so experimental set-ups, even from the strictest thresholds, contained scenes from all seasons. However, this is not the case for many other parts of the globe. Additionally, the more challenging task of crop monitoring and crop mapping necessitates multitemporal observations of high frequency in order to capture key periodic events in the dynamic lifecycle of the different species, and so others, less limiting than the threshold solutions, should be exploited.

In this regard, reliable cloud detection methodologies are essential for exploiting full optical information in satellite time series. However, accurate cloud and cloud shadow detection still remains a challenging task, mainly due to the significant diversity of cloud types/forms and the spectral similarity of shadows with other dark surfaces. There are numerous cloud detection algorithms available in the literature and the majority of them are based on physical rules, making use of spectral indices to separate clear-sky pixels from clouds (Tarrio et al. 2020). In the optical range, clouds are most often detected based on their very high reflectance values. Thermal bands are also considered a key data source for cloud detection, since clouds are colder than the Earths' surface, while shortwave infrared bands, aptly named "cirrus", have been specifically designed to improve the detection of cirrus clouds (Zhu et al. 2015; Tarrio et al. 2020). The most frequently cited masking methods used with S2 or/and L8 multitemporal data for LULC and crop mapping are F-mask, Sen2Cor and MAJA.

The Function of mask, in short F-mask, is a single-date, object-based method for cloud and cloud shadow detection, and it has been adopted by the USGS for operational processing of Landsat data (Zhu and Woodcock 2012). F-mask was initially designed for Landsat data and relies on the exploitation of thermal bands. Zhu et al. (2015) published an adapted version of F-mask to work in different configurations with options to omit thermal and/or cirrus bands and thus expand its usage to S2 data (Claverie et al. 2018). Frameworks for S2 and L8 ARD, as described in section 3.2.1, base their cloud detection step on the F-mask algorithm. The HLS project utilizes the adapted F-mask for the production of S2 S10 and S30 products (Zhu et al. 2015; Claverie et al. 2018), while the FORCE includes a version of the F-mask code that incorporates the latest updates (Frantz 2019).

Sen2Cor (Richter et al. 2012) is a single-date processor that is designed for land cover classification and atmospheric correction of Level-1C S2 data and used by ESA in Level-2A surface reflectance products generation. Sen2Cor employs a series of spectral reflectance thresholds, ratios and indices (e.g. NDSI and NDVI) to compute cloud probabilities for each pixel (Tarrio et al. 2020), resulting in a scene classification band, which includes labels such as clouds, cirrus clouds and cloud shadows. Multiple recent studies on LULC mapping relied on cloud masks derived from the Sen2Cor processor, applied before the main methodology steps (Malinowski et al. 2020; Preidl et al. 2020; Tassi and Vizzari 2020; Verde et al. 2020).

MAJA (Hagolle et al. 2010, 2017) is a spectro-temporal method for cloud detection and atmospheric correction, intended for use with Formosat-2, Landsat, VENμS and S2, and is currently used by the French Theia Land Data Center to deliver S2 surface reflectance products (Tarrio et al. 2020). Low-altitude cloud identification is based on the spectral differences of each scene, relative to a reference composite image created by the most recent cloud-free observation for each pixel. High-altitude clouds are detected using a single-date method based on thresholds that vary linearly with altitude on cirrus band. Cloud detection is performed at

240 m resolution, but after the recent version 4.2, MAJA is able to process cloud masks and aerosol detection at 120 m resolution. Many recent studies on LULC, crop and crop type mapping applied cloud masks derived from the MAJA/MACCS processor (Inglada et al. 2015, 2016, 2017; Matton et al. 2015; Pelletier et al. 2016; Defourny et al. 2019; Ienco et al. 2019; Stoian et al. 2019).

Few studies have compared the performance of the masks produced by different cloud and cloud shadow detection algorithms for S2 products. Baetens et al. (2019) developed a software program named Active Learning Cloud Detection (ALCD) to produce reference cloud masks in order to validate S2 cloud and shadow masks. The cloud masks were used to compare the performance of MAJA, Sen2Cor and F-mask processors. F-mask and MAJA gave very similar results with a small advantage for MAJA (overall accuracy (OA) of 90% and 91%, respectively), while the accuracy of Sen2Cor was lower by 3–6%. A recent study comparing cloud detection algorithms for S2 imagery (Tarrio et al. 2020) examined the relative performance of five different cloud-masking algorithms (Sen2Cor, MAJA, LaSRC, F-mask and T-mask) in S2 scenes distributed across the Eastern Hemisphere. The authors concluded that no single algorithm outperformed the others for both cloud and shadow detection, and proposed aggregating the results from multiple algorithms for an increased accuracy of up to 2.7%.

For the experiments in this chapter, we have selected the MAJA processor for the atmospheric corrections and the cloud/shadow detection procedures. This selection was based both on the analyzed literature and on a qualitative assessment we conducted, benchmarking the performance of four different cloud/shadow detection tools: (i) the python version of F-mask (Zhu and Woodcock 2012; Zhu et al. 2015), (ii) the Sen2Cor processor delivering the scene classification (SC) layer, (iii) the MAJA version 3.2 (processing at 240 m resolution) and (iv) the MAJA version 4.2. (processing at 120 m resolution). The qualitative evaluation on an annual time series for the year 2019, in various S2 tiles in Greece, indicated that MAJA in most cases provided the most balanced result in terms of omission and commission errors. The F-mask algorithm was generally sufficient in the detection of clouds and cloud-shadows, but there were many cases where high brightness objects (e.g. dense urban fabric and snow) were classified as clouds. Masks from the SC band of the Sen2Cor processor presented poor detection of shadows, while salt and pepper commission errors were observed in almost every scene studied.

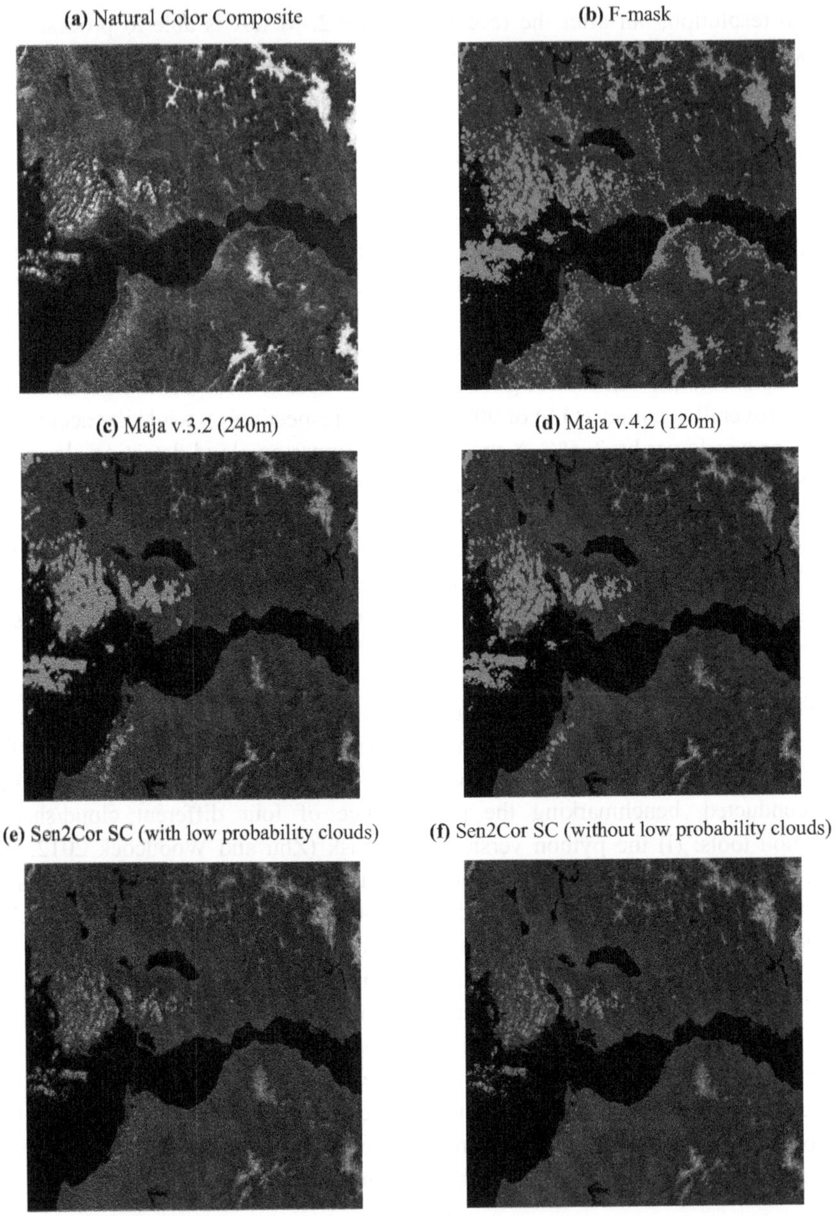

Figure 3.2. *The resulting cloud masks (in red) superimposed on the Red-Edge spectral band of the full (110 km x 110 km) S2 tile 34SEH, acquired on March 10, 2019. Higher altitude mountainous areas are covered with snow. For a color version of this figure, see www.iste.co.uk/mallet/multitemporal.zip*

More specifically, an indicative example is presented in Figure 3.2 for the S2 tile 34SEH over western continental Greece. The resulted cloud masks are presented in red, superimposed on the Red-Edge spectral band. The scene was acquired on March 10, 2019 and higher altitude mountainous areas are covered with snow. In particular, in Figure 3.2(b), extended commission errors can be observed for the F-mask output in snow covered areas and high-density urban areas. Both versions of MAJA, as observed in Figures 3.2(c) and (d), perform quite well, reporting few commission errors to snow, while v.4.2 provides an output of enhanced spatial detail. As is apparent in Figures 3.2(e) and (f), Sen2Cor presented a poor result for cloud's shadow detection and at the same time produced many salt and pepper commission errors, mainly for pixels of high brightness, even when excluding objects labelled as low probability clouds. Based on the performed analysis regarding the aspect of multitemporal cloud masking, our team utilized MAJA v.4.2, for the experiments showcased in this chapter.

3.2.5. *Geometric alignment and geolocation accuracy*

Observations of the Earth's surface need an accurate geometric alignment based on a dense network of ground control locations in order to be useful and reliable. Absolute geodetic accuracy refers to the accuracy of image registration relative to ground reference (Claverie et al. 2018). Additionally, geometric registration/alignment among the same scenes/tiles acquired on different dates (multitemporal co-registration), as well as different tiles or satellite data (co-registration) are crucial for time series analysis. This is especially important for ARD intended for automated processing, since thousands of scenes could be analyzed in a single run, without any or rather limited visual inspection of individual images (Wulder et al. 2019).

S2 data are provided in fixed tiles that have geodetic accuracy specifications of 12.5 m. These specifications were expected to be fully achieved with the incorporation of the ESA Global Reference Image (GRI), initially planned for 2017 (Gaudel et al. 2017; Yan et al. 2018). However, there have been significant delays in implementing this solution. Apart from the absolute geolocation accuracy, multitemporal mis-registration issues for S2 products have also been observed and documented. Yan et al. (2018) calculated the mean multitemporal misregistration between Level-1C tile acquisitions over southern Africa for a six-month period, at about 1.6 10-m pixels before ESA's software update (V2.04) on June 15, 2016 and after the update at about 0.4 pixels. Authors pointed out that this degree of misregistration is greater than the S2A 10-m band multitemporal registration accuracy specification and so users should exercise care when using multitemporal S2, acquired prior to June 15, 2016.

Despite the reported registration issues concerning mainly the multitemporal S2 imagery or the integration with L8 data, which can result in a misregistration of 38 m in some parts of the globe (Storey et al. 2016), the vast majority of relevant mapping works that exploit those data do not apply any step concerning further enhancing their geometric accuracy, registration or co-registration status. Exceptions to the latter are datasets and software that integrate the S2 and L8 data, such as the ARD provided by the HLS dataset that applies the Automated Registration and Orthorectification Package (AROP) (Gao et al. 2009), in order to align the two sensors' data to user-selected S2 Level-1C reference images of minimal cloud coverage (Claverie et al. 2018). Additionally, the FORCE software has integrated a co-registration step in the Level 2 Processing System, applying a modification of the Landsat Sentinel Registration (LSReg) algorithm to automatically co-register large volumes of S2 imagery with Landsat base images (Rufin et al. 2021).

In a similar manner, we have also implemented the AROP methodology to register multitemporal S2 data in recent publications (Karakizi et al. 2020, 2021), as well as for the experiments presented in this chapter. To this end, we have studied and analyzed the co-registration status between numerous S2 scenes, acquired during the year 2019, for the 53 tiles covering mainland and insular Greece. For each tile, we have selected one date of minimal cloud coverage that presented the optimal geometric co-registration status between its adjacent tiles in order to serve as the reference/base image for the AROP implementation. All other acquisitions for this tile were registered using the AROP software with the selected base image. Additional findings from our analysis highlighted that before the alignment procedure, 100% matching co-registration was ensured only for acquisitions of the same data taken (same orbit and date), while misregistration issues were greater and more frequent between data acquired by the different twin sensors (S2A and S2B).

3.2.6. *Temporal gap-filling, composites and interpolated time series*

The S2 constellation has a high revisit frequency; however, optimal exploitation of multitemporal data is still dependent on clear-sky conditions. At the same time, orbital cycles and varying acquisition schedules for different tracks also pose an irregularity factor for satellite time series toward mapping/monitoring extended areas. These factors can deteriorate the usability of such image time series and add significant complexity to the process of building an operational land cover mapping framework (Inglada et al. 2015). To alleviate these issues, temporal gap-filling and compositing procedures have been developed and applied in relevant mapping studies.

In brief, temporal gap-filling for image time series is the process of replacing missing or invalid pixel values (e.g. those masked as clouds) using values from valid dates/scenes of the time series for the same pixel. Temporal gap-filling has been employed in numerous recent land cover and crop type mapping studies (Matton et al. 2015; Inglada et al. 2016; Zhai et al. 2018; Ienco et al. 2019; de Oliveira Santos et al. 2019), and has also been combined with the production of synthetic images/dates on regular time intervals (Inglada et al. 2015, 2017; Defourny et al. 2019; Stoian et al. 2019). Despite a number of available sophisticated gap-filling methods, a recent study (de Oliveira Santos et al. 2019) highlighted that there was no clear improvement in using advanced gap-filling methods, such as linear interpolation or the missForest algorithm (Stekhoven and Bühlmann 2012), compared to using the band median value. Additionally, Inglada et al. (2015) indicated that a temporal linear interpolation of cloudy pixels gave similar results with a cubic spline, while being computationally less intensive for an operational application of crop type mapping.

Temporal irregularities in satellite image time series, as well as excessive computational demands that come with high dimensionality, have also been addressed in many mapping studies by producing image composites. Numerous recent studies have applied composites of various time intervals, that is, annual, seasonal, monthly/bi-monthly and weekly/bi-weekly, toward LULC and crop mapping (Huang et al. 2017; Xiong et al. 2017; Griffiths et al. 2019; de Oliveira Santos et al. 2019; Amani et al. 2020; Li et al. 2020; Phalke et al. 2020; Tassi and Vizzari 2020; Verde et al. 2020), using, in most cases, a median function. Relevant crop mapping studies (Matton et al. 2015; Waldner et al. 2017) have employed spectro-temporal composites, that is, composites of the spectral reflectances measured at a specific stage in the season (e.g. peak of vegetation defined as the max of NDVI annual values), to cope with satellite data heterogeneity and within-class spectral variability.

In two recent publications of ours (Karakizi et al. 2020, 2021), gap-filling on original acquisition dates of multitemporal S2 data has been addressed with a simple median function. Additional experiments were performed for the purposes of this chapter, by testing other approaches, for example, linear, cubic and Akima spline. A simple linear approach, where the linear piecewise function is defined by connecting all the available valid observations with straight lines, was chosen as the optimal trade-off between the output's quality and computational cost. The selection of the time interval was based on a highly acclaimed national-scale crop and land cover mapping study (Griffiths et al. 2019) that reported the highest accuracy rates, especially for crop classes, when using a 10-day composite time series, compared to monthly or seasonal composites. In summary, for the experiments in this chapter we adopted a piecewise linear interpolation to address concurrent gap-filling and the production of synthetic dates for one year with 10-day intervals from S2 of less than 80% cloud coverage.

As a last step in the pre-processing pipeline toward ARD time series, we downscale the medium resolution (20 m) S2 bands at 10 m in order to create an ARD product with a consistent high spatial resolution. Wang et al. (2016) have suggested that fusing the four 10 m and six 20 m bands is expected to increase the utility of these six downscaled bands, especially for urban classes mapping. Further supporting this, Gašparović and Jogun (2018) documented that fusing the S2 20 m bands with the 10 m bands, instead of downscaling with a simple resampling method, has improved the overall classification accuracy derived from various land cover classification schemes, especially for the pixel-based approaches. In this regard, we also opt for a sharpening method, by fusing the 20 m VNIR bands, using the spectrally nearest high resolution 10 m bands, applying the high-pass filter (HPF) algorithm. A simple resampling approach is adopted only for the 20 m SWIR bands that are spectrally further apart from the closest 10 m band, also based on our findings in a related publication (Vaiopoulos and Karantzalos 2016).

3.3. Classification and mapping

Once the consistent time series is formed, the next milestone toward LULC and crop type mapping is the selection of the basic aspects of the classification framework. As supervised classification is the most commonly applied approach adopted in relevant works (Chaves et al. 2020), we hereby discuss issues concerning the availability and creation of reference data, as well as the thematic analysis of the adopted nomenclatures. Analysis on classification units and algorithms' benchmarking is also performed, while further discussion is provided on the choice of classification features and the accuracy assessment strategies, also providing experimental results. We additionally elaborate on possible alternatives in constructing and evaluating classification models that are able to generalize over extended geographical regions of various climatic zones in Greece.

3.3.1. *Reference data*

Supervised classification techniques are considered superior to unsupervised ones for mapping tasks. Nonetheless, they require the availability of reference data of sufficient quality, adhering to various specifications; to be sufficient in number and well distributed in the study area, to contain all desired thematic classes and sub-classes, to be up-to-date and of course, accurate. Indeed, accurate reference data are crucial in order to train classification models and reliably evaluate the resulting land cover maps (Pelletier et al. 2017). However, it is highly difficult and in most cases unfeasible to collect reference data of optimal quality frequently, for example, on an annual basis. To this end, many LULC studies employ already available reference datasets, that is, global, continental or national land cover maps and products that are in many cases outdated and/or even of lower spatial resolution than the maps to be produced. Outdated land cover maps are subjected to two types of errors: classification errors and land cover changes since the production date (Waldner et al. 2017). To alleviate these issues, many studies employ some pre-processing steps in order to select reference samples that are considered more reliable toward the scope of their study.

Copernicus land monitoring products like CORINE land cover (CLC) and high-resolution layers (HRL) are some of the most commonly used datasets to serve as reference data for recent land cover mapping studies over Europe (Inglada et al. 2017; Stoian et al. 2019; Malinowski et al. 2020; Verde et al. 2020). In a similar manner, the large European-wide land survey LUCAS (Land Use/Cover Area frame Survey) data have been employed for national- and European-scale LULC mapping (Griffiths et al. 2019; Pflugmacher et al. 2019; Weigand et al. 2020; Venter and Sydenham 2021). Other products, like global (e.g. the LC-CCI, the MODIS MCD12Q1 and the GlobeLand30) and/or national land cover maps and geospatial datasets (e.g. the USGS NLCD products and the French National Land Cover database) have also been employed as the source of reference data for similar mapping studies using high-resolution multispectral data (Matton et al. 2015; Pelletier et al. 2016; Waldner et al. 2017; Venter and Sydenham 2021).

The availability of updated data for dynamic classes like crops comes as an even more challenging task. For Europe, information on agricultural lands and crop cultivation is included on the land parcel information system (LPIS) and the Geospatial Aid Application (GSAA) data as part of the Common Agricultural Policy (CAP). However, since those data are derived from individual claims for subsidies made by the farmers, they are prone to errors, while in many countries LPIS data are considered confidential and are not always timely and publicly available. Nonetheless, LPIS data have been employed in many recent LULC and crop type mapping studies (Matton et al. 2015; Pelletier et al. 2016; Inglada et al. 2017; Griffiths et al. 2019; Stoian et al. 2019; Giordano et al. 2020; Preidl et al. 2020; Verde et al. 2020). In the United States, the Cropland Data Layer (CDL), produced on an annual basis since 2008 by the Department of Agriculture, has been applied as the source of reference data in similar studies (Zhai et al. 2018; Nguyen and Henebry 2019).

The collection of new reference data customised for the scopes of each study, especially over extended geographical regions, is a time and cost intensive task, but it is the only way forward when there is no other available information of accepted quality. In situ data collection from field surveys, along with digitization and image interpretation from very-high resolution aerial or satellite imagery, have been practiced to create reliable reference data by a number of studies (Inglada et al. 2016; Huang et al. 2017; Xiong et al. 2017; Defourny et al. 2019; de Oliveira Santos et al. 2019; Ienco et al. 2019; Ghorbanian et al. 2020; Li et al. 2020; Phalke et al. 2020; Tassi and Vizzari 2020; Zhang et al. 2020).

In a similar manner, the reference data used for the experiments presented in this chapter derived from a database we created at a national scale for Greece. It is worth mentioning that we are aiming for a detailed land cover and crop type product with more than 30 classes, including numerous crop types, at a spatial resolution of 10 m. In this sense, the reference data to be used should address the corresponding requirements of this detailed nomenclature. In order to design the land cover classes' nomenclature, the nomenclatures of related land cover products and classification systems, such as the CLC, the LUCAS, the FAO Land Cover Classification System, the Land Cover CCI, the EAGLE framework, etc., were thoroughly studied. The scope, classes' definition, minimum mapping unit, production methods and specific characteristics were carefully examined for all these products and methods. Taking all these aspects into consideration, our nomenclature was designed based on the general axis of relative products/systems, but was customised to define spectrally and contextually homogeneous classes at the target 10 m spatial resolution and incorporate landscape' characteristics, specifically for the Greek territory. Additional parameters were taken into account, that is, representativeness, relative extent and distinctness of each land cover category, in order to decide the inclusion of each class to the nomenclature system. Crop nomenclature, in particular, was designed by also factoring in agronomical and spectral distinctness or similarity (grouping of classes, e.g. cereals), but also the cultivation extent and importance in the agricultural economy of Greece.

Our reference database was built through intensive photo-interpretation, in situ data collection and the integration of external data sources. Digitization was performed by photo-interpretation experts on very high-resolution imagery, that is, Google Earth and Bing satellite data, Planetscope data and aerial orthophotos of the National Cadastre and Mapping Agency. Various geospatial datasets were used as ancillary information, for example, geospatial data from field campaigns conducted by our laboratory, national geospatial information layers and Copernicus Global Land Service products like CLC and HRL. Reference data for crop type classes, whose identification and annotation via image interpretation is from challenging up to infeasible, were created based mainly on the Greek LPIS data and additional in situ data from field visits of the RSLab. The LPIS data were employed after some filtering procedures, conducted in order to exclude invalid geometries, outlier records and declaration errors, as described in Gounari et al. (2022).

3.3.2. *Classification units and classifiers*

Concerning classification units, although object-based approaches can be considered more sophisticated methods, by minimizing the within-class variability and integrating additional information in the procedure, the implementation of such frameworks for the multitemporal analysis of S2 and L8 is still limited, compared to pixel-based approaches. Indeed, only few of the recent LULC and crop type mapping studies cited in this chapter have adopted an object-based approach (Gbodjo et al. 2019; de Oliveira Santos et al. 2019; Amani et al. 2020; Tassi and Vizzari 2020; Verde et al. 2020). Recent studies have also indicated that the introduction of additional computing cost and complexity for the segmentation step did not result into a significant enhancement of the classification performance (Matton et al. 2015; Tassi and Vizzari 2020).

Concerning algorithmic frameworks, machine learning techniques are currently dominant in mapping studies. Among those, the most frequently used are random forest (RF), support vector machines (SVM) and deep learning frameworks based mainly on artificial NNs. RF classifiers prevail in the mapping domain and have been applied successfully in numerous recent studies for mapping LULC (Huang et al. 2017; Inglada et al. 2017; Zhai et al. 2018; Nguyen and Henebry 2019; Pflugmacher et al. 2019; Ghorbanian et al. 2020; Li et al. 2020; Malinowski et al. 2020; Preidl et al. 2020; Verde et al. 2020), crop type mapping (Inglada et al. 2015, 2016; Defourny et al. 2019; de Oliveira Santos et al. 2019) and joint LULC and crop type mapping (Griffiths et al. 2019; Karakizi et al. 2020). SVM have been also applied in a number of relevant studies, reporting similar accuracy levels to RF, but RF still presents the advantage of better computational time and easier parameterization (Inglada et al. 2015; Pelletier et al. 2016; Phalke et al. 2020).

Although the accuracy rates achieved by shallow machine learning algorithms like RF and SVM are quite high, the growing trend of deep learning in remote sensing has also been introduced for LULC and crop mapping tasks (Gbodjo et al. 2019; Ienco et al. 2019; Stoian et al. 2019; Amani et al. 2020; Turkoglu et al. 2021). Through deep learning, complexity and intra-class variability can possibly be tackled more efficiently, while the spatial context of pixels could be taken into account, especially for categories where the pixel context may be important (Stoian et al. 2019). However, deep NN requires a significantly higher computation cost. Indeed, based on a previous study of ours (Karakizi et al. 2018c), the benchmarking between SVM and deep convolutional NN (CNN) indicated that the OA and Cohen's kappa coefficient were quite similar, while the SVM was less computationally expensive, resulting also in geometrically more accurate class boundaries and relatively higher per-class accuracy metrics. A recent crop type mapping work (Sainte Fare Garnot et al. 2020) has also documented that CNN applied on S2 time series is cost-intensive in both processing time and memory requirements, and suggested an approach based on pixel-set encoders. Regarding operational applications, similar conclusions were drawn by Stoian et al. (2019), who expanded the RF operational mapping framework of Inglada et al. (2017), with the FG-Unet model. In particular, the authors concluded that the proposed model can yield improved results for certain classes, but its geometric accuracy was clearly lower and the computational demands significantly higher, compared to the pixel-based RF. In this perspective, for an operational application, the trade-off for the deep learning approach, regarding accuracy enhancement and computational cost, is still questionable (Mallet and Le Bris 2020).

Apart from the trade-off between computational requirements and accuracy, the current dominance of shallow techniques, like SVM and RF, in large-scale mapping studies can also be linked with the size of the available reference training data. Satellite image mapping, in most cases, involves the classification of an enormous number of pixels, and because of that, the reference data size is usually a very small percentage of the total of image pixels to be classified. At the same time, algorithms like SVM and RF are found to perform well, even when trained with a limited number of reference data, and they are also independent of data dimensionality, which is a critical characteristic when using numerous spectral bands such in hyperspectral imagery or multitemporal image analysis (Dixon and Candade 2008; Paneque-Gálvez et al. 2013; Ramezan et al. 2021). On the other hand, NN-based classifiers depend on dimensionality and thus training set sample size should increase in order to maintain classifier performance (Dixon and Candade 2008; Paneque-Gálvez et al. 2013). Taking all of the above into consideration, the mapping pipeline adopted for the experiments in this chapter employs a pixel-based RF classifier.

3.3.3. *Classification features*

Classification features, that is, the characteristics on which the discrimination of different classes will be based, are another major aspect of the supervised mapping pipeline. Undoubtedly, for multispectral satellite data, the available spectral bands, along with the corresponding spectral indices, compose the main set of classification inputs. Indeed, the vast majority of recent LULC and crop type mapping studies exploit spectral bands and indices from the visible spectrum (red, green, blue), the Red-edge and infrared regions (NIR, SWIR), usually excluding Coastal, Cirrus, thermal and panchromatic bands from the analysis. Indices, like NDVI, NDWI, NDBI and brightness are frequently used and documented as highly useful (Inglada et al. 2015, 2016, 2017; Pelletier et al. 2016; Defourny et al. 2019; Ienco et al. 2019; de Oliveira Santos et al. 2019; Stoian et al. 2019; Amani et al. 2020; Li et al. 2020; Phalke et al. 2020). This is in accordance with our previous studies (Karakizi et al. 2018c, 2021) in which the NDWI, NDVI and NDBI were the features with a major contribution and impact on the result, followed by NIR, SWIR and Red bands.

Temporal features are also widely applied in recent LULC mapping studies (Zhai et al. 2018; Pflugmacher et al. 2019; Tassi and Vizzari 2020; Zhang et al. 2020). These features refer to statistical derivations of the time series per pixel, for example, maximum, minimum, median and other percentile values of the spectral bands or indices. Temporal features or spectro-temporal composites, as described in section 3.2.6, can provide independence from the temporal shifts observed in crop classes, for example, discrepancies of crop calendars across extensive or different study areas. Hence, they have been employed and recommended in many crop mapping studies (Matton et al. 2015; Waldner et al. 2017; Defourny et al. 2019). At the same time, the use of these features as classification inputs comes with the advantage of a significantly lower data dimensionality, compared to an entire time series, and thus a decreased computational cost. Still, there is limited literature regarding the relative contribution of these advantages on classification accuracy, in comparison to the higher frequency of observations in a time series.

To this end, in two of our recently published works (Karakizi et al. 2020, 2021), we experimented with the employment of temporal features compared to time series of images for joint LULC and crop type mapping. In particular, in Karakizi et al. (2020, 2021), the performance of time series datasets was benchmarked against the use of temporal features, over one S2 tile for the year 2016 and over four S2 tiles for the year 2019, respectively. Quantitative analysis of the experimental results indicated that the use of temporal features produced slightly lower OA rates (up to 3%), in comparison to the time series datasets, that were from double to quadruple in dimension. For the purposes of this chapter, we further investigate the contribution of temporal features for building machine learning classifiers over larger areas.

Apart from spectral and temporal features, auxiliary data, for example, topographic, textural and climatic data have also been applied as additional classification inputs in order to provide information about the structure and morphology of the landscape context (Chaves et al. 2020). Textural information and surface elevation information are the most frequently used ancillary data for this type of mapping studies (Inglada et al. 2016; Huang et al. 2017; Xiong et al. 2017; de Oliveira Santos et al. 2019; Pflugmacher et al. 2019; Phalke et al. 2020; Tassi and Vizzari 2020; Verde et al. 2020; Zhang et al. 2020). Recent experiments performed in our laboratory advocate for the use of surface elevation information, in the form of a digital elevation model, as a good trade-off between added complexity (only one layer of additional information for the whole set of multitemporal data) and enhancement in accuracy metrics (Karakizi et al. 2021).

Regarding the experiments presented in this chapter, we have formed data cubes consisting of spectral bands and indices, by using all synthetic dates from the interpolated time series (section 3.2.6). From this point onward, we refer to these spectral features' datasets as synthetic time series or time series. We have also created data cubes consisting of temporal metrics derived from the synthetic time series. In the following sections, we refer to these datasets as temporal features.

3.3.4. *Accuracy assessment*

A key strength of mapping using remote sensing data is the enablement of wall-to-wall coverage of the study regions. Nonetheless, mapping results produced by classification procedures are far from being 100% accurate, since placing spatially and categorically continuous conditions into discrete classes may result in confusion at the categorical transitions (Olofsson et al. 2014). In an accuracy assessment of a produced map, the map is compared with higher quality reference data in order to derive estimations of its accuracy and reliability. In this sense, supervised classification techniques require reference data both for training and evaluating the resulting maps.

Literature on the subject highlights that samples used for accuracy assessment should be independent of the data used to train or develop the classification and that the validation samples should be derived from sources of higher quality and/or procedures of higher accuracy than those used to produce the map (Olofsson et al.

2014; Stehman and Foody 2019). Nonetheless, taking into consideration all the limitations and challenges regarding reference data availability and creation, it comes as no surprise that very few recent land cover mapping studies (Malinowski et al. 2020; Zhang et al. 2020) have documented an external independent validation of their results based on data from other sources than those of the training data. In many cases, the external validation procedure for a produced land cover map is the main goal of a separate study/work (Defourny et al. 2010; Manakos et al. 2017, 2018; Tsendbazar et al. 2018).

Indeed, the vast majority of LULC and crop type mapping studies assess the classification accuracy by splitting the available reference data into training and testing sets. Splitting the reference data may not be the best practice toward expressing the quality of the final product; however, some good strategies could be applied in order to reliably validate the accuracy of the produced classification model. One of them is to ensure the spatial independency of the training and validation samples. Several relevant studies have practiced the latter by splitting the reference data at the polygon level, instead of the pixel level (Inglada et al. 2016, 2017; Pelletier et al. 2016; Defourny et al. 2019; Stoian et al. 2019). We have experimented with both splitting strategies and have also assessed their impact on the derived accuracy metrics (Karakizi et al. 2019, 2020). In accordance with the literature, we have showcased that pixel level splitting, that is, allowing pixels of the same polygon to be used for both training and testing, may result in a 10%–30% overestimation of the overall and per-class accuracy rates, and thus it must be avoided.

We have also experimented with more rigorous strategies for splitting the reference data, that is, by partitioning the region/scene in spatially distinct regions that could be used separately for training and testing (Karakizi et al. 2018c). This strict geographical splitting scheme could provide a more reliable estimation of the robustness and generalization capacity of the classification model. However, there are a number of challenges to be addressed in such approaches, such as the attainment of a balanced training-testing dataset representing all classes and the incorporation of classes' spectral variability across different geographical regions, mainly due to different phenological patterns, crop cycles and diverse cultivation practices over the different eco-climatic zones. Therefore, beyond the optimal, completely independent external validation procedure on the map, the second best choice for expressing the accuracy of the classification procedure can be considered the splitting of the reference data at the polygon level with possible additional geographic distancing criteria.

3.3.5. *Benchmarking classification features and stratification strategies*

In order to further elaborate on scalability and geographic generalization issues, toward the systematic land cover mapping at the country scale, we hereby present indicative experiments over six S2 tiles in Greece, exhibiting different eco-climatic conditions. The study regions belong to different environmental zones based on the European stratification study (Metzger et al. 2005). More specifically, tiles 34SEG and 34SFH are located in South/central Greece and mostly correspond to the Mediterranean South zone, tiles 35TLF and 35TMF are located in northern Greece and correspond to the Mediterranean North zone and tiles 34SEJ and 34TEK are located in central/North Greece and mainly correspond to a combination of Mediterranean Mountains and Mediterranean North environmental zones. The geographically and environmentally different regions and corresponding tiles are presented in Figure 3.3.

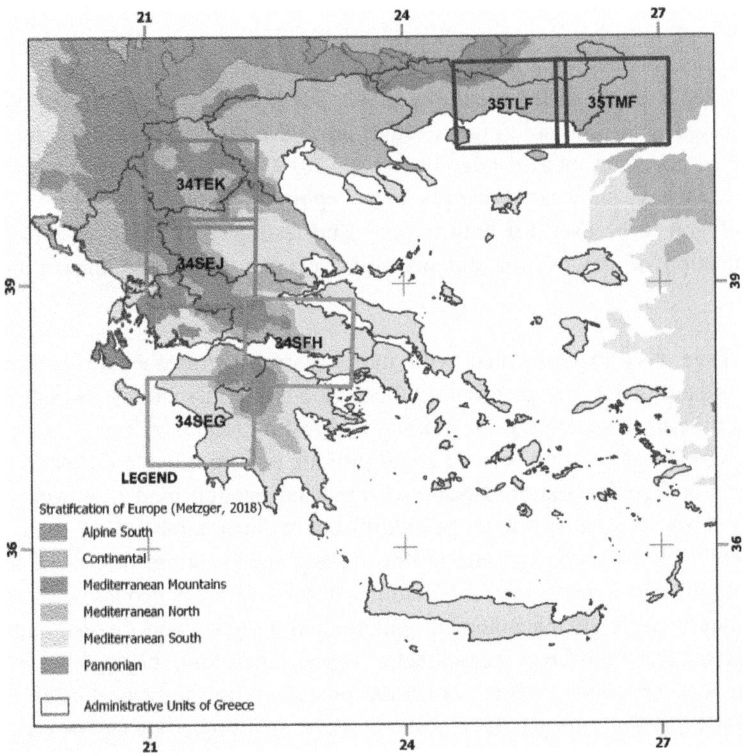

Figure 3.3. *Tiles 35SEG and 34SFH represent the South zone (cyan), 34SEJ and 34TEK the Mountains zone (red) and 35TLF and 35TMF the North zone (blue). For a color version of this figure, see www.iste.co.uk/mallet/multitemporal.zip*

Mapping is performed for the reference year 2019 toward the classification of 35 classes, including 18 crop types. To this end, multitemporal S2 Level-1C data, of less than 80% cloud coverage, acquired throughout 2019, were collected for the six studied S2 tiles. Pre-processing steps were employed, as described in the previous sections, to create a normalized synthetic time series dataset and the corresponding temporal features' one, for each tile. For the synthetic time series, a set of spectral features was formed for each tile, composed of selected spectral bands, that is, Blue (Band 2), Green (Band 3), Red (Band 4), Red-edge (Band 5), NIR (Band 8), SWIR 1 (Band 11), SWIR 2 (Band 12) and three spectral indices, that is, NDVI, NDWI and NDBI. For each tile, the synthetic time series was exploited in order to develop the temporal features, by calculating temporal metrics, that is, the 10th, 25th, 50th, 75th, 90th percentiles, minimum, maximum, mean values and standard deviation for all bands and indices. For both datasets, one additional feature was appended, consisting of a digital elevation model (*EU-DEM v1.1 – Copernicus Land Monitoring Service*, 2016) of 25 m spatial resolution, resampled to 10 m.

An RF classifier was implemented with Python 3.7.8 using scikit-learn and 100 trees. Reference data derived from the national scale database of our laboratory (section 3.3.1) and for each tile they were split randomly at the polygon level per class, using approximately 65% for training and 35% for testing. Different stratification strategies, to train and test the classification models, were examined in order to further evaluate their scalability and generalization capacity. In particular, two types of classification models were developed and validated (Figure 3.4) based on different splitting strategies and geographical criteria:

1) Regional models: For each region, a classification model was trained using 65% of the reference data from both tiles corresponding to this region (RM#1, RM#2, RM#3). The remaining 35% of reference data from both tiles was utilized for validation/testing purposes. Regional models were also created and tested with a more rigorous approach, that is, by using all of the reference data (100%) from one tile of each region for training, while testing on all of the reference data of the other same-region tile (RM#4, RM#5, RM#6). For each region, the tile used for training was the one that included comparatively more thematic classes. The developed regional models were trained and tested as follows:

– **RM#1**: Training (65%) and testing (35%) data from 35TLF and 35TMF.

– **RM#2**: Training (65%) and testing (35%) data from 34SEJ and 34TEK.

– **RM#3**: Training (65%) and testing (35%) data from 34SFH and 34SEG.

– **RM#4**: Training with all reference data from 35TLF. Testing with all reference data from 35TMF.

– **RM#5**: Training with all reference data from 34SEJ. Testing with all reference data from 34TEK.

– **RM#6**: Training with all reference data from 34SFH. Testing with all reference data from 34SEG.

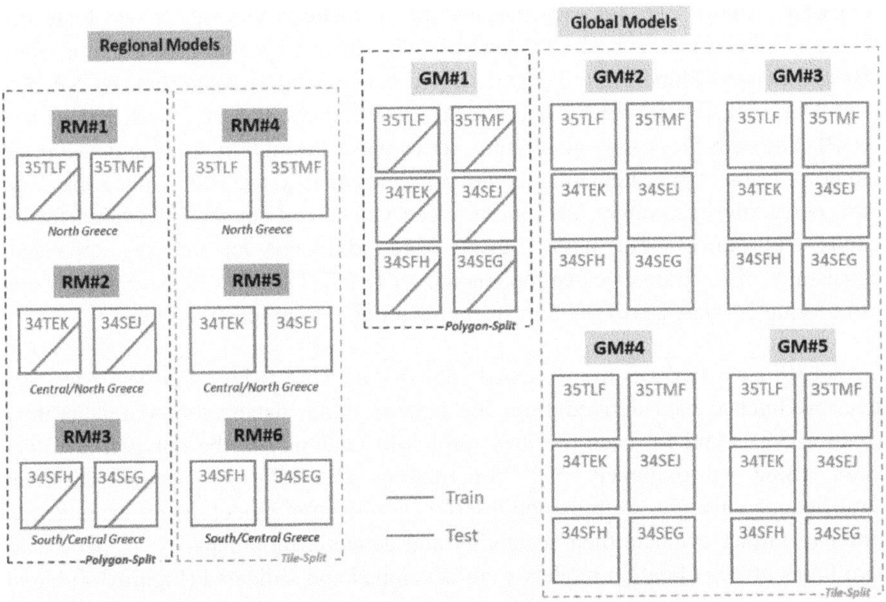

Figure 3.4. *Stratification strategies for training and testing the developed classification models for regional or global predictions. For a color version of this figure, see www.iste.co.uk/mallet/multitemporal.zip*

2) Global models: Global models, that is, models trained with reference data from different environmental regions, were also built to evaluate generalization capabilities at national scale. The first model (GM#1) was trained using 65% of the reference data from all six tiles and was tested on the remaining 35%. Global models were also created by employing all of the reference data (100%) from certain tiles for training, while testing was performed using all of the reference data from the remaining tiles (GM#2, GM#3, GM#4, GM#5). The developed global models were trained and tested as follows:

– **GM#1:** Training (65%) and testing (35%) data from all six tiles.

– **GM#2:** Training with all reference data from 35TLF, 34SEJ, 34SFH. Testing with all reference data from 35TMF, 34TEK, 34SEG (strategy: training with one tile from each geographical region and testing with the remaining).

– **GM#3:** Training with all reference data from 35TLF, 35TMF, 34SEJ, 34TEK. Testing with all reference data from 34SFH and 34SEG (strategy: training with north and North/central regions to predict the South/central).

– **GM#4:** Training with all reference data from 35TLF, 35TMF, 34SFH and 34SEG. Testing with all reference data from 34SEJ, 34TEK (strategy: training with North and South/central regions to predict the North/central).

– **GM#5:** Training with all reference data from 34SEJ, 34TEK, 34SFH, 34SEG. Testing with all reference data from 35TLF and 35TMF (strategy: training with North/central and South/central regions to predict the North).

These models were created based on (i) per-tile synthetic time series that correspond to 371 feature layers, that is, 37 virtual dates of 10 spectral layers each, including the DEM, and (ii) per-tile temporal features that consist of 91 feature layers, that is, nine temporal metrics for each of the 10 spectral layers, including the DEM. The performance of the land cover classification experiments was quantitatively assessed by calculating confusion matrices at the pixel level. The standard accuracy metrics of OA, User's and Producer's Accuracy (UA and PA) were utilized. The per class F-score (F1) measure was also computed as the harmonic mean between UA and PA. The accuracy metrics for the performed experiments are presented in Table 3.1., that is, OA rates, average F1 rates for all classes (avF1), average F1 rates only for the 18 crop classes (avF1 crops) and average F1 rates only for the 17 non crop classes (avF1 non-crops).

In Table 3.1, it can be observed that for all experiments, the results obtained utilizing the time series datasets achieved higher accuracy rates compared to those applying temporal features with approximately four times less dimension. Indeed, the difference for OA rates ranged from 1 to 15%, while the metric exhibiting the largest discrepancies (6–19%) was average F1 for crops. Less variant in time non-crop classes were significantly less affected when utilizing the temporal features, in most cases presenting differences in average F1 of less than 2%. For both datasets, but especially for the temporal ones, average F1 for non-crop classes was in most cases noticeably higher than average F1 for crop classes. Regarding validation approaches, experiments based on the polygon split (RM#1-3, GM#1), that allowed training and testing from the same tiles, presented the highest accuracy rates among all experiments, that is, in most cases OA was over 89%. In comparison, the more rigorous approach of training and testing in different tiles (RM#4-6, GM#2-5), resulted in lower accuracy rates for all metrics. The performance of all regional and global models will be further detailed in the next paragraphs.

	Feature Type	OA (%)	avF1 (%)	avF1 crops (%)	avF1 non-crops (%)
i) Regional models					
RM#1	Time series	94.99	82.82	74.73	90.91
	Temporal	92.23	76.10	62.60	89.59
RM#2	Time series	95.33	90.43	89.58	91.33
	Temporal	92.54	86.43	83.36	89.68
RM#3	Time series	89.10	84.99	80.62	89.61
	Temporal	85.13	80.00	71.39	89.12
RM#4	Time series	84.31	55.52	46.96	64.08
	Temporal	69.30	43.86	29.93	57.78
RM#5	Time series	70.62	55.13	47.68	62.57
	Temporal	69.60	48.72	36.16	61.27
RM#6	Time series	66.22	53.72	40.28	66.37
	Temporal	59.74	44.77	27.13	61.37
ii) Global models					
GM#1	Time series	93.15	88.03	85.20	91.03
	Temporal	89.61	83.06	76.73	89.76
GM#2	Time series	77.97	66.47	57.69	75.78
	Temporal	70.19	56.97	42.35	72.46
GM#3	Time series	73.52	62.00	50.85	73.80
	Temporal	66.70	53.70	35.47	72.99
GM#4	Time series	85.06	68.31	64.54	72.31
	Temporal	77.60	59.27	47.08	72.17
GM#5	Time series	77.60	57.40	52.30	62.50
	Temporal	65.28	46.92	33.39	60.46

Table 3.1. *Quantitative evaluation results for the performed experiments. Time series experiments were performed using 371 classification features, that is, 37 virtual dates of 10 spectral layers including the DEM, while temporal experiments were performed using 91 classification features, that is, nine temporal metrics for 10 spectral layers including the DEM*

Regional models with the polygon level splitting strategy, that is, RM#1, RM#2 and RM#3, resulted in high OA rates (85–95%) and avF1 rates (76–90%), regardless of feature type (Table 3.1). The stricter approach adopted with RM#4-6 yielded, as expected, significantly lower accuracy metrics, presenting OA rates of 66–84% for the time series and 60–70% for the temporal features (Table 3.1). For RM#4-6, the classifiers strive to incorporate the spectral variability across the whole study region, since part of this variability pertaining to one of the tiles, does not participate in training. In addition, the training to testing ratio in RM#4-6 experiments is closer to 50% than 65%, which is the case in RM#1-3. Regarding different regions' results, the pair of tiles corresponding to the South/central zone yielded in almost every case the lowest accuracy rates. This pair of tiles presented two challenges: the amount of reference data was lower and its tiles were further apart compared to the neighbouring tiles of the other two regions. Concerning feature type, for all regional models, the temporal features presented lower testing rates for all metrics. Especially for average crops' F1, differences compared to the time series approach were in the range of 6–17%.

Regarding global models, GM#1, which was built with training data from all six tiles of all environmental zones, reached very high OA rates of 93% and 90%, while average F1 rates were 88% and 83%, for the time series and temporal features, respectively (Table 3.1). To compare the global model with regional models RM#1-3, some additional testing was performed by predicting the testing data of each region separately, using GM#1. The quantitative results from this comparison did not exhibit remarkable differences, that is, regional models presented slightly higher rates, with differences in the range 0.41–1.76% for OA and 0.31–2.64% for average F1. No notable pattern concerning the impact of stratification strategy on specific classes was observed; however, higher differences were recorded for the temporal features' classifications.

In comparison to GM#1, the global models that were developed with the more rigorous splitting approach (GM#2-5) resulted in relatively lower accuracy rates. In particular, cross-regional model GM#2 resulted in an OA rate of 78% and average F1 of 67% for the time series, while for the temporal features, OA was at 70% and average F1 at 57% (Table 3.1). Nonetheless, for both datasets, the rates achieved by GM#2 were in most cases higher than those achieved when generalizing models on a per region basis (RM#4-6). In order to validate the latter in detail, we performed some additional experiments using the GM#2 model to predict the reference data of each testing tile separately, that is, 35TMF, 34TEK and 34SEG. In all cases and for all metrics, the use of the GM#2 model compared to the corresponding regional one, resulted in higher accuracy rates. In particular, OA presented increased rates up to 8%, avF1 up to 12%, avF1 crops up to 13% and avF1 non-crops up to 11%. These results highlight that the integration of additional training data, belonging to different regions, significantly contributed to the model's generalization capability.

For global model GM#3-5, training was performed using reference data from two different geographical regions and testing on the remaining one. Although these experiments were performed with a higher training to testing ratio (4/2 instead of 3/3), lower accuracy rates were reported in the cases of GM#3 and GM#5, when compared to GM#2. This, however, was not the case for GM#4, which presented OA rates of 85% for the time series and 78% for the temporal features (Table 3.1). More specifically, for almost all metrics, GM#4 exhibited the most accurate results, among experiments of the more rigorous splitting approach (RM#4-6, GB#2-5) presented in Table 3.1. The relative success of this experiment can be attributed to the specific tiles' combination that managed to more efficiently capture the variability of the 35 thematic classes across the different climatic zones' regions.

An indicative example on the increased efficiency of the classifier, when using the synthetic time series instead of temporal features, is presented in Figure 3.5. In particular, two maps produced with the RM#3 time series and temporal features classifiers are partially presented for comparison, corresponding to an agricultural area from tile 34SFH of the South/central region, near the town of Thiva. Comparing crops' ground truth (Figure 3.5(a)) with the classification maps (Figures 3.5(b) and (c)), it can be observed that time series classification gave enhanced results regarding crop type detection and geometrically more accurate parcel boundaries. At the same time, the urban/suburban area of the town of Thiva appears more accurately classified, mainly as Dense/Sparse Urban Fabric for the time series map, while for the temporal features' one, few commission errors to class soil, sand, dunes can be observed.

3.4. Data handling and computational challenges

Currently, more than 4 TB of new products are delivered and published every day on the Copernicus Portal. On the users' side, this unprecedented flow of imagery poses significant challenges, concerning data handling, storage and processing. For a country like Greece, 53 S2 tiles are required to cover the continental and island regions, which correspond to approximately 550,000 km^2. For annual country coverage, Level-1C products solely would require approximately 4TB of disk space for storage. When taking into account intermediate and final products, storage requirements are even more demanding. In this sense, annual land cover mapping at the large scales (e.g. country, continental and global) is significantly demanding on storage and computational resources.

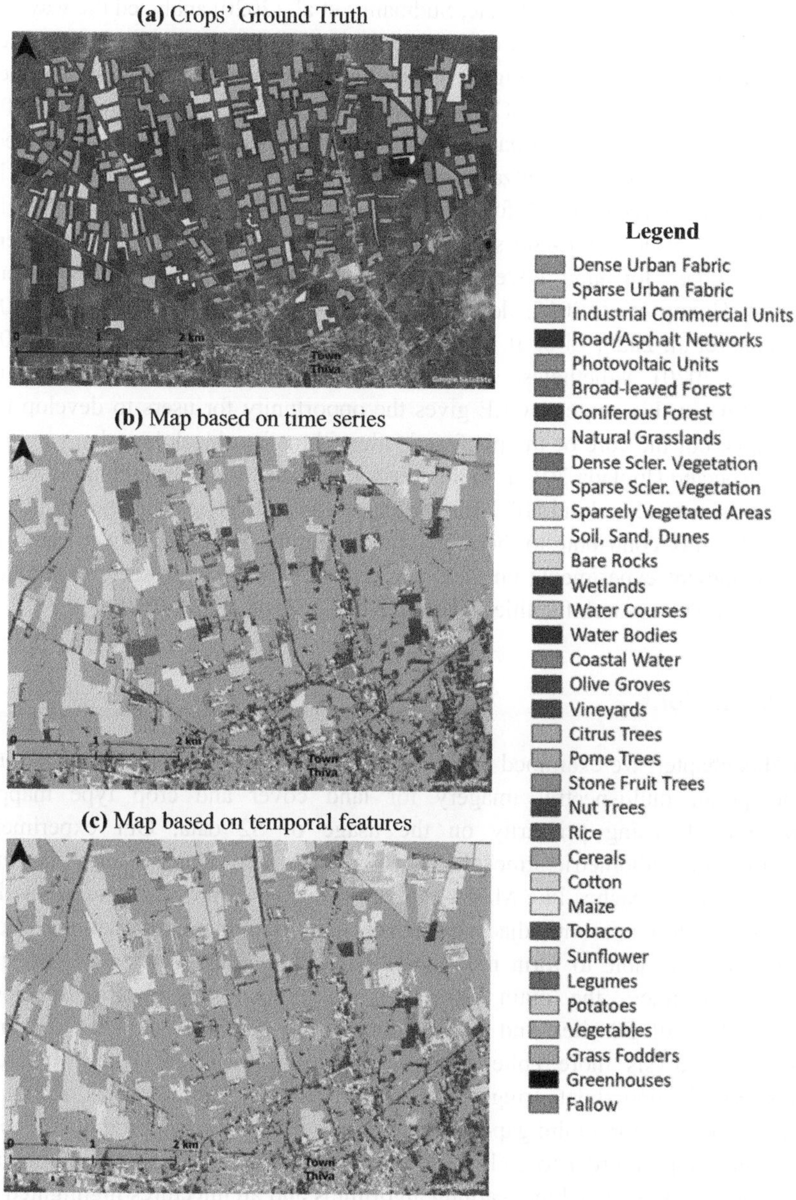

Figure 3.5. *Qualitative evaluation: comparing predicted maps for tile 34SFH from (b) the synthetic time series and (c) the temporal features datasets, using the regional model RM#3. For a color version of this figure, see www.iste.co.uk/mallet/multitemporal.zip*

To this end, in their recent article, Sudmanns et al. (2020) analyzed the way rapid development of cloud-service architectures, web technologies and related systems brought crucial changes in the highly demanding data analysis game. These systems include modules for handling EO datacubes via web-services (Giuliani et al. 2017; Lewis et al. 2017), geospatial data analytics components for applying cutting-edge machine learning tools (Karantzalos et al. 2015), as well as cloud platforms, like Copernicus DIAS (Data and Information Access Services), Amazon Web Services and GEE, that offer end-to-end solutions. Particularly based on GEE, a number of mapping studies for land cover mapping and monitoring have been recently presented (Huang et al. 2017; de Oliveira Santos et al. 2019; Amani et al. 2020; Ghorbanian et al. 2020; Li et al. 2020; Phalke et al. 2020; Tassi and Vizzari 2020; Verde et al. 2020). Nonetheless, controversial issues exist and should be taken into consideration in this context. GEE gives the opportunity for users to develop their own scripts, but the core of the platform's algorithms is a black box that cannot be further inspected (Inglada 2016). In this regard, national and international initiatives aiming at open-access cloud HPC infrastructures, as well as open science principles, will significantly contribute to the research community. In our case, we perform all our development experiments on local servers and deploy large-scale experiments on national HPC research facilities using free and open-source software.

3.5. Conclusions

In this chapter, we examined critical aspects toward the efficient exploitation of multitemporal, multispectral imagery for land cover and crop type mapping applications, focusing primarily on the usage of S2 data. Our experimental evaluation for radiometric/atmospheric corrections and cloud/shadow detection indicated that approaches like MAJA that are designed to process time series data, while taking into account adjacency effects and illumination variations due to topography, were able to form more consistent spatiotemporal cubes. Concerning geometric accuracy, the main issue that still remains is the multitemporal misregistration of S2 tiles and thus geometric alignment is required to render subsequent analysis more coherent and reliable at the pixel level. Regarding handling multitemporal data irregularities (cloud/shadows, orbital cycles, acquisition schedules), we opt for a joint gap-filling and synthetic time series production, with a regular time step, in order to build temporally normalized datasets. The comparison of various supervised machine learning algorithms and architectures highlighted that RF achieves a remarkable trade-off between computational complexity and product accuracy. Still, accurate reference data updated on an annual basis remains one of the most challenging issues for such supervised learning techniques.

Regarding geographical stratification and model generalization aspects, our experimental results targeting joint land cover and crop type mapping into a detailed nomenclature of 35 classes indicated that class variability across different geographical regions is crucial for the quality of the outcome, but also challenging to address. Global model GM#1 trained and tested with reference data from all environmental zones (polygon split approach), and reached into high OA (90–93%) and avF1 (83–88%) rates. Regional stratification adopted with models RM#1, RM#2 and RM#3 yielded similar or slightly higher quantitative results compared to GM#1 when testing on the same data. This observation was in accordance with the literature (Inglada et al. 2017). However, this was not the case for the more rigorous validation approach, that is, when the prediction was performed in tiles that did not participate in the training process. In these cases, regional stratification (RM#4, RM#5, RM#6) produced lower accuracy results for all metrics compared to the cross-regional model GM#2. This indicated that information from a relatively limited area (one tile) could not sufficiently represent the spectral variability, even when predicting across the same environmental zone. In support of the latter, the incorporation of additional reference data from different climatic regions, as in the case of GM#2, did strengthen the model's generalization capability and efficiency. In this sense, further assessment is required to interrogate the efficiency of both regional and global stratification approaches with regard to the amount of available reference data, the intra-class variability and the thematic classes' geographical distribution. There is an additional open question associated with the implementation of regional models, concerning the map's consistency along the boundaries of the climatic/environmental regions, since different models can possibly predict, in a different manner, the neighbouring pixels that may belong to the same thematic class.

Regarding the performance of the employed classification features, the experimental results indicated the effectiveness of the developed synthetic spectral time series, with a 10-day temporal interval, in various environmental zones of Greece. In addition, the use of temporal features' datasets that had significantly decreased dimensions compared to the synthetic time series, resulted in all cases in lower accuracy rates. However, differences concerning the general land cover classes (not crop types) were negligible. On the other hand, the classification of crop classes, which are, in most cases, dynamic in time and thus considered more challenging to discriminate, was noticeably worse when using the temporal features. These results highlighted that the approximately quadruple number of features in the time series has clearly played a significant role in the accurate classification of crop categories by successfully capturing their dynamic phenological patterns across the full extent of our study area.

3.6. Acknowledgments

This work was supported by computational time granted from the National Infrastructures for Research and Technology S.A. (GRNET S.A.) in the National HPC facility – ARIS. Part of this research was funded by the "ELKE" PhD Scholarship of the National Technical University of Athens. The authors would like to thank the Greek Paying Agency (OPEKEPE) for providing necessary geospatial data for this work.

3.7. References

Amani, M., Kakooei, M., Moghimi, A., Ghorbanian, A., Ranjgar, B., Mahdavi, S., Davidson, A., Fisette, T., Rollin, P., Brisco, B. et al. (2020). Application of Google Earth Engine cloud computing platform, sentinel imagery, and neural networks for crop mapping in Canada. *Remote Sensing*, 12(21), 3561.

Baetens, L., Desjardins, C., Hagolle, O. (2019). Validation of Copernicus Sentinel-2 cloud masks obtained from MAJA, Sen2Cor, and FMask processors using reference cloud masks generated with a supervised active learning procedure. *Remote Sensing*, 11(4), 433.

Borges, J., Higginbottom, T.P., Symeonakis, E., Jones, M. (2020). Sentinel-1 and Sentinel-2 data for Savannah land cover mapping: Optimising the combination of sensors and seasons. *Remote Sensing*, 12(23), 3862.

Chaves, M.E.D., Picoli, M.C.A., Sanches, I.D. (2020). Recent applications of Landsat 8/OLI and Sentinel-2/MSI for land use and land cover mapping: A systematic review. *Remote Sensing*, 12(18), 3062.

Claverie, M., Vermote, E., Franch, B., He, T., Hagolle, O., Kadiri, M., Masek, J. (2015). Evaluation of medium spatial resolution BRDF – Adjustment techniques using multi-angular SPOT4 (Take5) acquisitions. *Remote Sensing*, 7(9), 12057–12075.

Claverie, M., Ju, J., Masek, J.G., Dungan, J.L., Vermote, E.F., Roger, J.-C., Skakun, S.V., Justice, C. (2018). The harmonized Landsat and Sentinel-2 surface reflectance data set. *Remote Sensing of Environment*, 219, 145–161.

Copernicus Land Monitoring Service (2016). EU-DEM v1.1 [Online]. Available at: https://land.copernicus.eu/imagery-in-situ/eu-dem/eu-dem-v1.1.

Defourny, P., Bontemps, S., Obsomer, V., Schouten, L., Bartalev, S., Cacetta, P., de Wit, A., di Bella, C., Gerard, B., Giri, C. et al. (2010). Accuracy assessment of global land cover maps: Lessons learnt from the GlobCover and GlobCorine experiences. *Proceedings of ESA Living Planet Symposium*, 28 June-2 July, Bergen [Online]. Available at: https://ui.adsabs.harvard.edu/abs/2010ESASP.686E.224D/abstract.

Defourny, P., Bontemps, S., Bellemans, N., Cara, C., Dedieu, G., Guzzonato, E., Hagolle, O., Inglada, J., Nicola, L., Rabaute, T. et al. (2019). Near real-time agriculture monitoring at national scale at parcel resolution: Performance assessment of the Sen2-Agri automated system in various cropping systems around the world. *Remote Sensing of Environment*, 221, 551–568.

Dixon, B. and Candade, N. (2008). Multispectral land use classification using neural networks and support vector machines: One or the other, or both? *International Journal of Remote Sensing*, 29(4), 1185–1206.

Doxani, G., Vermote, E., Roger, J.-C., Gascon, F., Adriaensen, S., Frantz, D., Hagolle, O., Hollstein, A., Kirches, G., Li, F. et al. (2018). Atmospheric correction inter-comparison exercise. *Remote Sensing*, 10(2), 352.

Drusch, M., Del Bello, U., Carlier, S., Colin, O., Fernandez, V., Gascon, F., Hoersch, B., Isola, C., Laberinti, P., Martimort, P. et al. (2012). Sentinel-2: ESA's optical high-resolution mission for GMES operational services. *Remote Sensing of Environment*, 120, 25–36.

European Space Agency (2015). Sentinel-2 User Handbook [Online]. Available at: https://sentinel.esa.int/documents/247904/685211/Sentinel-2_User_Handbook.

Foga, S., Scaramuzza, P.L., Guo, S., Zhu, Z., Dilley, R.D., Beckmann, T., Schmidt, G.L., Dwyer, J.L., Joseph Hughes, M., Laue, B. (2017). Cloud detection algorithm comparison and validation for operational Landsat data products. *Remote Sensing of Environment*, 194, 379–390.

Frantz, D. (2019). FORCE – Landsat + Sentinel-2 analysis ready data and beyond. *Remote Sensing*, 11(9), 1124.

Gao, F., Masek, J.G., Wolfe, R.E. (2009). Automated registration and orthorectification package for Landsat and Landsat-like data processing. *Journal of Applied Remote Sensing*, 3(1), 033515.

Gašparović, M. and Jogun, T. (2018). The effect of fusing Sentinel-2 bands on land-cover classification. *International Journal of Remote Sensing*, 39(3), 822–841.

Gaudel, A., Languille, F., Delvit, J., Michel, J., Cournet, M., Poulain, V., Youssefi, D. (2017). Sentinel-2 global reference image validation and application to multitemporal performances and high latitude digital surface model. *ISPRS - International Archives of the Photogrammetry, Remote Sensing and Spatial Information Sciences*, XLII-1/W1, 447–454.

Gbodjo, Y.J.E., Ienco, D., Leroux, L., Interdonato, R., Gaetano, R., Ndao, B., Dupuy, S. (2019). Object-based multi-temporal and multi-source land cover mapping leveraging hierarchical class relationships. *arXiv:1911.08815*.

Ghorbanian, A., Kakooei, M., Amani, M., Mahdavi, S., Mohammadzadeh, A., Hasanlou, M. (2020). Improved land cover map of Iran using Sentinel imagery within Google Earth Engine and a novel automatic workflow for land cover classification using migrated training samples. *ISPRS Journal of Photogrammetry and Remote Sensing*, 167, 276–288.

Giordano, S., Bailly, S., Landrieu, L., Chehata, N. (2020). Improved crop classification with rotation knowledge using Sentinel-1 and -2 time series. *Photogrammetric Engineering & Remote Sensing*, 86, 431–441.

Giuliani, G., Chatenoux, B., Bono, A.D., Rodila, D., Richard, J.-P., Allenbach, K., Dao, H., Peduzzi, P. (2017). Building an earth observations data cube: Lessons learned from the Swiss Data Cube (SDC) on generating Analysis Ready Data (ARD). *Big Earth Data*, 1(1–2), 100–117.

Gounari, O., Karakizi, C., Karantzalos, K. (2022). Filtering LPIS data for building trustworthy training datasets for crop type mapping: A case study in Greece. *The International Archives of the Photogrammetry, Remote Sensing and Spatial Information Sciences*, XLIII-B3-2022, 871–877.

Griffiths, P., Nendel, C., Hostert, P. (2019). Intra-annual reflectance composites from Sentinel-2 and Landsat for national-scale crop and land cover mapping. *Remote Sensing of Environment*, 220, 135–151.

Hagolle, O., Huc, M., Pascual, D.V., Dedieu, G. (2010). A multi-temporal method for cloud detection, applied to FORMOSAT-2, VENμS, LANDSAT and SENTINEL-2 images. *Remote Sensing of Environment*, 114(8), 1747–1755.

Hagolle, O., Sylvander, S., Huc, M., Claverie, M., Clesse, D., Dechoz, C., Lonjou, V., Poulain, V. (2015). SPOT-4 (Take 5): Simulation of Sentinel-2 time series on 45 large sites. *Remote Sensing*, 7(9), 12242–12264.

Hagolle, O., Huc, M., Desjardins, C., Auer, S., Richter, R. (2017). MAJA algorithm theoretical basis document. *Zenodo*. doi: 10.5281/zenodo.1209633.

Hagolle, O., Colin, J., Coustance, S., Kettig, P., D'Angelo, P., Auer, S., Doxani, G., Desjardins, C. (2021). Sentinel-2 surface reflectance products generated by CNES and DLR: Methods, validation and applications. *ISPRS Annals of the Photogrammetry, Remote Sensing and Spatial Information Sciences*, V-1–2021, 9–15.

Huang, H., Chen, Y., Clinton, N., Wang, J., Wang, X., Liu, C., Gong, P., Yang, J., Bai, Y., Zheng, Y. et al. (2017). Mapping major land cover dynamics in Beijing using all Landsat images in Google Earth Engine. *Remote Sensing of Environment*, 202, 166–176.

Ienco, D., Interdonato, R., Gaetano, R., Ho Tong Minh, D. (2019). Combining Sentinel-1 and Sentinel-2 satellite image time series for land cover mapping via a multi-source deep learning architecture. *ISPRS Journal of Photogrammetry and Remote Sensing*, 158, 11–22.

Inglada, J. (2016). Is Google Earth Engine evil? [Online]. Available at: http://jordiinglada.net/sblog/geee.html.

Inglada, J., Arias, M., Tardy, B., Hagolle, O., Valero, S., Morin, D., Dedieu, G., Sepulcre, G., Bontemps, S., Defourny, P. et al. (2015). Assessment of an operational system for crop type map production using high temporal and spatial resolution satellite optical imagery. *Remote Sensing*, 7(9), 12356–12379.

Inglada, J., Vincent, A., Arias, M., Marais-Sicre, C. (2016). Improved early crop type identification by joint use of high temporal resolution SAR and optical image time series. *Remote Sensing*, 8(5), 362.

Inglada, J., Vincent, A., Arias, M., Tardy, B., Morin, D., Rodes, I. (2017). Operational high resolution land cover map production at the country scale using satellite image time series. *Remote Sensing*, 9(1), 95.

Karakizi, C., Antoniou, G., Karantzalos, K. (2018a). Towards joint land cover and crop type mapping with numerous classes. *IGARSS 2018 - 2018 IEEE International Geoscience and Remote Sensing Symposium*, Valencia, 2980–2983. doi: 10.1109/IGARSS.2018.8517473.

Karakizi, C., Gounari, O., Karantzalos, K. (2018b). Annual land cover and crop type mapping from Landsat-8 and Sentinel-2 multi-temporal data. *3rd EARSeL LULC and NASA LCLUC Joint Workshop*, Chania.

Karakizi, C., Karantzalos, K., Vakalopoulou, M., Antoniou, G. (2018c). Detailed land cover mapping from multitemporal Landsat-8 data of different cloud cover. *Remote Sensing*, 10(8), 1214.

Karakizi, C., Tsiokou, K., Karantzalos, K. (2019). Annual land cover and crop type mapping at Central Greece from Sentinel-2 data. *9th ESA Advanced Training Course on Land Remote Sensing: Agriculture*, Louvain-la-Neuve.

Karakizi, C., Tsiotas, I.A., Kandylakis, Z., Vaiopoulos, A.D., Karantzalos, K. (2020). Assessing the contribution of spectral and temporal features for annual land cover and crop type mapping. *ISPRS – International Archives of the Photogrammetry, Remote Sensing and Spatial Information Sciences*, XLIII-B3-2020, 1555–1562.

Karakizi, C., Kandylakis, Z., Vaiopoulos, A.D., Karantzalos, K. (2021). Joint land cover and crop type mapping using multi-temporal Sentinel-2 data from various environmental zones in Greece. *The International Archives of the Photogrammetry, Remote Sensing and Spatial Information Sciences*, XLIII-B3-2021, 319–326.

Karantzalos, K., Bliziotis, D., Karmas, A. (2015). A scalable geospatial web service for near real-time, high-resolution land cover mapping. *IEEE Journal of Selected Topics in Applied Earth Observations and Remote Sensing*, 8(10), 4665–4674.

Lewis, A., Oliver, S., Lymburner, L., Evans, B., Wyborn, L., Mueller, N., Raevksi, G., Hooke, J., Woodcock, R., Sixsmith, J. et al. (2017). The Australian geoscience data cube – Foundations and lessons learned. *Remote Sensing of Environment*, 202, 276–292.

Li, Q., Qiu, C., Ma, L., Schmitt, M., Zhu, X.X. (2020). Mapping the land cover of Africa at 10 m resolution from multi-source remote sensing data with Google Earth Engine. *Remote Sensing*, 12(4), 602.

Malinowski, R., Lewiński, S., Rybicki, M., Gromny, E., Jenerowicz, M., Krupiński, M., Nowakowski, A., Wojtkowski, C., Krupiński, M., Krätzschmar, E. et al. (2020). Automated production of a land cover/use map of Europe based on Sentinel-2 imagery. *Remote Sensing*, 12(21), 3523.

Mallet, C. and Le Bris, A. (2020). Current challenges in operational very high resolution land-cover mapping. *ISPRS - International Archives of the Photogrammetry, Remote Sensing and Spatial Information Sciences*, XLIII-B2-2020, 703–710.

Manakos, I., Karakizi, C., Gkinis, I., Karantzalos, K. (2017). Validation and inter-comparison of spaceborne derived global and continental land cover products for the Mediterranean region: The case of Thessaly. *Land*, 6(2), 34.

Manakos, I., Tomaszewska, M., Gkinis, I., Brovkina, O., Filchev, L., Genc, L., Gitas, I.Z., Halabuk, A., Inalpulat, M., Irimescu, A. et al. (2018). Comparison of global and continental land cover products for selected study areas in South Central and Eastern European region. *Remote Sensing*, 10(12), 1967.

Matton, N., Canto, G., Waldner, F., Valero, S., Morin, D., Inglada, J., Arias, M., Bontemps, S., Koetz, B., Defourny, P. (2015). An automated method for annual cropland mapping along the season for various globally-distributed agrosystems using high spatial and temporal resolution time series. *Remote Sensing*, 7(10), 13208–13232.

Mayer, B. and Kylling, A. (2005). Technical note: The libRadtran software package for radiative transfer calculations? Description and examples of use. *Atmospheric Chemistry and Physics Discussions*, 5(2), 1319–1381.

Metzger, M.J., Bunce, R.G.H., Jongman, R.H.G., Mücher, C.A., Watkins, J.W. (2005). A climatic stratification of the environment of Europe. *Global Ecology and Biogeography*, 14(6), 549–563.

Nguyen, L.H. and Henebry, G.M. (2019). Characterizing land use/land cover using multi-sensor time series from the perspective of land surface phenology. *Remote Sensing*, 11(14), 1677.

de Oliveira Santos, C.L.M., Lamparelli, R.A.C., Dantas Araújo Figueiredo, G.K., Dupuy, S., Boury, J., dos Santos Luciano, A.C., da Silva Torres, R., le Maire, G. (2019). Classification of crops, pastures, and tree plantations along the season with multi-sensor image time series in a subtropical agricultural region. *Remote Sensing*, 11(3), 334.

Olofsson, P., Foody, G.M., Herold, M., Stehman, S.V., Woodcock, C.E., Wulder, M.A. (2014). Good practices for estimating area and assessing accuracy of land change. *Remote Sensing of Environment*, 148, 42–57.

Paneque-Gálvez, J., Mas, J.-F., Moré, G., Cristóbal, J., Orta-Martínez, M., Luz, A.C., Guèze, M., Macía, M.J., Reyes-García, V. (2013). Enhanced land use/cover classification of heterogeneous tropical landscapes using support vector machines and textural homogeneity. *International Journal of Applied Earth Observation and Geoinformation*, 23, 372–383.

Pelletier, C., Valero, S., Inglada, J., Champion, N., Dedieu, G. (2016). Assessing the robustness of Random Forests to map land cover with high resolution satellite image time series over large areas. *Remote Sensing of Environment*, 187, 156–168.

Pelletier, C., Valero, S., Inglada, J., Champion, N., Marais Sicre, C., Dedieu, G. (2017). Effect of training class label noise on classification performances for land cover mapping with satellite image time series. *Remote Sensing*, 9(2), 173.

Pflugmacher, D., Rabe, A., Peters, M., Hostert, P. (2019). Mapping pan-European land cover using Landsat spectral-temporal metrics and the European LUCAS survey. *Remote Sensing of Environment*, 221, 583–595.

Phalke, A.R., Özdoğan, M., Thenkabail, P.S., Erickson, T., Gorelick, N., Yadav, K., Congalton, R.G. (2020). Mapping croplands of Europe, Middle East, Russia, and Central Asia using Landsat, random forest, and Google Earth Engine. *ISPRS Journal of Photogrammetry and Remote Sensing*, 167, 104–122.

Phiri, D., Morgenroth, J., Xu, C., Hermosilla, T. (2018). Effects of pre-processing methods on Landsat OLI-8 land cover classification using OBIA and random forests classifier. *International Journal of Applied Earth Observation and Geoinformation*, 73, 170–178.

Phiri, D., Simwanda, M., Salekin, S., Nyirenda, V.R., Murayama, Y., Ranagalage, M. (2020). Sentinel-2 data for land cover/use mapping: A review. *Remote Sensing*, 12(14), 2291.

Preidl, S., Lange, M., Doktor, D. (2020). Introducing APiC for regionalised land cover mapping on the national scale using Sentinel-2A imagery. *Remote Sensing of Environment*, 240, 111673.

Ramezan, C.A., Warner, T.A., Maxwell, A.E., Price, B.S. (2021). Effects of training set size on supervised machine-learning land-cover classification of large-area high-resolution remotely sensed data. *Remote Sensing*, 13(3), 368.

Richter, R. and Schlapfer, D. (2011). Atmospheric/topographic correction for satellite imagery. User guide, DLR – German Aerospace Center, Remote Sensing Data Center.

Richter, R., Louis, J., Müller-Wilm, U. (2012). Sentinel-2 MSI – Level 2A products algorithm theoretical basis document. *European Space Agency (Special Publication) ESA SP*, 49, 1–72.

Rosentreter, J., Hagensieker, R., Waske, B. (2020). Towards large-scale mapping of local climate zones using multitemporal Sentinel 2 data and convolutional neural networks. *Remote Sensing of Environment*, 237, 111472.

Roy, D.P., Zhang, H.K., Ju, J., Gomez-Dans, J.L., Lewis, P.E., Schaaf, C.B., Sun, Q., Li, J., Huang, H., Kovalskyy, V. (2016). A general method to normalize Landsat reflectance data to nadir BRDF adjusted reflectance. *Remote Sensing of Environment*, 176, 255–271.

Roy, D.P., Li, J., Zhang, H.K., Yan, L., Huang, H., Li, Z. (2017). Examination of Sentinel-2A multi-spectral instrument (MSI) reflectance anisotropy and the suitability of a general method to normalize MSI reflectance to nadir BRDF adjusted reflectance. *Remote Sensing of Environment*, 199, 25–38.

Rufin, P., Frantz, D., Ernst, S., Rabe, A., Griffiths, P., Özdoğan, M., Hostert, P. (2019). Mapping cropping practices on a national scale using intra-annual Landsat time series binning. *Remote Sensing*, 11(3), 232.

Rufin, P., Frantz, D., Yan, L., Hostert, P. (2021). Operational coregistration of the Sentinel-2A/B image archive using multitemporal Landsat spectral averages. *IEEE Geoscience and Remote Sensing Letters*, 18(4), 712–716. doi: 10.1109/LGRS.2020.2982245.

Sainte Fare Garnot, V., Landrieu, L., Giordano, S., Chehata, N. (2020). Satellite image time series classification with pixel-set encoders and temporal self-attention. *2020 IEEE/CVF Conference on Computer Vision and Pattern Recognition (CVPR)*, Seattle, 12322–12331. doi: 10.1109/CVPR42600.2020.01234.

Schug, F., Frantz, D., Okujeni, A., van der Linden, S., Hostert, P. (2020). Mapping urban-rural gradients of settlements and vegetation at national scale using Sentinel-2 spectral-temporal metrics and regression-based unmixing with synthetic training data. *Remote Sensing of Environment*, 246, 111810.

Stehman, S.V. and Foody, G.M. (2019). Key issues in rigorous accuracy assessment of land cover products. *Remote Sensing of Environment*, 231, 111199.

Stekhoven, D.J. and Bühlmann, P. (2012). MissForest – Non-parametric missing value imputation for mixed-type data. *Bioinformatics*, 28(1), 112–118.

Stoian, A., Poulain, V., Inglada, J., Poughon, V., Derksen, D. (2019). Land cover maps production with high resolution satellite image time series and convolutional neural networks: Adaptations and limits for operational systems. *Remote Sensing*, 11(17), 1986.

Storey, J., Roy, D.P., Masek, J., Gascon, F., Dwyer, J., Choate, M. (2016). A note on the temporary misregistration of Landsat-8 Operational Land Imager (OLI) and Sentinel-2 Multi Spectral Instrument (MSI) imagery. *Remote Sensing of Environment*, 186, 121–122.

Sudmanns, M., Tiede, D., Lang, S., Bergstedt, H., Trost, G., Augustin, H., Baraldi, A., Blaschke, T. (2020). Big Earth data: Disruptive changes in Earth observation data management and analysis? *International Journal of Digital Earth*, 13(7), 832–850.

Sütterlin, M., Schaaf, C.B., Stöckli, R., Sun, Q., Hüsler, F., Neuhaus, C., Wunderle, S. (2015). Albedo and reflectance anisotropy retrieval from AVHRR operated onboard NOAA and MetOp satellites: Algorithm performance and accuracy assessment for Europe. *Remote Sensing of Environment*, 168, 163–176.

Tarrio, K., Tang, X., Masek, J.G., Claverie, M., Ju, J., Qiu, S., Zhu, Z., Woodcock, C.E. (2020). Comparison of cloud detection algorithms for Sentinel-2 imagery. *Science of Remote Sensing*, 2, 100010.

Tassi, A. and Vizzari, M. (2020). Object-oriented LULC classification in Google Earth Engine combining SNIC, GLCM, and machine learning algorithms. *Remote Sensing*, 12(22), 3776.

Tsendbazar, N.-E., Herold, M., de Bruin, S., Lesiv, M., Fritz, S., Van De Kerchove, R., Buchhorn, M., Duerauer, M., Szantoi, Z., Pekel, J.-F. (2018). Developing and applying a multi-purpose land cover validation dataset for Africa. *Remote Sensing of Environment*, 219, 298–309.

Turkoglu, M.O., D'Aronco, S., Perich, G., Liebisch, F., Streit, C., Schindler, K., Wegner, J.D. (2021). Crop mapping from image time series: Deep learning with multi-scale label hierarchies. *arXiv:2102.08820*

Vaiopoulos, A.D. and Karantzalos, K. (2016). Pansharpening on the narrow VNIR and SWIR spectral bands of Sentinel-2. *The International Archives of the Photogrammetry, Remote Sensing and Spatial Information Sciences*, XLI-B7, 723–730.

Venter, Z.S. and Sydenham, M.A.K. (2021). Continental-scale land cover mapping at 10 m resolution over Europe (ELC10). *Remote Sensing*, 13(12), 2301.

Verde, N., Kokkoris, I.P., Georgiadis, C., Kaimaris, D., Dimopoulos, P., Mitsopoulos, I., Mallinis, G. (2020). National scale land cover classification for ecosystem services mapping and assessment, using multitemporal Copernicus EO data and Google Earth Engine. *Remote Sensing*, 12(20), 3303.

Waldner, F., Hansen, M.C., Potapov, P.V., Löw, F., Newby, T., Ferreira, S., Defourny, P. (2017). National-scale cropland mapping based on spectral-temporal features and outdated land cover information. *PLoS One*, 12(8), e0181911.

Wang, Q., Shi, W., Li, Z., Atkinson, P.M. (2016). Fusion of Sentinel-2 images. *Remote Sensing of Environment*, 187, 241–252.

Weigand, M., Staab, J., Wurm, M., Taubenböck, H. (2020). Spatial and semantic effects of LUCAS samples on fully automated land use/land cover classification in high-resolution Sentinel-2 data. *International Journal of Applied Earth Observation and Geoinformation*, 88, 102065.

Wulder, M.A., Loveland, T.R., Roy, D.P., Crawford, C.J., Masek, J.G., Woodcock, C.E., Allen, R.G., Anderson, M.C., Belward, A.S., Cohen, W.B. et al. (2019). Current status of Landsat program, science, and applications. *Remote Sensing of Environment*, 225, 127–147.

Xiong, J., Thenkabail, P.S., Tilton, J.C., Gumma, M.K., Teluguntla, P., Oliphant, A., Congalton, R.G., Yadav, K., Gorelick, N. (2017). Nominal 30-m cropland extent map of continental Africa by integrating pixel-based and object-based algorithms using Sentinel-2 and Landsat-8 data on Google Earth Engine. *Remote Sensing*, 9(10), 1065.

Yan, L., Roy, D.P., Li, Z., Zhang, H.K., Huang, H. (2018). Sentinel-2A multi-temporal misregistration characterization and an orbit-based sub-pixel registration methodology. *Remote Sensing of Environment*, 215, 495–506.

Zhai, Y., Qu, Z., Hao, L. (2018). Land cover classification using integrated spectral, temporal, and spatial features derived from remotely sensed images. *Remote Sensing*, 10(3), 383.

Zhang, M., Huang, H., Li, Z., Hackman, K.O., Liu, C., Andriamiarisoa, R.L., Ny Aina Nomenjanahary Raherivelo, T., Li, Y., Gong, P. (2020). Automatic high-resolution land cover production in Madagascar using Sentinel-2 time series, tile-based image classification and Google Earth Engine. *Remote Sensing*, 12(21), 3663.

Zhu, Z. and Woodcock, C.E. (2012). Object-based cloud and cloud shadow detection in Landsat imagery. *Remote Sensing of Environment*, 118, 83–94.

Zhu, Z., Wang, S., Woodcock, C.E. (2015). Improvement and expansion of the Fmask algorithm: Cloud, cloud shadow, and snow detection for Landsats 4–7, 8, and Sentinel 2 images. *Remote Sensing of Environment*, 159, 269–277.

4

Irrigation Monitoring Using High Spatial and Temporal Resolutions Remote Sensing Time Series

Hassan BAZZI[1] **and Nicolas BAGHDADI**[2]

[1]*Atos France, Technical Services, Bezons, France*

[2]*INRAE, UMR TETIS, Montpellier, France*

4.1. Introduction

Higher food demands and changing agricultural cultivation practices are natural effects of the growing population and it is expected, according to the United Nations, to reach 9 billion in 2050. With changing climatic conditions, both natural resources and food production are threatened. In the last decade, the agricultural intensification strategies are pushing toward increasing resource use, especially water, in order to meet the global food demand. As a result, several regions are facing rapid degradation of water resources due to the increasing irrigation rates. The continuing expansion of irrigated areas in an unsustainable manner has necessitated the implementation of different management strategies to mobilize water resources with the aim to reach the sustainable water management and water saving. One of the key elements required for better management of irrigation policies is the accurate spatial information about the irrigated area extents. Accurate estimation of irrigated surfaces could help to assess the food production, evaluate water consumption and accurately measure the water discharge from rivers and ground aquifers.

Multitemporal Earth Observation Image Analysis,
coordinated by Clément MALLET and Nesrine CHEHATA.

Large-scale irrigation quantification have been initially based on the use of national statistical data. For example, the FAO's (Food and Agriculture Organization of the United Nations) GMIA map (Siebert et al. 2005, 2015) and the MICRA map (Portmann et al. 2010) provides only census irrigation data derived from the national declaration resampled at 5 arc-minute spatial resolution. However, the low spatial resolution and the lack of credibility of these products remain an obstacle for irrigation management, especially in small and medium agricultural areas.

Currently, advanced technologies in remote sensing allowed the monitoring of irrigated areas at large scales. Large-scale monitoring of irrigated areas using satellite remote sensing can offer a global view with less time and money resources compared to terrain surveys. To perform irrigation mapping at different scales, optical remote sensing data have been extensively used. Optical remote sensing for irrigation mapping mainly relies on the difference in the spectral signature in optical bands between irrigated and rain-fed crops. In the last decade, traditional image classification techniques such as the spectral matching and maximum likelihood classifiers have been first employed to classify irrigated and rain-fed cropland using optical spectral bands (Thenkabail et al. 2009; Dheeravath et al. 2010). Particularly, coarse resolution optical data such as MODIS (Moderate Resolution Imaging Spectroradiometer) at 500 m spatial resolution has been extensively used to map irrigated areas. Recently, the availability of massive time series optical satellite data at higher spatial resolution, such as that offered by the Sentinel-2 and Landsat-8 satellites, has opened the way toward enhanced mapping of irrigated areas using optical time series spectral data.

SAR (Synthetic Aperture Radar) satellite data also presents new perspectives in irrigation mapping. The key point that links SAR data to irrigation detection is the surface soil moisture, which changes due to the application of water by irrigation. Microwave remote sensing is sensitive to the water content of soil due to the increase in the dielectric constant with the increase in the soil water content (Baghdadi et al. 2008, 2011a). This relation between irrigation, soil moisture and SAR data is the primary catalyst for irrigation mapping using SAR data. However, available SAR satellite data such as the RADARSAT-2, ALOS and TerraSAR-X are characterized by low temporal resolutions, which remained an obstacle to obtain sufficient time series data to perform irrigation mapping. Thus, the lack of operational SAR satellites in free and open access modes at high temporal resolution has limited the use of SAR data for irrigation mapping. Nowadays, the availability of the Sentinel-1 (S1) SAR data in open access mode and at high spatial and temporal resolutions encouraged researchers to explore the potential of the C-band (~5.4 GHz) SAR data offered by the S1 satellite for irrigation mapping.

In this chapter, the potential of optical and radar remote sensing data for mapping irrigated areas will be presented. Specifically, this chapter deals with the mapping of irrigated areas and monitoring of irrigation events at plot scale using the Sentinel-1 and Sentinel-2 time series data. The chapter first demonstrates the capability of discriminating irrigated and rain-fed crops using both optical and radar time series data at high spatial and temporal resolutions. Then, it presents several methodological applications recently developed for mapping irrigated areas and detecting irrigation events at plot scale using Sentinel-1 (S1) and Sentinel-2 (S2) time series data.

4.2. Fundamentals behind remote sensing for irrigation mapping

4.2.1. *Optical remote sensing for irrigation mapping*

Additional water for the crop caused by irrigation affects different biophysical properties of the plant. When the crop benefits from an additional amount of water from irrigation, higher levels of photosynthesis are achieved, leading to higher biomass and denser vegetation cover. Thus, compared to a rain-fed crop, an irrigated crop relatively shows higher biophysical properties such as vegetation density and biomass. As a result, the reflectance captured by optical remote sensing satellites differs between an irrigated agricultural plot and a rain-fed agricultural plot due to the difference in the biophysical properties. This difference in the spectral signature between irrigated and rain-fed plots is the main assumption used for mapping irrigated agricultural plots. Studies dealing with irrigation mapping using optical remote sensing have demonstrated that the use of vegetation indices derived from optical satellite images could be powerful for discriminating between irrigated and rain-fed crops. The most popular vegetation index used to describe the vegetation status is the NDVI (Normalize Difference Vegetation Index[1]). The NDVI is a proxy measure of the absorbed photosynthesis active radiation based on the difference between the infrared reflectance and the red reflectance. In the photosynthesis process, the plant absorbs the red radiation and emits back the infrared radiation. Higher levels of photosynthesis for the plants, due to additional amounts of water, increases the infrared emission and the red absorption. Consequently, irrigated crops could present higher NDVI values than rain-fed crops and reach maximum values exceeding the peak of rain-fed crops. Several studies have demonstrated that the NDVI could be used to discriminate between irrigated and rain-fed crops. Other studies have also demonstrated the use of the GI (greenness index[2]) for

---

1 NDVI $= \frac{\rho_{NIR} - Red\ \rho}{\rho_{NIR} + Red\ \rho}$, where ρ_{NIR} is the near infrared reflectance and $Red\ \rho$ is the red reflectance.

2 GI $= \frac{\rho_{NIR}}{Green\ \rho}$, where ρ_{NIR} is the near infrared reflectance and $Green\ \rho$ is the green reflectance.

discriminating between irrigated and rain-fed crops (Chen et al. 2018). The greenness index is a ratio between the near infrared reflectance and the green reflectance.

The temporal profile of vegetation indices derived from satellite optical time series data has proven better efficiency in discriminating irrigated and rain-fed crops compared to using single date image (Chen et al. 2018). Thus, time series data at high temporal resolution could allow for better mapping of irrigated areas. Previously, high temporal resolution optical images such as that of MODIS (500 m spatial resolution) did not provide appropriate spatial resolution for mapping irrigated areas at plot scale. For this reason, previous studies used vegetation indices derived from coarse spatial resolution optical satellites to map irrigated areas at grid scales of several hectares. Nowadays, the Sentinel-2 satellite offers optical images with 5 days revisit time and 10–20 m spatial resolution depending on the spectral band in free open access mode. These satellite specifications permit accurate monitoring of the vegetation temporal profile for irrigated and rain-fed crops at plot scale, which enables better detection of irrigated croplands.

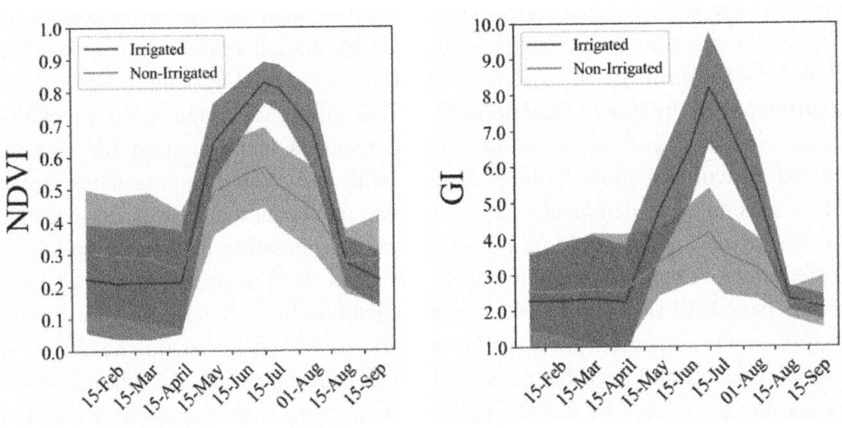

Figure 4.1. *Temporal profiles of (a) NDVI and (b) GI over irrigated (blue) and non-irrigated (red) plots. For a color version of this figure, see www.iste.co.uk/mallet/multitemporal.zip*

Figure 4.1 presents the temporal profiles of NDVI (Figure 4.1(a)) and GI (Figure 4.1(b)) for irrigated (blue curve) and rain-fed (red curve) plots, derived from Sentinel-2 time series images. The temporal profiles are constructed by averaging the vegetation index values over a sample of irrigated and rain-fed plots, of the same crop type, at each available S2 image. The bounding shaded area shows the standard deviation of the average NDVI. The irrigated and non-irrigated sample plots used to construct the temporal profiles are derived from an in situ terrain data collected over a study site near the city of Orléans in north central France. In both indices, the temporal evolution of the irrigated plots differs from that of the non-irrigated plots. Both indices demonstrate that irrigated plots show higher values than that of non-irrigated plots with higher maximum value for irrigated plots in both indices.

Several studies have already explored the use of the optical vegetation indices time series to map irrigated and rain-fed crops. Studies first tended to merge both remote sensing time series derived from coarse spatial resolution satellites such as MODIS and statistical data to obtain large scale irrigated area maps. Dheeravath et al. (2010) used MODIS data at 500 m spatial resolution to map irrigated areas in India between 2001 and 2003. In their study, they used seven optical reflectance bands of MODIS at 8 days temporal resolution combined with precipitation data and elevation data. The correlation coefficient (R^2) between their annual irrigation map over India and the national data reached 0.84. Finally, they reported that the uncertainty of the produced map is mainly linked to the limitation of the spatial resolution (500 m), the irrigated area fraction in the 500 m remote sensing pixel and the lack of robust irrigation declaration in the national surveys. The Global Rain-fed Irrigated and Paddy Croplands (GRIPC) product provided irrigated area extent at 500 m spatial resolution (Salmon et al. 2015). To perform irrigation mapping, Salmon et al. (2015) used the enhanced vegetation index (EVI[3]), the normalized differential water index (NDWI[4]), the land surface temperature (LST) and the NDVI derived from MODIS data jointly with auxiliary data related to climate (climate moisture index, annual moisture index) and agro-ecological zoning. Using terrain-training data, a univariate decision tree classification algorithm was applied on both MODIS satellite data and auxiliary data to obtain the irrigation classification map. In the assessment of the GRIPC map against other products, the results showed that the GRIPC irrigated cropland agrees closely with national inventory statistics from FAO (FAOSTAT).

To obtain a spatial resolution finer than that of MODIS, studies dealing with irrigation mapping have also exploited the potential of Landsat imageries at 30 m spatial resolution to map irrigated areas (Deines et al. 2017; Chen et al. 2018; Demarez et al. 2019; Ren et al. 2021). In Chen et al. (2018), irrigation extent was mapped by using the GI temporal series derived from Landsat 8 images combined

---

3 EVI = $2.5x \frac{(\rho_{NIR} Red\ Green - \rho)}{\rho_{NIR} + 6 * \rho_{NIR} - 7.5 * \rho + 1}$, where ρ_{NIR}, $Red\ \rho$, and $Blue\ \rho$ are the near infrared, red and blue reflectance, respectively.

4 NDWI = $\frac{Green\ \rho - \rho_{NIR}}{Green\ \rho + \rho_{NIR}}$, where ρ_{NIR} is the near infrared reflectance and $Green\ \rho$ is the green reflectance.

with MODIS (250 m) and precipitation data to detect the irrigation extent. They relied on the assumption that the GI index, which is the ratio between the near infrared and the green bands, is more sensitive to chlorophyll content than other indices such as the NDVI. This enhanced sensitivity of the GI with respect to the NDVI is due to the prior saturation of the red (used by NDVI) and blue absorptions compared to the green absorption, especially when the chlorophyll content is high. In their study applied over the Hexi corridor in China, the overall accuracy for detecting irrigation reached 87%.

Combining Landsat 8 and MODIS data for irrigation mapping has also been employed in a recent study by Ren et al. (2021) to map irrigated maize (corn) using time series optical satellite data. In their study, they tested three image classification algorithms including the neural network (NN), random forest (RF) and support vector machine (SVM) to map irrigated and rain-fed corn at 30 m spatial resolution in Nebraska (USA). First, they performed a MODIS based classification of irrigated and non-irrigated areas, which was then used to support the Landsat irrigation classification. The overall accuracy for irrigation mapping in 2012 over Nebraska reached 89.6% using the Neural Network classifier. Demarez et al. (2019) tested the potential of Landsat 8 images combined with elevation data to map irrigated summer crops in the south of France. They showed that the combined use of optical time series vegetation indices with elevation data produces an irrigation classification accuracy of 89%.

With the arrival of Sentinel-2 satellite that provides better spatial and temporal resolution compared to other commercial free satellites, studies have recently started to explore the potential of S2 data to map irrigated areas. Bazzi et al. (2019) used the NDVI temporal series to map irrigated areas in the Catalonia region of northeast Spain by applying a convolutional neural network (CNN). They achieved an overall accuracy of 91% for irrigated area mapping. In a recent study by Pageot et al. (2020), they showed that the vegetation indices NDVI, NDWI (Normalized Difference Water Index) and the NDRE (Normalized Difference Red-Edge) presents a good separability between irrigated and rain-fed maize and soybean crops. The use of these optical vegetation indices in a RF machine-learning algorithm allowed the classification of irrigated crops with an accuracy between 65% and 72%.

4.2.2. SAR remote sensing for irrigation mapping

Microwave data presents new perspectives in irrigation mapping. Unlike optical sensors, SAR data are not limited to good weather conditions and can be available during day and night. The key point that links SAR data to irrigation detection is the soil moisture of the agricultural plots. SAR remote sensing is sensitive to the water content of soil due to the increase in the dielectric constant with the increase in the soil water content (Baghdadi et al. 2011a, 2011b, 2016; Aubert et al. 2013). Therefore, the application of water in irrigation would obviously cause an increase in the soil moisture for an irrigated agricultural plot. Following this logic, El Hajj et al. (2014) analyzed the sensitivity of the TerraSAR X-band SAR data over irrigated grassland plots in the Crau plain of southeastern France. They reported that the SAR signal could be used to identify irrigations that took place at a maximum of 3 days before the SAR acquisition. In their study, they showed that after an irrigation event, the X-band SAR signal increased by approximately 1.4 dB due to an irrigation occurring 1 day before the satellite acquisition.

Figure 4.2 presents the temporal evolution of the SAR signal in VV polarization derived from Sentinel-1 C-band time series data for irrigated (black dashed line) and rain-fed (red dashed line) plots. The green solid line and the green dotted line show the NDVI temporal profiles of the irrigated plot and rain-fed plots, respectively. The blue bars represent the rainfall data.

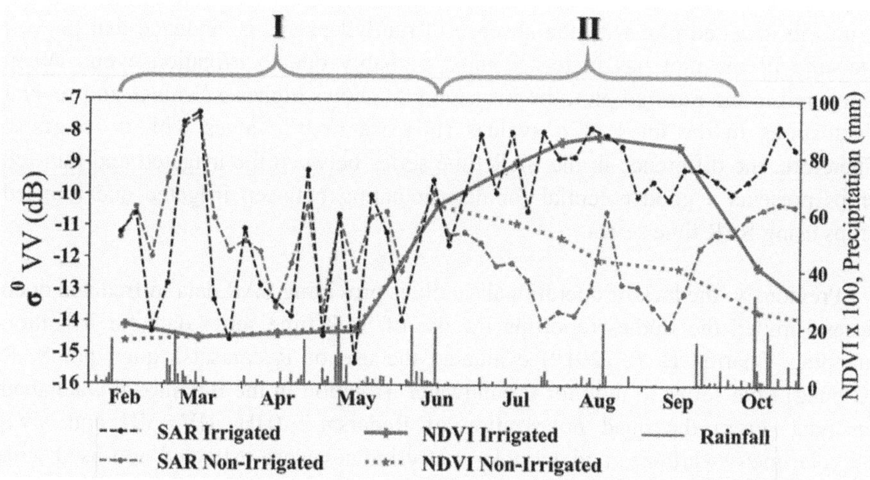

Figure 4.2. *S1 time series profiles for irrigated and rain-fed plots. For a color version of this figure, see www.iste.co.uk/mallet/multitemporal.zip*

In the NDVI temporal profile, it is clearly visible that the irrigated plots have different behavior than the rain-fed plots with higher NDVI values especially in the dry period (between June and September). The SAR signal also presents a difference between the irrigated and rain-fed plots. The sub-figure "I" shows the evolution of the SAR signal in the period between February and June (rainy period). In this period, both irrigated and rain-fed plot show similar behavior with simultaneous increase and decrease of the radar backscattering coefficient σ^0 mainly due to the presence and absence of rainfall, respectively. With a rainfall event, both plots show increase in the σ^0 value due to the increase in the soil moisture values. With the absence of rainfall, both plots encounter a decrease in the σ^0 values caused by the dryness of the soil. For this period, the increase in the σ^0 values for both irrigated and rain-fed plots is mainly linked to the change of soil moisture due to rainfall events.

The sub-figure "II" in Figure 4.2 shows the behavior of the σ^0 values for the period between June and September (the usual irrigation period for summer crops). During this period, the behavior of σ^0 for irrigated plot is different from that of the rain-fed plot. With the absence of rainfall (dry period), the rain-fed plot showed a gradual decrease in the σ^0 values between the end of June and mid-August. This is mainly due to the decrease in the soil water content for the rain-fed crops especially with high evaporation rates in summer and absence of any rainfall event. In this period, only one increase in the σ^0 value is observed in mid-August due to an important rainfall event. On the other hand, the irrigated plots show several fluctuations in the σ^0 values between June and August. The increase in the σ^0 values in irrigated plot with the absence of rainfall events is evidence that the soil moisture of the plot has increased most probably due to irrigation event. When compared to the rain-fed plot, the irrigated plot shows higher σ^0 values and several fluctuations in the level of σ^0-values (increase in σ^0 values with no rainfall). Therefore, the difference in the SAR time series between the irrigated and rain-fed crops presents a good potential for discriminating between irrigated and rain-fed crops using SAR time series.

Previously, the lack of operational satellites providing SAR data in free and open access limited the studies reporting the use of SAR time series data for irrigation mapping. Sharma et al. (2019) evaluated the use of Radarasat-2 quad pol SAR (C-band) time series to monitor groundwater irrigation in the Berambadi watershed in India. From the quad polarization of Radarsat-2 (HH, HV, VH and VV), 15 polarimetry variables and 11 polarimetry indices were extracted and used with ground-collected data to perform a SVM classification for irrigated and non-irrigated crops. They report a classification accuracy of 91.4% and 90.4% for winter and summer irrigated crops, respectively.

The Sentinel-1 (S1) satellite in C-band provided new opportunities to map irrigated areas as it offers over several territories better than 6-days revisit time images at 10 m pixel spacing. Thus, better spatiotemporal monitoring of irrigated and rain-fed plots could be achieved recently using S1 data. Gao et al. (2018) first reported the use of statistical metrics derived from S1 time series in VV and VH polarizations to map irrigated areas in a semi-arid region in Catalonia, Spain. They assume that due to high persisting soil moisture values in irrigated plots compared to rain-fed plots during the year, the S1 temporal signal over irrigated plot must have high average value, low variance value and high signal correlation length. Thus, they used these three statistical metrics (mean, variance and correlation length) in a RF classifier with collected in situ data to map irrigated areas at plot scale. They achieved an overall accuracy of 81%. Bousbih et al. (2018) followed the same methodology of statistical metrics derived from S1 time series to map irrigated and rain-fed wheat crops in an arid area of Tunisia. The validation of the irrigation mapping showed an overall accuracy up to 77%. Demarez et al. (2019) also reported the importance of jointly using S1 SAR and Landsat optical data for irrigated summer crop mapping in south France. The study showed that the joint use of SAR and Optical data provided the best accuracy for irrigation classification.

4.3. New methodologies for irrigation mapping using S1 and S2 time series

4.3.1. *Decision tree algorithm for detecting irrigation events*

Remote sensing applications in irrigation mainly focused on the mapping of irrigation extent without giving significant attention to the detection of the irrigation frequency and timing. In the context of irrigation water management, the detection of irrigation events is important to quantify the water discharge, better schedule the irrigation episodes and improve the water use efficiency. As discussed in section 4.2.2, SAR time series data could be used to detect the wetness information caused by irrigation episodes. Nevertheless, irrigation is a time dynamic activity and may require dense time series data in order to be detected. After the irrigation episode, the surface soil moisture could increase from a value of 15–20 vol.% to reach 25–40 vol.% and then may dry out to reach the same value before irrigation two to three days after the irrigation (El Hajj et al. 2014). Therefore, an irrigation event occurring four to five days before the SAR acquisition could be hardly detected.

The availability of S1 data has encouraged studies to discover the potential of the S1 time series to detect irrigation events at the plot scale. A study performed by Bazzi et al. (2020a) proposed a decision tree algorithm that detects the irrigation episodes occurring on the plot scale in a quasi-real time scenario. To build the decision tree algorithm, the sensitivity of the S1 SAR signal following irrigation events was studied over several study sites located in Catalonia in Spain, West Occitanie and East Occitanie in France and the PACA region in France. The decision tree called the IEDM (Irrigation Event Detection Model) detects the change in the σ^0 values at plot scale (σ^0_{plot}) and compares it to the σ^0 values derived at basin or grid scales σ^0_{Grid}. It assumes first that when the surface soil moisture at the plot increases between two consecutive S1 acquisitions, the σ^0_{plot} value between these two dates increases (as shown in Figure 4.2). However, the increase in the σ^0_{plot} value could be due to either rainfall or irrigation. Thus, the challenge was to discriminate the increase in the σ^0_{plot} due to irrigation from that due to rainfall events. To overcome this limitation, Bazzi et al. (2020a) proposed to jointly use the σ^0_{Grid} at 10 km scale with the σ^0_{plot}. They suppose that when a rainfall event occurs between two S1 images, all of the bare soil agricultural plots within a given radius (e.g. 10 km × 10 km) must show an increase in their soil moisture values and consequently in their σ^0_{plot} value. Thus, due to a rainfall event between two S1 images, the average σ^0 calculated for all of the bare soil plots within a 10 km grid cell (σ^0_{Grid}) would increase. This correlation between soil moisture at grid scale and rainfall was also demonstrated by Bazzi et al. (2019b). In the IEDM, the increase in the soil moisture values at an agricultural plot due to irrigation was detected by comparing the increase in the σ^0_{plot} to that of the σ^0_{Grid}. When both σ^0_{plot} and σ^0_{Grid} increase simultaneously between the two S1 images, then a rainfall event occurred and if an irrigation event had also occurred it would not be detected. When the σ^0_{plot} increases between two S1 images and the σ^0_{Grid} value decreases or remains stable at low levels (stable and low σ^0_{Grid} values mean persistence of dry conditions), then an irrigation event most probably occurred at the plot.

For each agricultural plot and at each S1 image, the IEDM gives an irrigation likelihood of 0%, 25%, 50% and 100% (values empirically established). The four irrigation possibility values are directly related to the change in the σ^0_{plot} (plot scale) and σ^0_{Grid} (grid scale) values between two dates t_i and t_{i-1} : $\Delta\sigma^0_{plot} = \sigma^0_{plot}(t_i) - \sigma^0_{plot}(t_{i-1})$ and $\Delta\sigma^0_{Grid} = \sigma^0_{Grid}(t_i) - \sigma^0_{Grid}(t_{i-1})$.

– The value 0% corresponds to the absence of any irrigation chance between t_i and t_{i-1} due to either decrease in the σ^0_{plot} ($\Delta\sigma^0_{plot} \leq -0.5$ dB) or increase in the $\Delta\sigma^0_{Grid}$ ($\Delta\sigma^0_{Grid} \geq 1$ dB rainfall event occurred).

– The low irrigation possibility weight (value = 25%) corresponds to the absence of rainfall events between t_i and t_{i-1} ensured by the stability or the decrease in σ^0_{Grid} ($\Delta\sigma^0_{Grid} \leq 0.5$ dB) and a slight modification in the σ^0_{plot} between t_i and t_{i-1} ($-0.5 \leq \Delta\sigma^0_{plot} < 0.5$ dB).

– The medium irrigation possibility weight (value = 50%) is associated with a moderate increase in the σ^0_{plot} ($0.5 \leq \Delta\sigma^0_{plot} < 1$ dB).

– The high possibility of irrigation (value = 100%) is ensured when the σ^0_{plot} strongly increases ($\Delta\sigma^0_{plot} \geq 1$ dB) with the absence of rainfall events between t_i and t_{i-1} ensured by the stability or the decrease of σ^0_{Grid}($\Delta\sigma^0_{Grid} \leq 0.5$ dB).

Besides the major filters discussed above, the IEDM uses some auxiliary filters to confirm the existence of an irrigation event or to remove falsely detected irrigation events. One of the additional filters considers the surface soil moisture estimations at plot scale estimated using the neural network proposed by El Hajj et al. (2017). For example, when the σ_{plot}^{0} slightly increases between t_i and t_{i-1}, the IEDM uses the soil moisture estimation at plot scale to confirm the existence of low irrigation possibility weight (value = 25%). If the plot's soil moisture estimation at time t_{i-1} is high (≥20 vol.%) due to either rainfall or irrigation and remains with high value to time t_i without rainfall event detected by σ_{Grid}^{0} (humid soil conditions persisted from time t_{i-1} to t_i), then the irrigation at low possibility is confirmed. Moreover, a filter using the NDVI time series is applied to reduce some falsely detected irrigation events. These false detections could be due to the increase in the σ_{plot}^{0} values related to the change of surface roughness or vegetation cover. The NDVI filter proposes that if an event is detected with low NDVI value at date t_i (NDVI < 0.4) and the NDVI value 1 month later at t_{i+30} decreases or remained stable (NDVI(t_{i+30}) − NDVI(t_i) ≤0.1), then the event is discarded (crop cycle in decreasing stage or persistent bare soil conditions).

Figure 4.3(a) shows the temporal profile of σ_{plot}^{0} (in dashed black) and σ_{Grid}^{0} (in dashed red) for an irrigated grassland plot located in the Crau plain of southeast France. The NDVI shows two growth cycles occurring on the plot between July and October (green line). Six irrigations were applied on this plot with a frequency of approximately one irrigation each 10 days. Blue points show the irrigation events detected by the IEDM (blue point) on each acquired S1 image. In fact, after each irrigation event, the σ_{plot}^{0} value increases between 1.3 dB and 1.7 dB, whereas the σ_{Grid}^{0} values remained stable and at low level during the dry summer season (no rainfall events). Figure 4.3(b) shows also the temporal profile of σ_{plot}^{0} (in dashed black) and σ_{Grid}^{0} (in dashed red) for an irrigated maize plot located in the irrigated basin of Catalonia region northeast Spain. The real in situ irrigation events are not available for the region but the irrigation episodes are clearly detected by the IEDM. Between February and May, both σ_{plot}^{0} and σ_{Grid}^{0} increases simultaneously following rainfall events and then decreases a couple of days later. With the beginning of the summer and the start of the maize cycle, σ_{plot}^{0} shows several increases (between 1.4 and 1.8 dB) with a stability of the σ_{Grid}^{0} values at low level (–13 dB) indicating dry conditions at grid scale and absence of rainfall events. Thus, the IEDM will identify these increases in the σ^0_{plot} values which are mainly related to the increase in the soil moisture of the plots due to irrigation events. Consequently, nine irrigation episodes were detected by the IEDM on the irrigated maize plot in Figure 4.3(b).

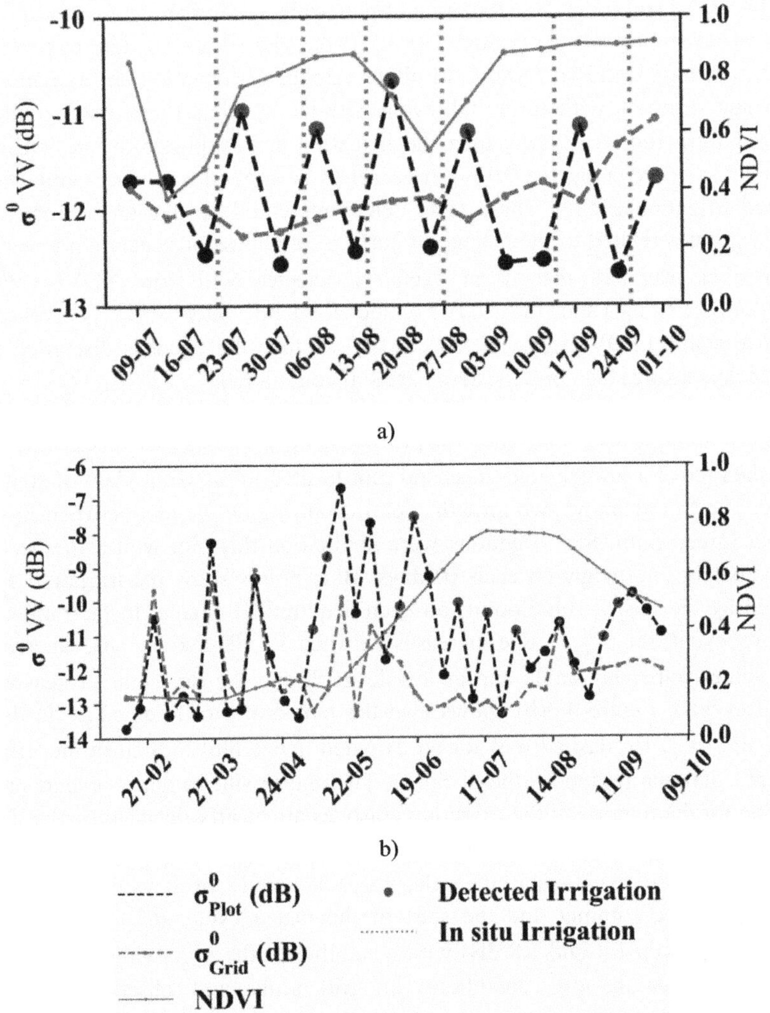

Figure 4.3. *Example of the detection of irrigation events using the IEDM over (a) irrigated grassland plot and (b) irrigated maize plot. For a color version of this figure, see www.iste.co.uk/mallet/multitemporal.zip*

The IEDM was validated on four different study sites having different climatic properties. One site located in the Catalonia region (Spain) showed that the IEDM did not detect any irrigation event over 90% of the non-irrigated plots and detected more than two irrigation events on 72% of the irrigated plots (overall accuracy). In another study area located in a humid region in west Occitanie of southern France, the application of the IEDM detected more than one irrigation event on 91% of the irrigated plots and more than two irrigation events on 75% of the irrigated plots. In a third study site located near Montpellier city in southeast France, the application of the IEDM detected 28 out of 33 irrigation events registered over three irrigated maize plots (Bazzi et al. 2020a) with a detection accuracy reaching 85%. Over a fourth study site in the Crau plain in southeast France, the IEDM was applied on 45 irrigated grassland plots having a detailed in situ irrigation calendar showing the exact irrigation dates at each plot. The comparison between the detected irrigation events and the in situ recorded irrigation events showed that the IEDM is capable of detecting 82.4% of the irrigation events with an F_score value reaching 73.8% (Bazzi et al. 2020b).

4.3.2. *Machine learning algorithms for mapping irrigated areas*

The capability of discriminating irrigated from rain-fed crops using optical and SAR time series data have encouraged researchers to develop classification approaches able to spatially map irrigated and rain-fed plots. Studies lastly focused on the Sentinel-1 and Sentinel-2 data as a main source for optical and SAR imageries due to the availability of the data in free open access modes, the high revisit time of the satellites and the high spatial resolution well adapted to plot scale irrigation mapping. To classify irrigated and rain-fed crops, machine-learning algorithms were deployed to obtain classifiers capable of judging each plot as either irrigated or not using optical and/or SAR time series data. Machine learning models are based on training an algorithm by a sufficient amount of optical and/or SAR data for in situ reference irrigated and rain-fed plots in order to classify the different agricultural plots of the zone as irrigated/rain-fed crops. These called supervised classifications are widely reported in literature for land cover mapping applications including the irrigated area mapping. Recently, Bazzi et al. (2019a) tested several approaches for mapping irrigated areas using two machine-learning algorithms (RF and NN) and Sentinel-1/2 time series data. Over a study site located in Catalonia region of northeastern Spain (Figure 4.4), a sufficient amount of in situ irrigated/rain-fed sample plots were available over the study site to build supervised classification models.

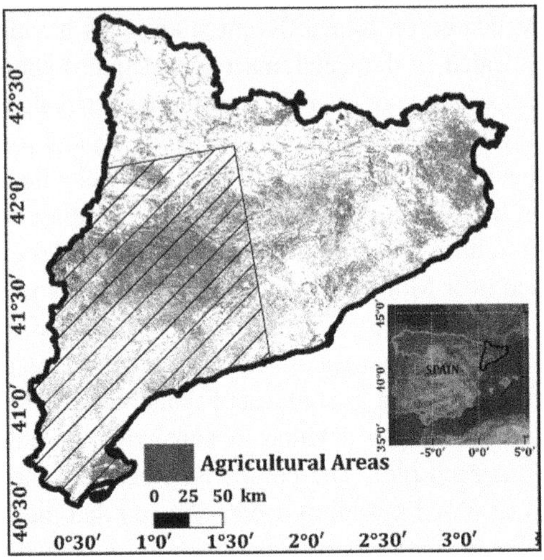

Figure 4.4. *Location of the Catalonia study site. Hatched areas represent the studied area. For a color version of this figure, see www.iste.co.uk/mallet/multitemporal.zip*

Using 82 S1 images acquired between August 2017 and December 2018, a dense S1 backscattering coefficient time series was obtained for each plot in VV and VH polarizations. As discussed in the previous section, the σ^0 value at grid scale σ^0_{Grid} could be a good representative of rainfall events and helps remove the ambiguity between rainfall and irrigation events. Thus, the σ^0_{plot} was used jointly to σ^0_{Grid} (at 10 km × 10 km) by substracting the grid scale σ^0_{Grid} from plot scale σ^0_{plot} to remove the effect of rainfall events present in the σ^0_{plot} time series. Later, the principal component analysis (PCA) and the wavelet transformation (WT) were applied to S1 temporal series obtained in both VV and VH polarizations for each plot. The PC values and the WT values were then integrated into an RF classifier to map irrigated areas. In addition to the use of the PC and WT values in a RF classifier, Bazzi et al. (2019a) tested the use of the VV and VH S1 time series in a convolutional neural network to map irrigated areas.

In addition to S1 time series, Bazzi et al. (2019a) tested the use of NDVI time series derived from S2 images for irrigation mapping using RF and CNN approaches. Table 4.1 summarizes the overall accuracies obtained for the several tested configurations using S1 and/or S2 data.

The results derived from the RF classifiers using the S1 time series show high overall accuracy with the PC values (90.7%) and the WT coefficients (89.1%). By applying the CNN approach on the raw S1 time series, a significant overall accuracy of 94.1% was obtained. The overall accuracy obtained using RF classifier with the PC values calculated on NDVI time series reached 89.5%, while that obtained using the CNN with the raw S2-NDVI time series reached 91.6%. The combined use of S1 and S2 data slightly enhanced the classification by RF classifier (using PC components) to obtain an accuracy of 92.3%. However, no significant change in the accuracy is obtained in the CNN approach using S1 and S2 time series compared to that using only S1.

	Data	S1 time series (%)	S2 NDVI time series (%)	S1 + S2 time series (%)
RF	PC values	90.7	89.5	92.3
	WT coefficients	89.1	–	–
CNN	Raw time series	94.1	91.6	94.5

Table 4.1. *Irrigation classification accuracies using S1 and S2 time series data over Catalonia site of northeastern Spain*

Supervised machine-learning models depend, in general, on the studied geographical zone and considered year, which may lead to unsatisfactory results if used over different regions or across several years with different climatic conditions. Consequently, a model built on one specific area may not lead to the same accuracy when applied over another region or over the same region for different years. In order to overcome the limitation of the supervised machine-learning models on the studied region and year, Bazzi et al. (2020c) proposed a framework capable of transferring the supervised classification model from one region to other region with different climatic conditions. The CNN model built over Catalonia using the Catalonian in situ data was chosen to be transferred to a humid region located in the southwest of France (in the Occitanie region) to map the irrigated area in the latter. The region adaptation called "Distilling before refine" consists of transferring the model from Catalonia (source area) and refining it with small amount of in situ data in the French site (target area) to map irrigated areas. First, the CNN model built on Catalonia is characterized by large volume of labelled samples. This model is denoted by the "teacher model". The teacher model is then distilled into a lighter model following a distillation procedure developed by Hinton et al. (2015). This lighter model, called the "student model" behaves in the same way as the teacher model but has less parameters than the teacher model. With less parameters, the student model is refined with small amount of in situ data collected over the French site (300 plots). Then the refined student model is used to perform the irrigation mapping. The main objective was that if small amounts of in situ data are available over a study site, which does not permit the creation of a well-trained supervised model, transferring an already existing model over another study site could give good classification accuracy. The accuracy of the irrigation mapping on the French study site reached 83% using the proposed transfer-learning framework.

Figure 4.5 shows the irrigation maps obtained over Catalonia using the source CNN (Figure 4.5(a)) and the irrigation map over southwest France obtained using the regionally adapted CNN of Catalonia (Figure 4.5(b)).

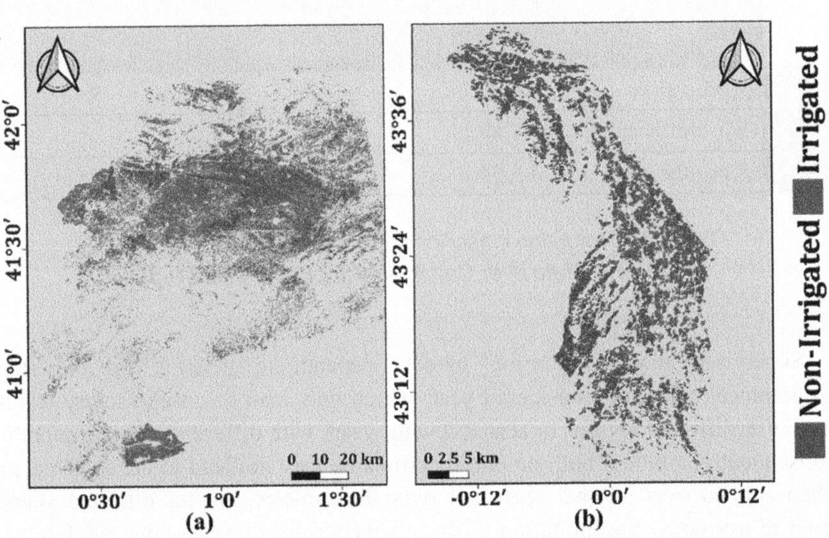

Figure 4.5. *Irrigation maps over (a) Catalonia using CNN and (b) southwest France using the CNN transferred from Catalonia with the transfer-learning framework. For a color version of this figure, see www.iste.co.uk/mallet/multitemporal.zip*

4.3.3. *Operational semi-supervised method for mapping irrigated areas*

Machine and deep learning algorithms such as the RF and the CNN have proven their efficiency in mapping irrigated areas using S1, S2 or S1 and S2 time series data. The only debatable issue regarding the use of these algorithms is the availability of in situ terrain data. In fact, developing such models requires an extensive in situ database of irrigated/rain-fed plots to train the model, especially when using deep learning methods. Acquiring an in situ database for irrigated/rain-fed plots on a yearly basis could be computational and time consuming. Therefore, the absence of terrain campaigns can block the possibility of obtaining yearly irrigation maps.

In trying to overcome this blocking issue, a new study performed by Bazzi et al. (2021) has proposed an automated semi-supervised classification model that maps irrigated areas at plot scale with no need for in situ data. The proposed methodology, referred to as "S^{2}IM" (Sentinel-1/2 Irrigation Mapping), generates automatically its own reference data to train the model without using in situ data. Selecting reference data depends on two metrics, one derived from the S1 time series and the other derived from the S2-NDVI time series. The reference data are then implemented in a RF classifier to classify irrigated/rain-fed plots using S1 and S2-NDVI time series. The method was tested over a study site located near Orléans city of northcentral France (Figure 4.6).

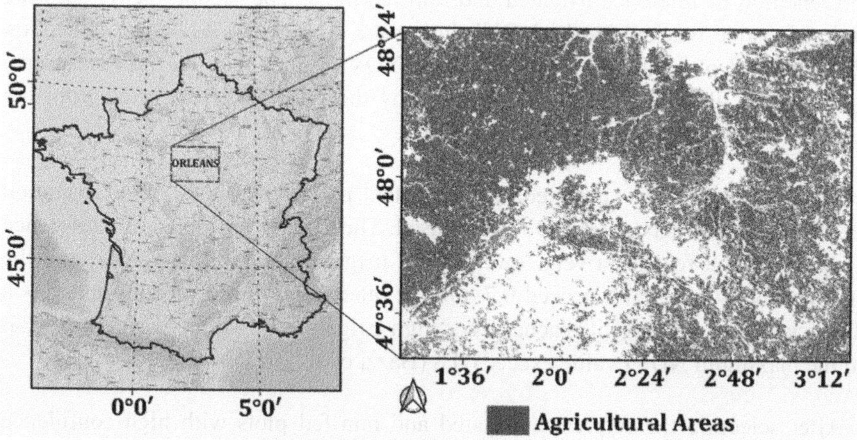

Figure 4.6. *Location of the study site near Orléans in northcentral France. For a color version of this figure, see www.iste.co.uk/mallet/multitemporal.zip*

Two criteria were used to select the reference dataset. The first criteria are based on the use of the previously discussed IEDM (section 4.3.1). All of the available S1 images from different S1 orbits over the study site were first downloaded. The overlapping of different S1 images originating from different S1 orbits acquisitions allows for obtaining four different S1 temporal series. Each S1 temporal series is at 6 days temporal resolution (S1 revisit time is 6 days in order to obtain the same image in the same orbit acquisition). Each of the four temporal series was treated separately due to the difference in the incidence angles between the orbit acquisitions as well as the diurnal effect between the morning and evening acquisitions (Brisco et al. 1990). The IEDM was applied on the four S1 temporal series in VV and VH polarizations. The IEDM gives at each image of each temporal series an irrigation possibility value (0%, 25%, 50% and 100%). Then, the irrigation possibilities obtained at each image of each temporal series were accumulated using both VV and VH polarizations for each plot (Bazzi et al. 2021). This cumulative irrigation possibility represents a metric for selecting the plots for reference data. If the plot is not irrigated, the IEDM would most probably detect no irrigation possibilities. Thus, a low value of the cumulated irrigation possibilities from the four temporal series could be evidence to deem that the plot is not irrigated during the season. If the plot is irrigated, the IEDM would be capable of detecting several irrigation events and thus the cumulative irrigation possibility for the plot would be high.

Another metric is used to ensure that the selected reference plots are a good representation of reference irrigated and rain-fed plots. The second metric uses the maximum value reached by the NDVI during the crop cycle. As shown in section 4.2.1, irrigated plots show higher maximum NDVI values than the rain-fed crops. For each plot, the maximum value reached by the NDVI was extracted from the S2-NDVI temporal series.

To select reference irrigated and rain-fed plots, the cumulative irrigation possibility and the maximum NDVI were used. The plot is considered as a reference rain-fed plot if it reaches a very low value of irrigation possibility (<25%) and the maximum NDVI does not exceed 0.7. On the other hand, a plot is considered as a reference irrigated plot if it reaches very high irrigation possibility value (>250%) and the maximum NDVI value exceeds 0.8 (Bazzi et al. 2021).

After selecting the reference irrigated and rain-fed plots with high confidence degree using both metrics, an RF classifier is used to build the classification model. The classifier uses the S1 time series (in VV and VH polarizations) and the NDVI time series of the selected reference data. The method was tested for 4 years (2017, 2018, 2019 and 2020). To assess the accuracy of irrigation mapping using the proposed method, Bazzi et al. (2021) used in situ terrain data to compare the produced irrigation map to in situ observed irrigated/rain-fed plots.

Figure 4.7 summarizes the accuracies obtained for the 4 years when applying the proposed S^{2}IM. The accuracies are reported as a function of the overall accuracy, the overall F_score, the F_score of the irrigated class and the F_score of the non-irrigated class. The validation of the irrigation map performed by S^{2}IM using the selected 2020 reference data generally shows very good accuracy. The overall accuracy obtained using S^{2}IM reaches 84.3% with a similar F_score value (84.1%). The irrigated class appears to have higher accuracy (86.4%) than the non-irrigated class (81.3%). In 2019, the results show that high accuracy was achieved for the irrigation mapping. Indeed, validation of the irrigation map with in situ data shows that all four accuracy metrics achieved high values between 92.5% and 93%. An overall accuracy of 81.8% was obtained for the year 2018 with an F_score of 82.2%. While the irrigated class showed good accuracy (86.8%), the non-irrigated class showed lower accuracy than the irrigated class (70.0%). When validating the irrigation map using in situ data in 2017, an overall accuracy of 72.8% was achieved with an F_score value of 74.0%. However, the accuracy between both classes was different. The irrigated class showed good accuracy (78.1%), whereas the non-irrigated class showed moderate accuracy reaching 62%.

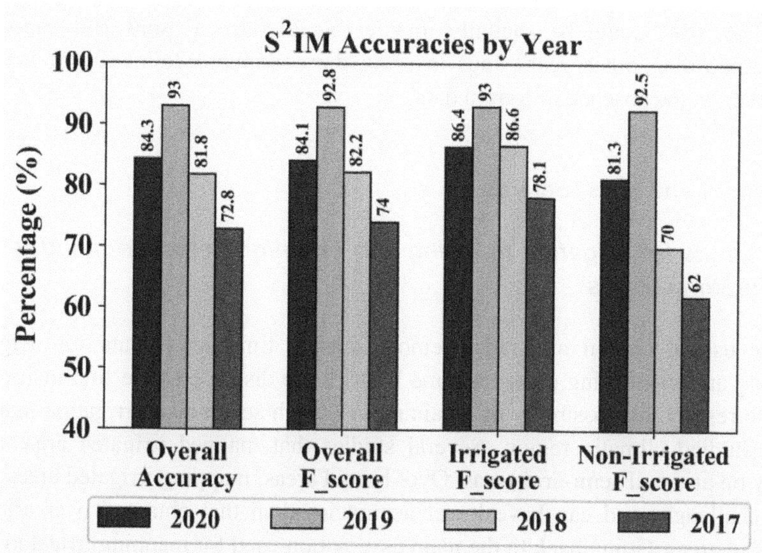

Figure 4.7. *Accuracies for irrigated area mapping using the S²IM. For a color version of this figure, see www.iste.co.uk/mallet/multitemporal.zip*

The difference in the obtained accuracies between the studied years was mainly correlated with the climatic conditions. In their study, Bazzi et al. (2021) noticed that the years attaining the highest overall accuracies (93.0% in 2019 and 84.3% in 2020), encountered the least amount of cumulative rainfall during summer season reaching 150 and 180 mm for 2019 and 2020, respectively. Nevertheless, years with less accuracies (81.8% and 72.8% in 2018 and 2017, respectively) encountered higher rainfall amounts in the summer reaching 210 mm and 320 mm for 2018 and 2017, respectively. Years with higher cumulative rainfall during summer had also the least accuracy for the non-irrigated class (78.1% and 62% for 2018 and 2017, respectively).

To analyze the effectiveness of the proposed S^2IM method, Bazzi et al. (2021) also compared the maps produced from the S^2IM using the selected reference data to maps produced using directly the in situ data in the RF classifier for the 4 years. They reported that the S^2IM had nearly the same performance as the traditional RF that uses in situ data for the training phase. The difference between the overall accuracy obtained using the S^2IM and the RF using in situ data varied between 1.7% and 6.2% only. Results of both classification approaches using automatically generated reference data and in situ data ensures that the proposed S^2IM could be powerful to map irrigated areas especially in the absence of terrain collected in situ data. The S^2IM could be spatially transferable to other regions and temporally transferable over other years, thus insuring the continuous follow-up of irrigated areas even in the absence of terrain data.

4.4. Limits and perspectives

4.4.1. *Limits for irrigation mapping and irrigation detection using remote sensing time series*

The use of optical and radar remote sensing time series data for irrigation mapping and monitoring presents some limits to be discussed. The first major limit that can restrict the possibility of obtaining very high accuracy in irrigation mapping is the studied climatic region. Several studies that mapped irrigated areas focus mainly on arid and semi-arid areas. Over humid areas, mapping irrigated areas faces more challenges and can have lower accuracies than that obtained over arid and semi-arid areas. Going back to the main assumption used for mapping irrigated areas using optical data, we can find that the mapping technique relies on the difference in the spectral signature (thus vegetation indices) between irrigated and rain-fed plots. Indeed, when the studied area is humid, the frequent rainfall events insure nearly the same development of the vegetation for both irrigated and rain-fed plots, which creates less difference in the vegetation index between the irrigated and rain-fed plots. The similarity between irrigated and rain-fed plots in humid areas is not only present on optical derived vegetation indices but also on the SAR temporal series. With frequent rainfall events, the difference between irrigated and rain-fed plots in the SAR time series also decreases. This is mainly caused by the fact that frequent rainfall events can decrease the chance of detecting irrigation events especially when using the S1 time series at 6 days temporal resolution. During these 6 days (time delta between two consecutive S1 images), if an irrigation event occurs and then was followed by a rainfall event, it is impossible to distinguish if the increase in the soil moisture values (thus the σ^0_{plot}) is due to rainfall or to irrigation. The effect of abundant rainfall events on the accuracy of mapping irrigated areas has been demonstrated by several studies including and not limited to the two studies performed by Pageot et al. (2020) and Bazzi et al. (2021). In the study performed by Pageot et al. (2020), it has been demonstrated over a humid study site in southwestern France that the difference between irrigated and rain-fed crops in the vegetation indices was low. In their study, despite of using climatic data (precipitation) in addition to the S1 and S2 time series in the RF classifier, the overall accuracy for irrigation mapping did not exceed 78%. Bazzi et al. (2021) showed that years affected by abundant rainfall events in the summer show less accuracy than years having drier summers.

Another major concern when using the C-band SAR data for irrigation events detection is the limited penetration of the C-band signal (wavelength ~6 cm) in very well-developed vegetation covers. Over soil with low vegetation cover, the emitted C-band SAR signal (wavelength ~6 cm) has no difficulty to interact with the surface layer of the soil (top 5 cm), which makes the radar backscattered signal highly correlated to soil moisture. In such case of soil, the detection of irrigation events based on the increase in soil moisture values between two S1 acquisitions could be easily done. When the canopy cover develops, the penetration of the SAR signal in the canopy becomes dependent on the crop type and canopy growth stage (Ulaby et al. 1982). In such case, the emitted SAR signal in C-band may not be able to reach the soil layer and thus the soil moisture contribution in the backscattered SAR signal will be negligible. As a result, the detection of the irrigation based on the change in the soil moisture values could be difficult due to the absence of the soil moisture information in the backscattered SAR signal.

The penetration of the C-band SAR signal and the vegetation contribution could also be different among several crop types. For cereal crops such as wheat and some grassland crops, the crop growth until a NDVI value of 0.7 does not strongly attenuate the C-band SAR backscattered signal. Beyond a NDVI of 0.7, the sensitivity of the backscattered signal becomes negligible to soil moisture content due to very high attenuation of the C-band SAR backscattered signal in such crop types (cereals and similar grassland). The low penetration of the C-band signal over cereals in the case of very well-developed vegetation cover has been demonstrated by several studies (El Hajj et al. 2018; Nasrallah et al. 2019). This low sensitivity of the C-band signal to soil moisture values for developed vegetation cover has limited the capability to detect the irrigation events in the study performed by Bazzi et al. (2020b) over irrigated grassland plots. They showed that in one of the grassland cycles that was rich in grasses such as coarse hay, which are similar in geometric structure and type to cereal crops, the detection of the irrigation events was extremely hard for NDVI greater than 0.7, due to the low soil moisture contribution to the backscattered radar signal. For maize crop, the penetration of the

C-band SAR data into the canopy cover is different from the wheat or grassland. In two studies by El Hajj et al. (2018) and Joseph et al. (2010), it has been showed that surface soil moisture in maize plot could still contribute to the C-band SAR backscattering signal even at maximum biomass stage (thus high NDVI values). This soil moisture contribution even in developed maize vegetation cover is mainly explained by the significant soil-vegetation scattering pathway that includes soil moisture information. Thus, the scattering along the soil-vegetation pathway present in dense maize cover allows for obtaining soil moisture information (Macelloni et al. 2001) and thus detect irrigation events.

The third major concern for the use of S1 time series data in detecting irrigation events is the temporal resolution of the satellites. Currently, the S1 satellites are providing images at 6 days revisit time. Benefiting from several S1 orbits, this temporal resolution can be enhanced to obtain two to four images within these 6 days. However, joining the S1 images from different S1 orbits can lead to unsatisfactory results in irrigation detection due to the effect of having several incidence angle and the morning/evening diurnal effect on the σ^0 values. Irrigation is a time dynamic activity where its detection is also affected by the other factors such as the evaporation rate, temperature and soil properties. With high temperatures and high evaporation rates, the effect of the irrigation on the soil moisture value for a given S1 image could not be clear if the irrigation took place a long time before the acquisition date of the S1 image. Thus, irrigation detection using SAR data could be limited to the time laps between the irrigation date and the SAR image date. A study performed by Bazzi et al. (2020b) analyzed the effect of the 6 days revisit time of the S1 on the detection of irrigation events. They showed that for low vegetation cover (NDVI < 0.7), the irrigation event could be detected, using the change in soil moisture values, until 2–3 days after the irrigation event. Beyond three days, the detection of the irrigation event becomes difficult. For NDVI greater than 0.7, they showed that it is possible to detect the irrigation if it occurs on the same day of the S1 acquisition or 1 day before. However, more than 2 days between the irrigation and the S1 acquisition makes it difficult to detect the irrigation event with the presence of well-developed vegetation cover. Therefore, the temporal resolution of the SAR satellite plays an important role in the detection of irrigation events.

4.4.2. *Perspectives in irrigation mapping using remote sensing time series data*

Based on some limits previously discussed for irrigation mapping using remote sensing time series data, this section presents some perspectives. First, the most important limit concerning the penetration of the C-band SAR data in developed vegetation covers necessitates the use of higher wavelength SAR data to reduce the effect of vegetation. Previously, it has been demonstrated that the L-band SAR signal (wavelength ~24 cm) penetrates the vegetation cover well more than the C-band for several crop types (El Hajj et al. 2018; Joseph et al. 2010). The vegetation attenuation in the L-band SAR signal is less than that of the C-band (Ulaby et al. 1982). Therefore, the change in soil moisture values due to irrigation could be better detected using L-band in the presence of well-developed canopy cover. Using L-band can lead to better accuracy in both irrigation event detection and irrigation mapping. Nevertheless, the L-band satellites currently operating still lack the high temporal resolution and the free data accessibility. Satellites in L-band such as PALSAR-2 are not in an open access mode and have very low temporal resolutions (very often one or two images per month). As a future perspective, several L-band satellites are to be launched soon such as the NISAR NASA-ISRO SAR (planned in 2022), the Tandem-L (German Aerospace Center, planned in 2022) and ALOS-4 (JAXA planned in 2022). Some of these missions (NISAR and Tandem-L) aim at providing open access images on a weekly basis temporal resolution. Thus, the open access L-band satellite data opens the way towards better mapping of irrigated areas because (1) L-band will allow better detection of irrigation events if the vegetation is well developed and (2) the next L-band data combined with C-band data will give several radar acquisitions per week, which allows for reducing the time laps between irrigation events and SAR acquisitions and thus detecting almost all of the irrigation events.

Another perspective for irrigation mapping includes the cartography of irrigated fruit trees. Mapping irrigated trees using satellite imagery is still an unexplored subject. Specifically, irrigating vineyards is becoming an imperative issue in several regions in Europe and especially in the south of Europe. Irrigation mapping for vineyards presents many challenges. Notably the use of micro-irrigation and low applied doses in vineyards make it difficult to detect irrigation by the approaches developed for irrigated crops by sprinklers or gravity irrigation techniques. Methods developed for soil moisture estimation or irrigation mapping (detection) at major crops (cereals and grassland) cannot be directly applied on vineyards because of the complexity of vineyard plots (presence of metal and wooden stakes, metal wires, etc.). Thus, one of the future perspectives is to develop a method capable of estimating soil moisture and detecting irrigation events at vine plots.

4.5. Conclusions

In this chapter, a synthesis of irrigation mapping and monitoring using high-resolution remote sensing data is presented. Optical and SAR time series data are effective data sources for discriminating between irrigated and rain-fed areas. Specifically, with Sentinel-1 (S1) and Sentinel-2 (S2) data, this chapter demonstrates the capability of obtaining high-resolution irrigation mapping using these high-resolution satellite data. Optical vegetation indices time series derived from the S2 data shows important differences between irrigated and rain-fed plots especially in arid and semi-arid areas. The S1 time series data, which is mainly correlated to the soil moisture values, also show great potential for mapping irrigated plots. Several machine-learning models are tested and used through literature to obtain irrigation maps at high spatial resolution using S1 and S2 data. With the absence of in situ terrain data to train machine-learning models and map irrigated plots, recent studies proposed unsupervised and semi-supervised models for both irrigation event detection and irrigated area mapping using S1 and S2 time series data. The main limitations for irrigation detection and irrigation mapping are the temporal resolution of the SAR satellite and the penetration capability of the SAR signal (case of C-band using S1). Future perspectives include the use of L-band SAR data as it is well known for having higher penetration depth than the C-band data. The combined use of S1 C-band data and future planned L-band data can enhance the detection of irrigation events at plot scale.

4.6. References

Aubert, M., Baghdadi, N.N., Zribi, M., Ose, K., El Hajj, M., Vaudour, E., Gonzalez-Sosa, E. (2013). Toward an operational bare soil moisture mapping using TerraSAR-X data acquired over agricultural areas. *IEEE Journal of Selected Topics in Applied Earth Observations and Remote Sensing*, 6, 900–916.

Baghdadi, N., Zribi, M., Loumagne, C., Ansart, P., Anguela, T. (2008). Analysis of TerraSAR-X data and their sensitivity to soil surface parameters over bare agricultural fields. *Remote Sensing of Environment*, 112, 4370–4379.

Baghdadi, N., Camus, P., Beaugendre, N., Issa, O.M., Zribi, M., Desprats, J.F., Rajot, J.L., Abdallah, C., Sannier, C. (2011a). Estimating surface soil moisture from TerraSAR-X data over two small catchments in the Sahelian part of Western Niger. *Remote Sensing*, 3, 1266–1283.

Baghdadi, N., Abou Chaaya, J., Zribi, M. (2011b). Semiempirical calibration of the integral equation model for SAR data in C-band and cross polarization using radar images and field measurements. *IEEE Geoscience Remote Sensing Letters*, 8, 14–18.

Baghdadi, N., Choker, M., Zribi, M., El Hajj, M., Paloscia, S., Verhoest, N., Lievens, H., Baup, F., Mattia, F. (2016). A new empirical model for radar scattering from bare soil surfaces. *Remote Sensing*, 8, 920.

Bazzi, H., Baghdadi, N., Ienco, D., El Hajj, M., Zribi, M., Belhouchette, H., Escorihuela, M.J., Demarez, V. (2019a). Mapping irrigated areas using Sentinel-1 time series in Catalonia, Spain. *Remote Sensing*, 11, 1836.

Bazzi, H., Baghdadi, N., El Hajj, M., Zribi, M. (2019b). Potential of Sentinel-1 surface soil moisture product for detecting heavy rainfall in the South of France. *Sensors*, 19, 802.

Bazzi, H., Baghdadi, N., Fayad, I., Zribi, M., Belhouchette, H., Demarez, V. (2020a). Near real-time irrigation detection at plot scale using Sentinel-1 data. *Remote Sensing*, 12, 1456.

Bazzi, H., Baghdadi, N., Fayad, I., Charron, F., Zribi, M., Belhouchette, H. (2020b). Irrigation events detection over intensively irrigated grassland plots using Sentinel-1 data. *Remote Sensing*, 12, 4058.

Bazzi, H., Ienco, D., Baghdadi, N., Zribi, M., Demarez, V. (2020c). Distilling before refine: Spatio-temporal transfer learning for mapping irrigated areas using Sentinel-1 time series. *IEEE Geoscience Remote Sensing Letters*, 17(11), 1909–1913. doi: 10.1109/LGRS.2019.2960625.

Bazzi, H., Baghdadi, N., Amin, G., Fayad, I., Zribi, M., Demarez, V., Belhouchette, H. (2021). An operational framework for mapping irrigated areas at plot scale using Sentinel-1 and Sentinel-2 data. *Remote Sensing*, 13, 2584.

Bousbih, S., Zribi, M., El Hajj, M., Baghdadi, N., Lili-Chabaane, Z., Gao, Q., Fanise, P. (2018). Soil moisture and irrigation mapping in a semi-arid region, based on the synergetic use of Sentinel-1 and Sentinel-2 data. *Remote Sensing*, 10, 1953.

Brisco, B., Brown, R., Koehler, J., Sofko, G., Mckibben, M. (1990). The diurnal pattern of microwave backscattering by wheat. *Remote Sensing of Environment*, 34, 37–47.

Chen, Y., Lu, D., Luo, L., Pokhrel, Y., Deb, K., Huang, J., Ran, Y. (2018). Detecting irrigation extent, frequency, and timing in a heterogeneous arid agricultural region using MODIS time series, Landsat imagery, and ancillary data. *Remote Sensing of Environment*, 204, 197–211.

Deines, J.M., Kendall, A.D., Hyndman, D.W. (2017). Annual irrigation dynamics in the U.S. northern high plains derived from Landsat satellite data: Satellite-derived irrigation dynamics. *Geophysical Research Letters*, 44, 9350–9360.

Demarez, V., Helen, F., Marais-Sicre, C., Baup, F. (2019). In-season mapping of irrigated crops using Landsat 8 and Sentinel-1 time series. *Remote Sensing*, 11, 118.

Dheeravath, V., Thenkabail, P.S., Chandrakantha, G., Noojipady, P., Reddy, G.P.O., Biradar, C.M., Gumma, M.K., Velpuri, M. (2010). Irrigated areas of India derived using MODIS 500 m time series for the years 2001–2003. *ISPRS Journal of Photogrammetry and Remote Sensing*, 65, 42–59.

El Hajj, M., Baghdadi, N., Belaud, G., Zribi, M., Cheviron, B., Courault, D., Hagolle, O., Charron, F. (2014). Irrigated grassland monitoring using a time series of TerraSAR-X and COSMO-SkyMed X-Band SAR data. *Remote Sensing*, 6, 10002–10032.

El Hajj, M., Baghdadi, N., Zribi, M., Bazzi, H. (2017). Synergic use of Sentinel-1 and Sentinel-2 images for operational soil moisture mapping at high spatial resolution over agricultural areas. *Remote Sensing*, 9, 1292.

El Hajj, M., Baghdadi, N., Bazzi, H., Zribi, M. (2018). Penetration analysis of SAR signals in the C and L bands for wheat, maize, and grasslands. *Remote Sensing*, 11, 31.

Gao, Q., Zribi, M., Escorihuela, M., Baghdadi, N., Segui, P. (2018). Irrigation mapping using Sentinel-1 time series at field scale. *Remote Sensing*, 10, 1495.

Hinton, G., Vinyals, O., Dean, J. (2015). Distilling the knowledge in a neural network. *arXiv:1503.02531*.

Joseph, A.T., van der Velde, R., O'Neill, P.E., Lang, R., Gish, T. (2010). Effects of corn on C- and L-band radar backscatter: A correction method for soil moisture retrieval. *Remote Sensing of Environment*, 114, 2417–2430.

Macelloni, G., Paloscia, S., Pampaloni, P., Marliani, F., Gai, M. (2001). The relationship between the backscattering coefficient and the biomass of narrow and broad leaf crops. *IEEE Transactions on Geoscience and Remote Sensing*, 39, 873–884.

Nasrallah, A., Baghdadi, N., El Hajj, M., Darwish, T., Belhouchette, H., Faour, G., Darwich, S., Mhawej, M. (2019). Sentinel-1 data for winter wheat phenology monitoring and mapping. *Remote Sensing*, 11, 2228.

Pageot, Y., Baup, F., Inglada, J., Baghdadi, N., Demarez, V. (2020). Detection of irrigated and rainfed crops in temperate areas using Sentinel-1 and Sentinel-2 time series. *Remote Sensing*, 12, 3044.

Portmann, F.T., Siebert, S., Döll, P. (2010). MIRCA2000-Global monthly irrigated and rainfed crop areas around the year 2000: A new high-resolution data set for agricultural and hydrological modeling: Monthly irrigated and rainfed crop areas. *Global Biogeochemical Cycles*, 24. doi: 10.1029/2008GB003435.

Ren, J., Shao, Y., Wan, H., Xie, Y., Campos, A. (2021). A two-step mapping of irrigated corn with multi-temporal MODIS and Landsat analysis ready data. *ISPRS Journal of Photogrammetry and Remote Sensing*, 176, 69–82.

Salmon, J.M., Friedl, M.A., Frolking, S., Wisser, D., Douglas, E.M. (2015). Global rain-fed, irrigated, and paddy croplands: A new high resolution map derived from remote sensing, crop inventories and climate data. *International Journal of Applied Earth Observation and Geoinformation*, 38, 321–334.

Sharma, A.K., Hubert-Moy, L., Sriramulu, B., Sekhar, M., Ruiz, L., Bandyopadhyay, S., Mohan, S., Corgne, S. (2019). Evaluation of Radarsat-2 quad-pol SAR time-series images for monitoring groundwater irrigation. *International Journal of Digital Earth*, 12, 1177–1197.

Siebert, S., Döll, P., Hoogeveen, J., Faures, J.-M., Frenken, K., Feick, S. (2005). Development and validation of the global map of irrigation areas. *Hydrology and Earth System Sciences*, 9, 535–547.

Siebert, S., Kummu, M., Porkka, M., Döll, P., Ramankutty, N., Scanlon, B.R. (2015). A global data set of the extent of irrigated land from 1900 to 2005. *Hydrology and Earth System Sciences*, 19, 1521–1545.

Thenkabail, P.S., Biradar, C.M., Noojipady, P., Dheeravath, V., Li, Y., Velpuri, M., Gumma, M., Gangalakunta, O.R.P., Turral, H., Cai, X. et al. (2009). Global irrigated area map (GIAM), derived from remote sensing, for the end of the last millennium. *International Journal of Remote Sensing*, 30, 3679–3733.

Ulaby, F.T., Moore, R.K., Fung, A.K. (1982). *Microwave Remote Sensing: Active and Passive. Volume 2: Radar Remote Sensing and Surface Scattering and Emission Theory*. Addison-Wesley, Reading, MA.

5

Trends in Satellite Time Series Processing for Vegetation Phenology Monitoring

Santiago BELDA[1], Luca PIPIA[2] and Jochem VERRELST[3]

[1] *University of Alicante, Applied Mathematics, Space Geodesy Group, Alicante, Spain*

[2] *Institut Cartogràfic i Geològic de Catalunya (ICGC), Barcelona, Spain*

[3] *Image Processing Laboratory (IPL), Universitat de València, Spain*

Time flies over us, but leaves its shadow behind

– Nathaniel HAWTHORNE, 1860.

Optical remotely sensed data are typically discontinuous in time, with missing values due to cloud cover or noisy observations, which can be problematic for many applications. Enduring clouds over agricultural fields can mask key stages of crop growth, leading to unreliable monitoring practices such as yield predictions. Consequently, gap-filling solutions are needed for accurate crop phenology characterization. In this chapter, we will review and evaluate common gap-filling methods as well as flexible machine learning (ML) methods for the reconstruction of cloud-free Sentinel-2 maps based on a multi-year time series of established vegetation indicators, such as leaf area index (LAI) and Normalized Difference Vegetation Index (NDVI). The fitting methods Whittaker smoother and the Bayesian ML algorithm Gaussian processes regression (GPR) have proven to fill in the gaps in time series the most accurately when comparing a reconstructed LAI map against a

Multitemporal Earth Observation Image Analysis,
coordinated by Clément MALLET and Nesrine CHEHATA.

reference LAI map. Processed cloud-free LAI images were subsequently used to quantify phenology indicators, such as start- and end-of-season. Nevertheless, fitting methods tend to fail in correctly reconstructing time series trends when the gaps are too long in time, for example, due to persistent cloud cover for multiple weeks or months. Then the only solution is to retrieve some information concerning the missing dates from other data sources, such as from all-weather radar data. Progressing along this line, a time series fusion approach has been presented, whereby optical data are complemented with radar data. A multi-output GPR (MOGPR) model was proposed for this fusion task, which has the advantage that the added value of complementary datasets can be quantified. MOGPR was proven to be effective in fusing Sentinel-2 LAI and radar Sentinel-1 time series data: radar data turned out to be essential for realistic gap filling in case of long time gaps. Outcomes from this analysis underlines that fusion of optical with radar data will lead to improved time series analysis and so more reliable detection of key seasonal trends. As a second fusion application, MOGPR enabled to reconstruct NDVI time series for almost two decades on the basis of S2 and historical Landsat data. Practically, all of the methods explored in this chapter have been implemented into a new in-house developed toolbox, called DATimeS (Decomposition and Analysis of Time Series Software). DATimeS is a stand-alone GUI toolbox for time series analysis that brings together conventional interpolation techniques and versatile ML algorithms, and can then derive a suite of phenology indicators.

5.1. Introduction

From our beginnings, humans were only able to perceive a few square kilometers of their immediate landscape. Since the 1960s, that local and small-scale perception has changed thanks to the satellite observations, which have contributed toward the global monitoring and assessing of Earth's dynamic systems (e.g. global climate and land surface environmental change) (Townshend 1994; Ramachandran et al. 2011). The main goal of Earth Observation (EO) is to monitor the Earth and its interaction with the atmosphere. Satellite sensors provide useful data to tackle the problem and address many important applications with deep societal, environmental and economical implications. Among the objectives of the multiple EO missions launched in the last decades, primary importance has been given to natural and managed vegetation land covers (Chevrel et al. 1981; Huete et al. 2002; Zhang et al. 2003; Boles et al. 2004; Dierckx et al. 2014). The strong relationship between the response of vegetation in the visible and near-infrared spectrum and its biophysical activities led the preference toward optical satellite sensors, even if on average clouds prevent measuring more than 70% of Earth surface at any time over the globe (Chelton and Wentz 2005). Besides, only to some extent, the revisit time was considered a demanding constraint for these satellite missions, since its increase goes at the expense of the imagery's spatial resolution. This trade-off long narrowed down the analysis to global vegetation dynamics, with processing limitations over areas with persistent cloud covers.

Yet, the usage of repetitive satellite observations over a region, further referred to as time series, has opened the door to the monitoring of seasonal changes in vegetated areas through their spectral properties (Reed et al. 2009). In fact, the possibilities to monitor seasonal dynamics from satellite time series has never been as rich as in the current era of EO. We are witnessing an ever increasing amount of optical EO satellites, endowed with a either a spatial or temporal resolution, with current and upcoming EO satellite missions, such as the Sentinel constellation on behalf of the joint ESA/European Commission initiative Copernicus and the NASA A-Train satellite constellation. With the operational super-spectral Copernicus' Sentinel-2 (Drusch et al. 2012) and Sentinel-3 missions (Donlon et al. 2012) as well as the recently launched Hyperspectral Precursor and Application Mission (PRISMA) (Labate et al. 2009) and Environmental Mapping and Analysis Programme (EnMAP) (Guanter et. al. 2015), Hyperspectral Infrared Imager (HyspIRI) (Roberts et al. 2012) and Fluorescence Explorer (FLEX) (Kraft et al. 2012) imaging spectrometer missions, an unprecedented data stream for land monitoring becomes available. For instance, the unprecedented frequency of Sentinel-2 observations (every 5 days), with a nominal spatial resolution of 10-to-20 m, captures rapid changes of agricultural land-cover from national to field scale, serving as a major support for environmental monitoring and agricultural subsidy control (Drusch et al. 2012). Hence, Sentinel-2 time series allows for high-resolution coverage of large areas with systematic data acquisition with high-frequency sampling during critical phases of the crop growth cycle (Bolton and Friedl 2013). This constant data influx covering every corner of the globe requires adaptive processing techniques, for example, to enable dealing with cloud covers and noises. At the same time, the processing should also be accurate, robust, reliable and fast, when envisaging the quantification of crop dynamics such as plant growth or health status.

Given the current viability of multiple EO missions, a growing number of monitoring agencies and Earth System sciences depend on the influx of continuous satellite data, that is, gap-free spatiotemporal data. In reality, however, the need for continuous data is often unfulfilled, for multiple causes: (1) inadequate climatic conditions (clouds, snow, dust and aerosols); (2) instrumentation errors; (3) losses of image data during data transmission; or (4) low temporal resolution (i.e. high time needed to revisit and acquire data for the exact same location), among others. They degrade the availability/quality of spectral/temporal information required to retrieve land surface properties. Therefore, the impact of missing data on quantitative research can be serious, leading to biased estimates of parameters, loss of information, decreased statistical power, increased standard errors and weakened of findings (Dong and Peng 2013). For this reason, spatiotemporal reconstruction of gapped areas from satellite imagery is becoming a crucial step for monitoring purposes (Schwartz 2013), and for the knowledge of the lifecycle of vegetation, that is, vegetation phenology (White et al. 2005).

Accordingly, when aiming to process time series data, a critical aspect to deal with is that EO products are spatially and temporally discontinuous. This implies that the ability to process irregular time series becomes indispensable for studying land surface seasonal patterns. In this respect, the following challenges arise:

1) How can we achieve accurate gap-filled data given irregular remote sensing acquisitions?

2) How can we preserve sufficient prediction accuracy in case the temporal gaps become too large?

3) To what extent can these filled time series data be used for the detection of phenological indicators?

4) Can we reconstruct gap-filled time series data back in time?

This chapter is dedicated to tackle and answer these questions. It will do so by reviewing latest trends on image time series processing for the studying and quantifying of vegetation dynamics. In particular, the chapter will (1) address trends in fitting methods for filling up cloud-based gaps or for the reconstruction of regular cloud-free composite maps, (2) outline a novel fusion method that blends multi-source time series data and (3) present some methods and tools for further processing steps, such as the reconstruction of multiple years satellite products (e.g. LAI and NDVI), and the calculation of phenological indicators (i.e. crop key growth stages) along multiple seasons.

5.2. Time series processing for gap filling

5.2.1. *Gap-filling methods*

Essentially, time series gap-filling refers to the prediction of missing values in time. Mostly these missing values are within the time series dataset, so in principle interpolation methods would suffice. It is therefore no surprise that interpolation methods are commonly used in time series data processing as a first processing step. According to the recent review by Zeng et al. (2020), gap-filling methods can be categorized into (1) empirical smoothing methods, (2) data transformations and (3) curve fitting methods. In short, these three family methods are defined as follows:

1) *Empirical smoothing methods* operate over a local temporal window within the time series based on empirical assumptions. Often by making use of moving filters, time series are smoothed. A popular method is the Savitzky-Golay filter (Chen et al. 2004), which can be thought of as a generalized moving average. These methods can be used as first step, and then further processed with a subsequent gap-filling method so that noise artifacts are removed. Apart from moving window methods, all kind of smoothing methods can be used to fill up gaps. The most commonly used is smoothing splines, but also interpolation methods such as linear, polynomial and nearest neighbor can be used. Although these methods apply a local interpolation,

they are not following the general trends. Curve fitting methods would be better suited when it comes to irregular, noisy datasets.

2) *Data transformation methods* decompose the time series into cyclical, trend, seasonal and irregular (e.g. noise) components based on mathematical transformations (Zhang et al. 2005). Fourier transforms (Moody and Johnson 2001; Hermance 2007) and wavelet analysis (Ganguly et al. 2010; Sakamoto et al. 2005) are the most widely used data transformation methods to characterize recurrent patterns, such as the phenology of vegetation from satellite observations. However, these methods are only suitable for stationary data, and without aperiodic extreme behaviors.

3) *Curve fitting methods* apply mathematical functions to fit the time series curves to a specified function. Curve fitting is the most common phenology detection method. Widely used approaches include logistic models and variations to it (Zhang et al. 2003; Elmore et al. 2012; Cao et al. 2015; Beck et al. 2006; Fisher et al. 2006), asymmetric Gaussian functions (Jönsson and Eklundh 2004), quadratic function (De Beurs and Henebry 2004) and the nonlinear spherical model (De Beurs and Henebry 2005). The model fitting methods can effectively suppress the noise of data. In addition, they are expected to be more objective approaches and easier to adapt to a wide range of situations, as mathematical functions are used to approximate the time series trajectory of vegetation growth and no predefined thresholds or empirical constraints need to be applied.

Although this overview summarizes the commonly used gap-filling methods in remote sensing time series data processing, recently a new family of adaptive curve fitting methods emerged as alternative to the above-mentioned methods, being the family of machine learning (ML) algorithms. While for the last decade ML methods have become standard in classification and regression tasks (DeFries and Chan 2000; Abburu and Golla 2015; Verrelst et al. 2012), lately ML regression methods have been introduced in time series data processing as flexible curve fitting methods for gap-filling purposes (Belda et al. 2020). Earlier, the following ML methods were proven to perform highly adaptive in regression applications: (1) artificial neural networks; (2) decision trees (e.g. random forests); and (3) kernel-based MLs (e.g. support vector regression) (Verrelst et al. 2015, 2019). Hence, these methods would be a logical choice for testing in gap-filling applications of time series datasets, and they are only at the beginning of being exploited in time series processing. For a complete overview and more in-depth discussions of ML algorithms promising in gap-filling processing, we refer to the following publications: Smola and Schölkopf (2004); Camps-Valls et al. (2006); Rasmussen and Williams (2006); Thayananthan et al. (2006); Lázaro-Gredilla et al. (2014); Verrelst et al. (2019). In the following section, these methods will be quantitatively evaluated.

Among the multiple ML regression algorithms applicable to time series curve fitting tasks, special attention is paid to the Gaussian processes regression (GPR) (Rasmussen and Williams 2006). GPR belongs to the family of kernel-based regression methods, but carries out curve fitting modeling in a Bayesian framework, that is, it provides mean estimates along with uncertainty intervals. This distinct feature of uncertainty provision can open a unique source of information to assess the robustness of the predictions at various temporal scales (Verrelst et al. 2013). For instance, when the uncertainty becomes too large, for example, passing a predefined threshold, it can be decided not to consider the prediction as valid.

5.2.2. *Assessment of gap-filling methods*

Given the existence of multiple gap-filling methods, our interest is to evaluate their performances quantitatively against a common dataset. This section provides a demonstration case of reconstructing leaf area index (LAI) time series from Sentinel-2 (S2) acquisitions over an agricultural region in northwestern Spain. LAI is defined as the total one-sided area of leaf tissue per unit ground surface area, and is amongst the most important vegetation properties to quantify vegetation density (Watson 1947). Being a key component of biogeochemical cycles in ecosystems, LAI drives canopy microclimate and influences rainfall interception, amount of intercepted radiation and thus carbon gas exchange (Ewert 2004; Bréda 2003). Within agricultural monitoring LAI can also be related to crop biomass or crop yield (Zhou et al. 2019; Bai et al. 2019).

The S2 acquisitions cover an unevenly spaced time series starting from late November 2015 until September 2019. Several of the recorded S2 images are partly covered by clouds. It means that gap-filling methods are indispensable for subsequent phenology analysis. The selected area for the study is a crop region in Castile and Leon, dominated by wheat, maize, barley, beet, potato, rape and alfalfa. To speed up the whole of the processing, a single parcel consisting of 21,291 pixels was selected. The most commonly used gap-filling methods are evaluated, that is, smoothing methods, data transformation methods and curve fitting methods, and furthermore some promising ML regression methods. See Table 5.1 for an overview. It is therefore of particular interest to evaluate how well these ML fitting methods behave in reconstructing cloud-free time series datasets as opposed to the commonly used methods.

The subset shown in Figure 5.1 was selected for demonstration cases as part of a wider validation region of Sensagri Horizon2020 Project (Amin et al. 2018), for which a highly detailed land-cover map is yearly retrieved. The land cover map distinguishes between 50 specific crop types, 35 of them being arable crops, seven being irrigated crops and eight being for permanent crops (Gómez et al. 2018). The scene is mainly characterized by an intensive dryland agricultural system where the arable land comprises up to 80% of the available area. For the experiments, the atmospherically corrected S2 time series images were first converted to green leaf area index (LAI) using the GPR model developed in the framework of this project (Amin et al. 2018).

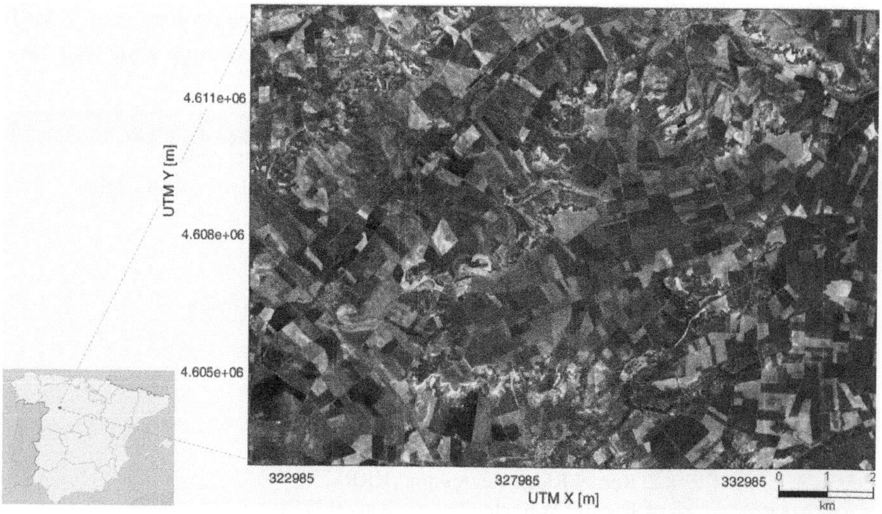

Figure 5.1. *RGB image of the test site in Castile and Leon region, northwestern Iberian peninsula, from Sentinel-2 capture of 2017, April 12. For a color version of this figure, see www.iste.co.uk/mallet/multitemporal.zip*

In order to evaluate the capabilities of the method to accurately reconstruct cloud-free time series LAI maps, a sufficiently large gap was artificially created. This was done by removing the LAI maps within the window during the period of 13 March until May 2, 2017, that is, artificially creating a data gap during most of the growing season. Within this gap, one LAI map (April 12, 2017) was taken as reference map. Subsequently, the reconstructed LAI map was quantitatively compared against the LAI map by calculating goodness-of-fit estimates. To do so, it is important to point out that the gap-filling methods have different modeling assumptions. By definition, parametric regression algorithms assume an explicit relationship between response (dependent) and explanatory (independent) variables using a finite number of parameters. On the contrary, for the ML fitting algorithms, the information is derived from the training data without any assumption about their statistical distribution or variable interrelations, making the number of parameters (theoretically) infinite. Parametric models such as linear regression, logistic regression and harmonic algorithms are typical examples of "learners" with a fixed size of parameters. Conversely, algorithms for nonparametric methods such as K-nearest neighbor, decision trees or GPR, the number of parameters grows with the size of the training set. For these reasons, the use of the traditional Akaike information criteria (AIC) (Akaike 1998) and Bayesian information criteria (BIC) (Schwarz 1978) that normalize the amount of parameters have not been taken into consideration for the analysis. Therefore, the statistical assessment measures used in this study are as follows: coefficient of variation R^2, root mean square error RMSE and relative RMSE (RRMSE (%)). At the same time, processing time was also recorded.

Gap-filling methods	
Smoothing and interpolation methods	Linear, Polynomial, Nearest, Next, Previous, Spline
	Piecewise Cubic Hermite Interpolating Polynomial (PCHIP)
Data transformation	Offset + harmonic analysis
	Offset + harmonic analysis + linear term
	Offset + harmonic analysis + linear term
	Offset + harmonic analysis using sliding window
Fitting methods (e.g. machine learning)	Bagging trees (BAGTREE)
	Adaptive Regression Splines (ARES)
	Boosting trees (BOOST)
	k-Nearest neighbors regression (KNNR)
	Gaussian Process Regression (GPR)
	Kernel Ridge Regression (KRR)
	Locally Weighted Polynomials (LWP)
	Support Vector Regression (SVR)
	Neural networks (NN)
	Random forests (RF2)
	Boosting random trees (RF1)
	Structured Kernel Ridge Regression with linear Kernel (SKRRlin)
	Relevance Vector Machine (RVM)
	Sparse Spectrum Gaussian Process Regression (SSGPR)
	Structured Kernel Ridge Regression with RBF kernel (SKRRrbf)
	Decision trees (TREE)
	Variational Heteroscedastic Gaussian Process Regression (VHGPR)
	Double logistic curve
	Whittaker

Table 5.1. *Interpolation methods used for gap-filling*

An overview of validation results against the reference image and processing time is provided in Table 5.2. Based on these results, the following trends are worth noting.

– The Whittaker fitting method and the ML regression algorithm GPR are evaluated as most accurately reconstructing time series data. Both methods reconstructed the reference LAI map with an R^2 accuracy of 0.89 (Whittaker) and 0.88 (GPR). Moreover, these two methods substantially outperformed the most commonly used method in reconstructing time series data, that is, Sigmoid, with a R^2 of 0.78. By taking additional processing speed into account, Whittaker is then clearly the most attractive method: it runs 56 times faster than GPR in fitting data. This method is thus particularly attractive to reconstruct cloud-free composite maps, for example, at regular time intervals (typically every 10 days). However, GPR, as being designed in a Bayesian context, provides associated uncertainties next to the predictions, which is a strong advantage as opposed to any other gap-filling method: the uncertainty intervals provide information as to whether the predictions are trustworthy.

Method	RMSE	R2	RRMSE	Processing time	
				Total (min)	Per pixel (s)
Next	0.46	0.77	14.93	0.26	0.0007
Polyfit	1.38	0.04	44.90	0.37	0.0010
Previous	0.82	0.52	26.70	0.26	0.0007
Spline	0.47	0.76	15.36	0.26	0.0007
Fourier1	0.82	0.26	26.65	1.27	0.0036
Fourier2	0.78	0.31	25.25	1.26	0.0036
Fourier3	1.43	0.05	46.36	2.78	0.0078
Sigmoid	0.60	0.78	19.54	758.05	2.1363
Whittaker	0.35	0.89	11.48	0.58	0.0016
ARES	0.95	0.43	30.71	31.51	0.0888
BAGTREE	0.85	0.63	27.75	148.97	0.4198
GPR	0.36	0.88	11.68	32.78	0.0924
KNNR	0.78	0.50	25.25	3.20	0.0090
KRR	0.50	0.76	16.09	5.60	0.0158
NN	0.78	0.56	25.26	1232.74	3.4740
RF1	0.50	0.80	16.11	46.91	0.1322
RF2	0.46	0.77	14.93	74.62	0.2103

Table 5.2. *Goodness-of-fit statistics and processing time for the reference versus LAI-reconstructed map as produced by the most promising gap-filling methods from Table 5.1 for 21,291 pixels*

– The commonly used gap-filling methods such as Sigmoid, Fourier, Spline performed considerably poorer in the order of R^2 0.70–0.80. Moreover, the Sigmoid method is computationally expensive and its run time is therefore slow. For instance, it takes about 23 times longer than GPR caused mainly by the tough tolerance and stopping criteria imposed in the iterative mathematical procedure. Conventional interpolation methods such as Next and Spline are probably more attractive for cloud-free time series generation, especially when it comes to large image processing: these two methods yielded good performances with a R^2 of 0.76–0.77 and reconstructed the reference map fast, that is, twice faster than Whittaker.

– It is also noteworthy that other popular ML methods did not perform better. Only RF1 reaches high accuracies with a R^2 of 0.80. Also KRR reached rather good accuracies, with an R^2 of 0.76 with reasonable time (map reconstruction in 5.6 min). But the other ML methods turned out to be exceedingly slow. This is especially the case for NN, which runs a factor 4.7 thousand times slower as opposed to the conventional interpolation functions (e.g. Spline). Its reason for slow processing lies in that for each pixel the model is trained again with the given time series data, which makes its processing burdensome and unappealing for large image processing. To accelerate the processing speed, it would be worth exploring training for multiple pixels at once, for example, at the parcel scale, so that training is done only once.

Overall, this quantitative comparison exercise suggests that Whittaker and GPR are the most appealing for the generation of gap-free time series data, with the difference that Whittaker is much faster, but GPR provides additional uncertainty estimates. Hence, these two methods would be the first choice for the generation of a clean, cloud-free time series dataset for subsequent calculation of phenology metrics. It should however be pointed out that these results are only valid for this dataset. Although in other time series reconstruction exercises we have observed top performances for these two methods, it is not necessarily the case that these methods are always top performing. For instance, the methods may be sensitive to local fluctuations and noises. Typically, gap-filling methods are combined with data smoothing methods, as they are expected to maintain the integrity of vegetation dynamics while removing the noise component (Beck et al. 2006; Hird and McDermid 2009). For most of the fitting methods, this may not be needed, as the fitting algorithm already functions as kind of a smoothing. It should also be remarked that care is required while using smoothing methods, as the smoothed profiles may deviate from the actual vegetation growth trajectory and have overcorrection problems (dampen the actual vegetation fluctuations). Altogether, it is difficult to determine the "best" method for all situations, and the selection of the most appropriate method should consider the biogeographical characteristics of the study region, potential noise sources in the used data and the general shape of time series vegetation indices curves (Hird and McDermid 2009; Zeng et al. 2020). Hence, an initial analysis of all factors playing a role in the time series processing is recommended. Such a systematic analysis is possible and streamlined in the here presented DATimeS toolbox (see section 5.5 for further information).

To illustrate the performance of the most interesting gap-filling methods for time series data reconstruction, an example is provided in Figure 5.2. This figure provides the original observations averaged over the studied area (in black), and then the reconstructed data for each of the 25 days. Apart from Whittaker and GPR, the most common gap-filling methods are Sigmoid, Nearest Neighbor, Spline, Sliding Windows Fourier and the ML regression algorithm KRR. It can be observed that Whittaker, GPR and KRR provide similar fittings and the smoothest profiles along the observations. Moreover, GPR provides additional uncertainty estimates, which tend to be small when data are available and grows when gaps become larger. Note also that these methods are very capable of reconstructing the seasonal trends over multiple years and this for irregular data. The conventional Fourier methods are not able to do so (not shown), as these methods provide systematic, harmonic trends. This limitation can be mitigated when introducing a sliding window. For instance, the Sliding Windows Fourier method provides sufficient flexibility to reconstruct irregular, multi-season time series. Yet, it can also be noted that this method still behaves in a more unstable way in comparison to the other fitting methods, that is, with small fluctuations in-between the main trends.

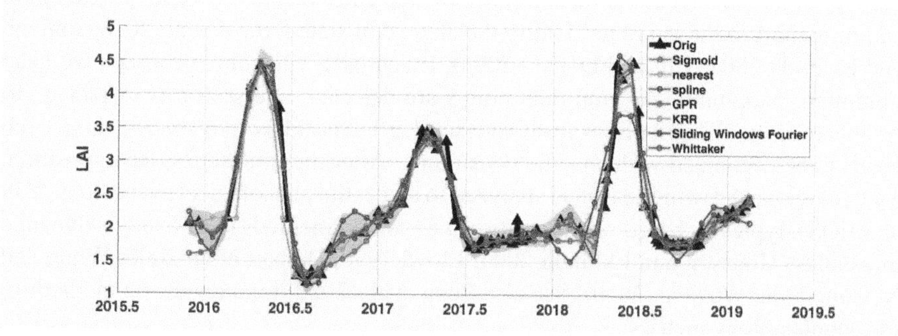

Figure 5.2. *Original and reconstructed time series of LAI using several gap-filling techniques. Interpolated values of time series at a higher sampling frequency (each every 25 days). The GPR uncertainties are shown in red shade areas. For a color version of this figure, see www.iste.co.uk/mallet/multitemporal.zip*

5.3. Time series processing for phenology indicators estimation

Once a gap-filled time series is generated, it can be used for the quantification of seasonal phenology indicators. Such spatiotemporally explicit knowledge of vegetation phenology is critical for understanding the change trend of natural seasonal phenomena and serve for agricultural production and global change studies (Yu et al. 2017; Tang et al. 2015; Ren et al. 2017). Moreover, the temporal and spatial variability of vegetation phenology variations help distinguish different vegetation types (Xue et al. 2014; Wang et al. 2013), especially crops (Wardlow and Egbert 2008). The following phenology metrics are commonly calculated:

1) start of season (SOS);

2) end of season (EOS);

3) day of maximum value: that is, when the largest value per season occurs (between SOS and EOS);

4) maximum value: largest value per season;

5) amplitude: difference between the maximum and the average of the left and right minimum values per season;

6) seasonal integral: area under the curve between SOS and EOS;

7) length of season: difference between SOS and EOS.

When aiming to calculate phenology indicators, a first step is to determine a season within a time series dataset. In fact, depending on the land cover type, multiple seasons can be identified within time series data (e.g. as a result of crop rotation), as well no season at all (e.g. over non-vegetated surfaces). The determination of the number of seasons (seasonal decomposition) typically makes use of local maximum and minimum of the curve to identify the change of state (Araya et al. 2018; Jönsson and Eklundh 2004; Hill and Donald 2003). Essentially, when three consecutive local minimum, maximum and minimum points are detected, one season is extracted. To be able to reduce contributions from undesired artifacts at low frequencies, this search needs to be optimized with specific constraints. After the time series decomposition, each individual growing season is analyzed to detect the phenological events (e.g. SOS and EOS), typically based on conventional threshold methods (e.g. fixed or dynamic thresholds) (Jönsson and Eklundh 2004; Lloyd 1990; Delbart et al. 2006; White and Nemani 2006; Wu et al. 2010). See also Zeng et al. (2020) for a review on methods to extract phenology metrics.

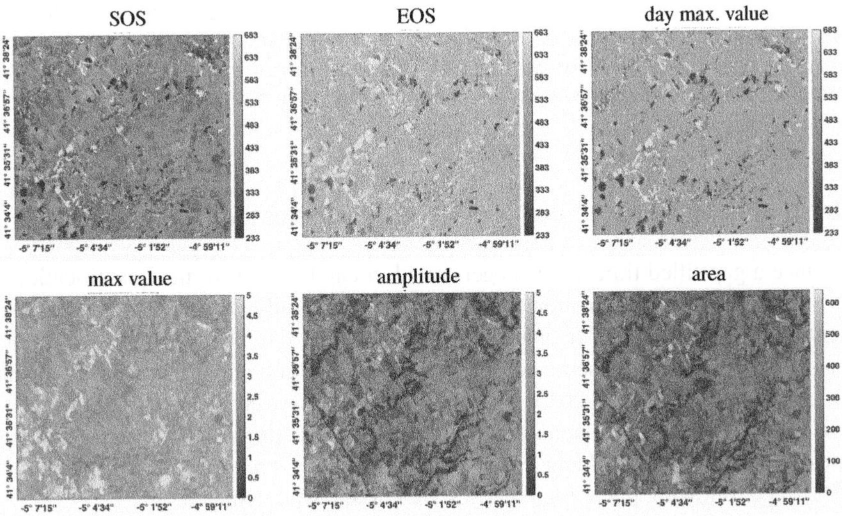

Figure 5.3. *Maps of phenological indicators estimated by using reconstructed Sentinel-2 LAI images with GPR. The spatial interpolation was applied for pixels that became impossible to calculate the indicators. SOS, EOS and Day of Maximum value (DOM) referred to 20160101 (day1). For a color version of this figure, see www.iste.co.uk/mallet/multitemporal.zip*

As a demonstration case, maps of the phenology indicators for the same subset as in the gap-filling section are calculated for a time series starting from January 1, 2016. Obtained maps are shown in Figure 5.3. The maps of SOS and EOS have been put within the same scale. Overall, for the large majority of the area, the SOS is consistent and starts beyond 300 days. This implies that the analysis did not identify the seasonality of the first year. Its reason lies in that the green-up already took place before the first of January 2016, and the software was therefore not able to identify the full cycle in the first year. One solution to enable detecting incomplete seasons is to relax the thresholds in the seasonality identification, for example, to allow a higher baseline for the SOS. For a few parcels, even no SOS was encountered in the second year, and SOS was only identified beyond 600 days. A piece of advice to ensure adequate identification of seasonal cycles is to make sure to have more consecutive years of data, taking into account the first and last years may be imprecise. Also the EOS map shows consistency across the landscape; for the majority of parcels, EOS takes place in the autumn of the second year. For a few parcels, EOS is a year earlier, because a full season is detected within the first year. In turn, some parcels also showed late EOS, that is, in the third year, in spatial agreement with the patterns of late SOS. The difference between EOS and SOS leads to the Length of Season map. As a result of the spatial variation of both SOS and EOS, this map is more heterogeneous, with parcels lasting only a few months, while seasonality for other parcels lasts more than 7 months. Probably of more interest are the maximum value and the day when this maximum value is obtained. These maps provide information when most vegetation cover (here: LAI) is created. Likewise, amplitude and area are other phenology metrics that can be related to crop productivity. Hence, these growth-related indicators can be valuable for crop productivity monitoring.

5.4. Fusion of time series products for improved gap filling

While the above-presented fitting methods are typically able to deal with even multiple medium-length data gaps (up to 2 months), often the situation occurs where gaps are too long for realizing the correct reconstruction of optical satellite data with a single source. Persistent clouds over agricultural fields can mask key stages of crop growth, for example, during the period of rapid spring green-up or during autumn senescence. Consequently, in long-lasting cloud-dominated areas, optical satellite imagery tends to fail in correctly reconstructing time series trends, which can be problematic, for example, for yield prediction. The only solution is to retrieve some information concerning the missing dates from other sources. An alternative remote-sensing monitoring solution is represented by synthetic aperture radar (SAR). SAR provides all-weather imagery which can potentially overcome this limitation (Mattia et al. 2003; Dusseux et al. 2014; Wiseman et al. 2014). Yet, given its high and distinct sensitivity to different surface properties, radar data alone face limitations to monitor directly phenological changes of crops, due to its relationship to also soil moisture and roughness (Ulaby et al. 1981; Ferrazzoli et al. 1992; Cloude 2010). Ideally, optical data with radar data can be fused over vegetation as long as just their common information is used (i.e. have a similar pattern in common), and thereby minimizing SAR response dependency on soil properties. It means that having a sound method able to extract shared information from multi-source data for the retrieval of a desired parameters becomes indispensable to tackle time series analysis in case of long data gaps. Nevertheless, the fusion of SAR and optical time series data still remains an open challenge (Schmitt et al. 2017; Kulkarni and Rege 2020). From this point of view, Multi-Output Gaussian Process regression (MOGPR) represents a promising solution, being a cutting-edge ML technique able to capture the dependencies among any kind of multiple, yet related, data collections (Álvarez et al. 2012). Essentially, MOGPR learns the relationship between even totally different sensors, builds a cross-domain kernel function able to transfer information across the time series and provides predictions (i.e. gap-filled intervals) on the regions where no optical data are available.

To demonstrate the potential of MOGPR to cope with long data gaps, we select a small subset within the above-mentioned study area and consider the time series of the S2 LAI time series and Sentinel-1 Radar Vegetation Index (RVI), which was shown to be the SAR descriptor correlated to LAI the most (Pipia et al. 2019). After a brief description of the theoretical background of MOGPR, we remove some of the dates from the LAI time series and show how (1) even the advanced gap-filling method fails over long data gap, whereas (2) MOGPR fulfills the reconstruction of LAI dynamics thanks to its correlation with RVI.

5.4.1. *MOGPR modeling*

One limitation of the standard GP formulation (Rasmussen 2004) is that it applies only to scalar functions, so that developing as many individual GP models as target variables does not take into account the relationships among outputs. This shortcoming is overcome by Multi-output Gaussian Process (MOGP) regression, based on the *linear model of coregionalization* (LMC) (Álvarez et al. 2012), also known as *co-kriging* in the field of geostatistics (Journel and Huijbregts 1978). See Álvarez et al. (2012) for details of the mathematical formulation of this supervised ML technique.

Here, we propose an adaptation of MOGPR formulation to model the active-passive synergy. We assume only one parameter is available from each Sentinel imagery, that is, the number of outputs $D = 2$. Referring to them as P_{S1} and P_{S2}, the collections $\mathcal{D}_{P_{S1}} = \{t_i, P_{S1,i}\}_{i=1}^{N_{P_{S1}}}$ and $\mathcal{D}_{P_{S2}} = \{t_i, P_{S2,i}\}_{i=1}^{N_{P_{S2}}}$ represent the information to be used for training. Moreover, we posit that the two outputs are obtained as a linear combination of latent functions from two independent GPs ($Q = 2$). By setting $R_q = 1$, we use only one latent function to describe each GP. This simplification, also known as *semiparametric latent factor model* (SLFM) (Álvarez et al. 2012), leads to the definition of the reference model as follows:

$$\mathbf{f}(t) = \begin{bmatrix} f_{P_{S1}}(t) \\ f_{P_{S2}}(t) \end{bmatrix} = \begin{bmatrix} \sum_{q=1}^{Q} \sum_{r=1}^{R_q} a_{1,q}^r u_q^r(t) \\ \sum_{q=1}^{Q} \sum_{r=1}^{R_q} a_{2,q}^r u_q^r(t) \end{bmatrix}$$
$$= \begin{bmatrix} a_{1,1}^1 u_1^1(t) + a_{1,2}^1 u_2^1(t) \\ a_{2,1}^1 u_1^1(t) + a_{2,2}^1 u_2^1(t) \end{bmatrix} = \mathbf{a}_1^1 u_1^1(t) + \mathbf{a}_2^1 u_2^1(t), \qquad [5.1]$$

where $a_{1,q}^r$ and $a_{2,q}^r$ are scalar coefficients, and $u_q^r(t)$ are latent functions sampled from the GP q with zero mean and covariance $\mathbf{K}_q$ (Rasmussen and Williams 2006). Accordingly, the two *coregionalization* matrices are rank-1 and related to the model's coefficients as:

$$[B_1] = \mathbf{a}_1^1 \mathbf{a}_1^{1^T} \qquad [B_2] = \mathbf{a}_2^1 \mathbf{a}_2^{1^T}. \qquad [5.2]$$

The expressions in [5.1] and [5.2] constitute the basic structure of the multi-output model. On the one hand, the two independent GPs account for details of the two parameters time series with a magnitude and at a time scale defined by the hyperparameters of each $\mathbf{K}_q$, optimized after the training step[1]. On the other hand, whether or not these details are shared between the two outputs throughout time is determined by the components of $\mathbf{a}_1$ and $\mathbf{a}_2$. Note that the ideal case is represented by a trained model with a dominant GP with all-ones vector $\mathbf{a}$, that is, completely shared by the two outputs, and a second GP with a null-vector $\mathbf{a}$. This corresponds to the case of two perfectly correlated parameters P_{S1} and P_{S2}. Finally, it is worth stressing that the only input to the trained MOGPR model is the time t at which the parameters must be estimated, as described visually in the sketch in Figure 5.4. Each output is predicted at any time along with the corresponding uncertainty estimation (σ_{PSi} in Figure 5.4).

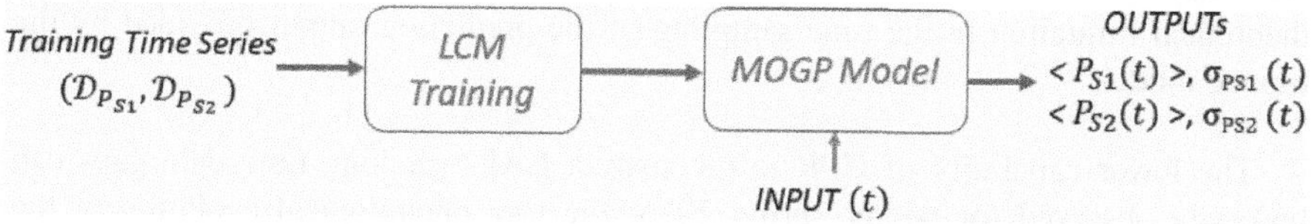

Figure 5.4. *Sketch of MOGP modelling based on the Linear Corregionalization Model (LCM). For a color version of this figure, see www.iste.co.uk/mallet/multitemporal.zip*

1 The hyperparameters are related to the kernel function to be use to define the distance criteria among samples. Here, we chose the anisotropic squared exponential (SE), a very flexible one that typically suffices to tackle most of the EO problems (Mateo et al. 2015).

5.4.2. *MOGPR assessment*

In order to assess the predictive power of the fusion method, we propose a quantitative assessment of the MOGPR gap-filling approach with the so-called leave-one-out technique applied to the image collection. Taking a group of LAI time samples out of the training collections, MOGPR predictions are assessed using these extracted values as reference. For the sake of clarity, we focus the analysis onto a single parcel selected within the study area, shown in Figure 5.5(a). Eliminating three samples in April 2016, three samples from April to May 2017 and one sample in February 2018 generates three artificial gaps nominally ranging from 25 up to 109 days, which exemplify realistic-to-extreme high-latitude S2 sampling conditions due to the prominent cloud cover. Concerning S1 collection, all of the captures from ascending and descending orbits covering the same time span were considered. First, speckle effects were reduced in space using a 7x7 Lee-Sigma filter, and in time using a 2-week Gaussian filter. Then, RVI was calculated from ascending and descending collections, and the images taken the same day from the two observation geometries were finally averaged (Pipia et al. 2019).

The assessment is performed both along time and space dimensions. First, the value of pixels within the parcel is averaged for each acquisition date. Then, the corresponding LAI time series was split into training-validation subset (section 5.4.1) to assess performance of MOGPR with respect to all of the methods in Table 5.2. Note that the validation data were not selected randomly according to a specific percentage, but in a way so as to create gaps of different length along the time series, from a few weeks up to almost 3 months. To highlight the benefits of the fusion approach, in Figure 5.5(b) we compare MOGPR retrievals to the estimations of GPR, which provided the best assessment performance along with Whittaker (see Table 5.2). The two predictions have been carried out every 10 days, but there is no theoretical limitation in the time sampling of the prediction output provided by the trained models.

The lower capability of GPR to reconstruct LAI over long time data gaps can be easily observed for winter–spring 2016; the true phenological evolution of the crop (magenta circles) is completely lost because all missing values are significantly underestimated. Conversely, the simultaneous use of SAR-based information allows a more correct retrieval of crop phenological stages, within the long data gap.

Similar results are instead obtained over shorter gaps, even if the quantitative assessment of the seven predictions carried out by GPR and MOGPR seems to indicate a slightly better performance of the latter approach ($R^2 = 0.93$, $RMSE = 0.36\text{m}^2/\text{m}^2$) with respect to the former one ($R^2 = 0.83$, $RMSE = 0.55\text{m}^2/\text{m}^2$). The images in Figure 5.6 offer a direct comparison of the

LAI reference map (a) obtained from S2 acquisition of April 27, 2016, over the selected parcel, and the maps predicted by the GPR (b) and MOGPR (c) models trained at pixel level. In (d), we show also the LAI histograms of the three maps. As already obtained for the time-dimension analysis, the spatial results confirm the higher error of GPR when attempting to estimate LAI over a long data gap, with significant underestimations of the real values (red line) over the studied area. Conversely, the use of an alternative source of information, such as RVI, with a degree of relationship with LAI, leads to a better recovery of the missing data (green line) and hence a more reliable reconstruction of the spatial patterns. Concerning the slight overestimation shown by MOGPR histogram, its interpretation is not straightforward. On the one hand, it may be due to spurious LAI-independent SAR information leaking into the retrieval (soil moisture changes, residual soil roughness effects, etc.). On the other hand, the smoothing process performed by the regression (see Figure 5.6) is expected to describe more likely dynamics of LAI evolution in time, meaning the overestimation may also represent an improvement of LAI actual values. In fact, the residual atmospheric correction errors affecting the reflectance estimation and the uncertainty of Sensagri LAI model (Amin et al. 2018) make each S2-based reference map an approximated description of LAI. In order to characterize the dynamics of vegetation phenology, the continuity condition among consecutive samples of the time series must be imposed over time, with the exception of crop sudden changes at the end of the season caused by harvest. These particular events should be dealt with caution, for example, probably by using specific covariance estimators that allow sudden discontinuities.

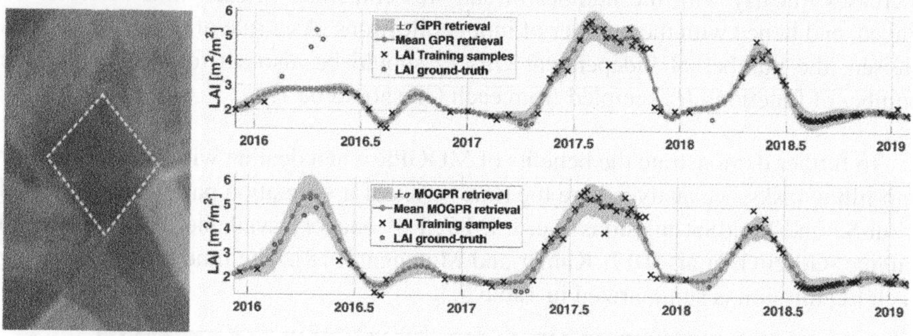

(a) (b)

Figure 5.5. *Assessment of standard GPR(a) and MOGPR(b) predictions (red circles) for data gap-filling on S2 cloud-free captures eliminated from training information (black crosses) and here used as reference (blue circles). The $\pm\ \sigma$ prediction uncertainty is represented by the boundary shade (pink) at any estimate. For a color version of this figure, see www.iste.co.uk/mallet/multitemporal.zip*

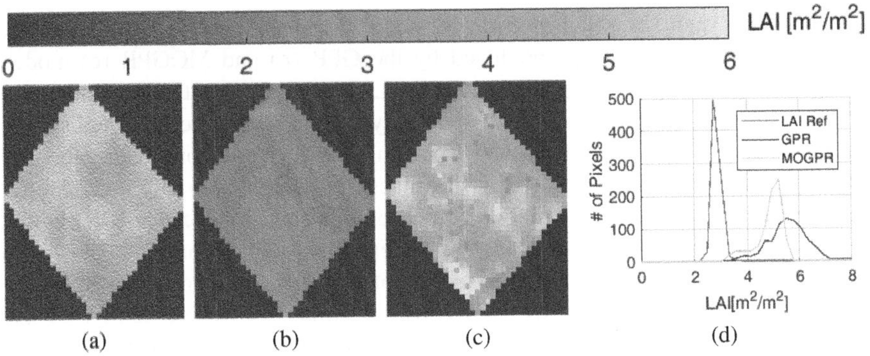

Figure 5.6. *Reference (a), GPR (b) and MOGPR (c) LAI maps for April 27, 2016. Retrievals obtained over a long data gap (from January 18 to June 6). For a color version of this figure, see www.iste.co.uk/mallet/multitemporal.zip*

5.4.3. *MOGPR generalization*

The mathematical factorization presented in section 5.4.1, as well as the performance assessment in section 5.4.2 for the case of LAI (optical) and RVI (SAR) time series, must be considered a demonstrative example of MOGPR potential. In fact, the number of outputs D is not limited to 2. The only drawback to be kept in mind is the computational cost, since the size of the final matrix $\mathbf{K}_q$ to be inverted increases linearly with the number of samples contained in each time series to be linked, and hence with the number of input collections. As a rule of thumb, once D is chosen, the number of independent GPs Q should be also set to D, whereas the number of functions R_q sampled from each GP should be 1.

To further demonstrate the benefits of MOGPR when dealing with multiple sensor gap-filling tasks, we analyze here the time series of a vegetation parameters obtained from S2 and LandSat missions (i.e. L5, L7, L8) imagery available in Google Earth Engine (Gorelick et al. 2017; Kumar and Mutanga 2018). The time span covered by each EO mission is summarized in Table 5.3.

Sensor	Landsat5 (L5)	Landsat7 (L7)	Landsat8 (L8)	Sentinel-2 (S2)
Start	1994/2/10	2000/01/18	2013/03/23	2017/04/12
Stop	2004/10/26	2016/02/22	2020/12/12	2021/01/21

Table 5.3. *Multi-sensor imagery for NDVI time series generation*

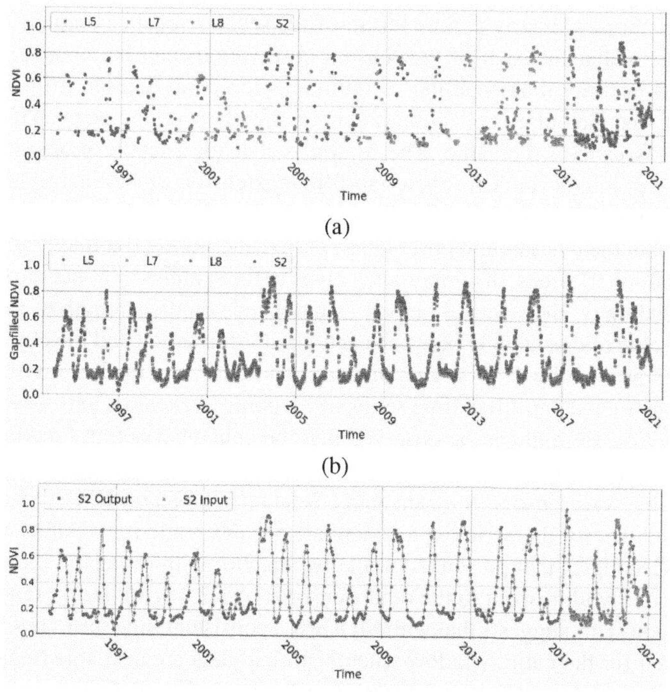

(a)

(b)

(c)

Figure 5.7. *4-sensor NDVI Input time series to MOGP (a), 4-sensor NDVI Gap-filled time series provided by MOGP (b), S2 input/gap-filled time series (c) based on multi-sensor information. For a color version of this figure, see www.iste.co.uk/mallet/multitemporal.zip*

For a meaningful comparison, we selected the green sector of the pivot area in the south-east part of the AOI (Figure 5.1). To avoid band and spatial resolution discrepancy across time, we calculated the NDVI vegetation parameter from each imagery and averaged it over the pivot area. The corresponding four input time series are shown in Figure 5.7(a). The different crop seasons can be detected along the almost 30-year observation period covered by the four sensors. As expected, the denser the sample distribution, the clearer the crop phenological evolution. In case a unique NDVI collection is needed for a specific sensor, two main issues must be taken into account: (1) the irregular sampling of each sensor due to cloud presence as well as its orbit/field-of-view specific limitation and (2) possible radiometric bias among the different sensors. MOGPR offers the opportunity to tackle these two issues at once, along with the possibility to define any desired time-step for the gap-filled output time series. The results obtained with MOGPR with $D = 4$ and $Q = R_q = 1$ with a sampling step of 15 days are described by the plots in Figure 5.7(b). All of the crop-season are now clearly visible for all of the sensors for the complete time span. Thus, it means that for each of the sensors observations can be reconstructed back and forward in time, thus enabling a continuous temporal dataset. Still, it is worth pointing out that the output collections are not perfectly overlapping even if the output time vector is exactly the same. The reason is that the specific properties of each sensor are kept into account for their gap-filling/extension to the final time coverage. For instance, possible tendency of L7 to provide lower NDVI values with respect to L8 is learned from the overlapping observation period and imposed throughout the whole output collection. Besides, the importance given to the samples of a specific time series varies depending on the selected output, as it depends on the strength of the link that MOGPR is able to disentangle between each input information and the chosen output, which is represented by the coregionalization matrices. As a final example, the input versus reconstructed gap-filled time series of S2 going backwards until 1994 is shown in Figure 5.7(c). Here, it can be observed that the generated output function respects the main temporal behavior of the input data, enclosing the overall observation period covered by the four sensors. It means that S2 data can be reconstructed back in time. At the same time, for the period when there are S2 observations available, all of those observed S2 samples that present an unlikely behavior with respect to the main trend (i.e. outliers) or with respect to the other sensors (L5,L7,L8) are nicely ignored in the reconstruction. This suggests that with MOGPR a consistent time series product can be reconstructed for the entire window when historical data are available (e.g. Landsat). This consistency opens the door for trends analysis such as phenology and long-term change detection.

5.5. Time series processing toolbox: DATimeS

The time series analysis was undertaken with a newly in-house developed toolbox named DATimeS (Decomposition and Analysis of Time Series software) (Belda et al. 2020). DATimeS is freely available as a powerful time series software that generates cloud-free composite maps and captures seasonal vegetation dynamics from regular or irregular satellite time series. It is a stand-alone Microsoft Windows application developed in Matlab. In comparison to exiting time series software packages (e.g. TIMESAT (Jönsson and Eklundh 2004), BFAST (Verbesselt et al. 2010), TIMESTATS (Udelhoven 2011), SPIRITS (Eerens et al. 2014) or EO Time Series Viewer (Jakimow et al. 2020)), DATimeS expands established time series interpolation methods with a diversity of advanced machine learning fitting algorithms.

In short, DATimeS is developed as a modular toolbox that can be applied to both images as well point time series data as stored in a text file. An overview of the DATimeS' modules contained in this version (v.1.10) is shown in Figure 5.8.

1) The first module, "Inputs", is mainly responsible for reading the data and masking cloudy pixels using a thresholding filter (i.e. user must provide a minimum

and maximum value to filter out pixels that fall outside these boundaries). Time series processing for spatiotemporal analysis can take place either on a single pixel, or on a stack of images. For single pixels, a .txt file is required. Imagery can be entered in TIFF or ENVI format from a specific folder. When loaded, the data will appear sorted in the DATimeS main window (Figure 5.9(a)).

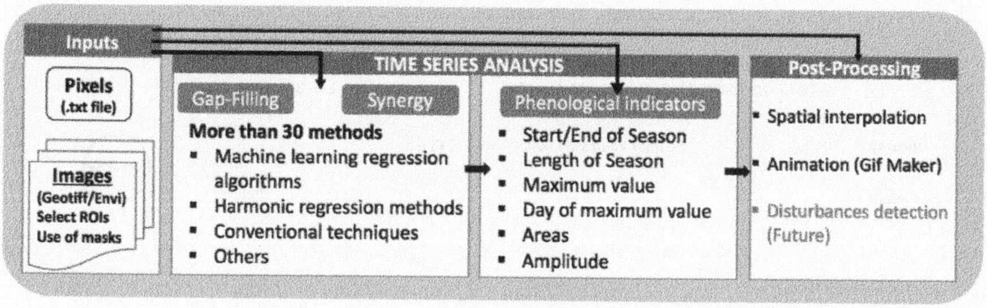

Figure 5.8. *Hierarchical design of DATimeS (v.1.10). For a color version of this figure, see www.iste.co.uk/mallet/multitemporal.zip*

2) The second module, "Time Series Analysis", is in charge of constructing composite images with any time step by applying the gap-filling (e.g. due to clouds) using a broad variety of advanced interpolation methods or establishing the synergy between different sensors (e.g. from active-passive sensors or from S2, L5, L7 and L8). Subsequently, it enables to calculate all kinds of phenological indicators. The GUIs corresponding to this module are shown in Figures 5.9(b) and 5.9(c).

3) The third module, "Post-processing", improves the obtained maps by spatial interpolation and gives the possibility to create videos, thus enabling easy visual comparison of different dates (not shown).

The most important module is the "Time Series Analysis" where the gap-filling and synergy methods can be selected and subsequent phenology indicators be computed. The user is also given the freedom of choosing whether or not to incorporate smoothing prior to the parameter estimation. This is commonly done, however it must be remarked that the ML methods perform a time series smoothing along with prediction. The phenological indicator estimation is recommended to be applied after the interpolation step so that cloud-free composite images are created, and trends become evident for easy phenological metrics derivation. Yet, it is also possible to go directly to this step. In this module, the growing seasons are identified and each individual season is then analyzed to estimate the phenological indicators (e.g. SOS and EOS) based on conventional threshold methods, analogous to some previous studies (Jönsson and Eklundh 2004; Lloyd 1990; Delbart et al. 2006; White and Nemani 2006; Wu et al. 2010).

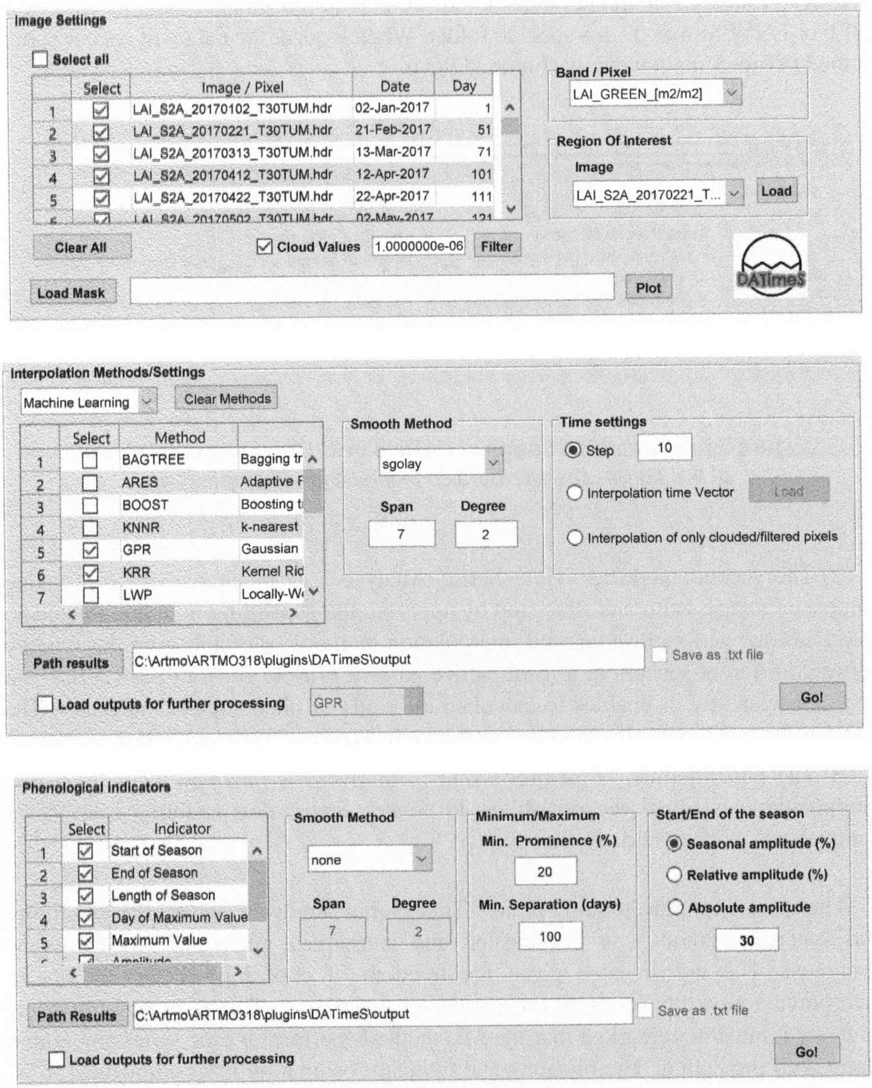

Figure 5.9. *DATimeS graphical user interface. (a) Main window of DATimeS. (b) Panel designed for gap-filling. It is divided into three logical areas: interpolation algorithms, smoothing methods and time settings. (c) Panel for phenological indicator estimation*

The computational strategy follows multiple steps. When running the phenological module for each pixel within an image, it (1) extracts time series data, (2) identifies automatically individual growing seasons within each year, (3) locates specific points within the growing season (e.g. SOS, EOS and peak), (4) computes integrals and cumulative integrals, and (5) stores the estimates in output ENVI or Tiff files. Because the time series data may consist of multiple cycles (e.g. seasons), the extracted features are added as layers (bands) in the output maps. Typically, for the majority of pixels, one seasonal cycle is encountered within a year. But due to noise or crop rotating practices, some pixels may have two or multiple cycles within a time series. Hence, those pixels will appear in the second or consecutive bands.

Three different methods have been implemented in DATimeS to calculate the SOS/EOS: (1) seasonal, (2) relative and (3) absolute amplitude. In the former case, the SOS/EOS are identified where the left/right part of the curve reaches a fraction of the seasonal amplitude along the rising/decaying part of the curve. The second approach is similar to the previous one, but now a mean amplitude is estimated considering the minimum/maximum values of all seasons. Consequently, the SOS/EOS correspond to dates where the curve reaches a percentage of this relative amplitude. In the latter method, the SOS/EOS is determined when each growing season reaches an absolute value.

5.6. Discussion

This chapter started with reviewing conventional gap-filling methods and proposed ML fitting methods as alternative to conventional gap-filling methods. Not only do ML regression methods lead to accurate reconstructions (see Table 5.2), it is also adaptive toward unevenly spaced data over multiple seasons. Particularly, the fitting functions Whittaker and GPR were proven to be of interest: Whitakker because of not only accurate in reconstructing time series trends but also fast, and GPR because of accuracy and providing associated uncertainty estimates, which provides per-pixel information of the gap-filling confidence. Typically, the longer the gap between two consecutive input samples, the higher the uncertainty. However, the provided gap-filling techniques offer different performances. Each method has advantages and drawbacks, which depend strongly on the characteristics of the input time series (Atzberger and Eilers 2011; Zeng et al. 2020), that is, a method that fits well with some data can be unsuited for a different set of data points. This implies that Whitakker and GPR are not always top performing in all situations, and in other situations alternative gap-filling methods may turn out performing superior. For that reason, it is recommended to analyze first the behavior of the fitting methods on single pixels, for example, as can be easily done in DATimeS. For instance, conventional harmonic analysis is only suitable for stationary data and without aperiodic extreme behaviors. Contrarily, Whittaker and most of the ML algorithms are effective to account for the amplitude and phase time variability.

Based on the gap-filled time series, data multiple vegetation phenology metrics can then be derived such as SOS and EOS. Here, it should be remarked that, just as the gap-filling methods, these metrics can be calculated in various ways, that is, determining where to put the thresholds (Sakamoto 2018). This ambiguity generates a wide discrepancy in the timings estimated for these metrics among these methods, making them difficult to compare and determine the most applicable method for a specific use (de Beurs and Henebry 2010; Zeng et al. 2020). As a result, careful consideration should be taken in comparing and interpreting the results from different methods. The choice of the most appropriate model depends upon the purpose of the study, the growth trajectory to be analyzed and the targeted land cover type(s) (Zeng et al. 2020). In addition, Zeng et al. (2020) further noted that a variety of factors can also impact the phenology detection for both general and specific land cover types that include (1) noise artifacts in remotely sensed time series data, (2) intrinsic limitations of method selected, (3) spatial scale differences between the phenological observation dates of individual plants/species and the mixed spectral signal observed from space in moderate to coarse spatial resolution imagery commonly used for this application, and (4) long-lasting cloud cover during specific phenological stages (e.g. peak greenness) or during a key transition stage in the vegetation growth trajectory such as SOS or EOS.

With respect to the latter, one of the crucial aspects in time series analysis is how to balance the requirement of high temporal and spatial resolution. Specifically, when the number of cloud-free observations is below the minimum quantity needed to characterize the basic trajectory of vegetation growth and key phenological transition stages, temporal resolution becomes the most important factor and the uncertainty in phenology detection increases substantially (Melaas et al. 2013; Pipia et al. 2019). In case of long-lasting cloud cover, a solution to fill in gaps would be to make use of all-weather radar data. While earlier studies have attempted to combine both sources, for example, by adding the data layers into classification schemes (Veloso et al. 2017; Dusseux et al. 2014), what remained yet to be developed is a method that transparently merges both time series data sources by fusing the complementary information of radar data with cloud-prone optical data for achieving accurate cloud-free predictions.

In this regard, among the most exciting progress in ML methods for time series analysis involves the multi-output (MO) version of GPR (MOGPR), as it allows to fuse multi-source time series datasets. Here, the utility of MOGPR has been demonstrated as a fusion method to combined active (S1) and passive (S2) time series data over vegetated areas. This approach has proven to be particularly advantageous for long gapped time series, such as prolonged cloud cover, where the S2-based fitting alone notoriously fails. Hence, the data from cloud-penetrating radar technology kick in as complementary information, although the relationship between radar and vegetation phenology is less obvious and strongly depends on structural properties. An absolute novelty of the solution we proposed is that the parameters of the trained model implicitly predict the meaningfulness of any fusion approach: they quantify the amount of information shared between the two time series and rule the interaction of low- and high-frequency GPs for output reconstruction. The LAI data gap-filling described in this work is only one example of MOGPR possible applications. The strength of the two time series synergy strictly depends on the crop type, with some crops being more prone to be well characterized by radar data (such as oat, wheat or maize), but also on crop-stage (Pipia et al. 2019). All the same, the radar information extracted by MOGPR has been shown to be key to reduce prediction uncertainties over long time data gaps. Outcomes from this analysis underlines that fusion of optical with radar data will lead to improved time series analysis, and so more reliable detection of key seasonal trends. Moreover, MOGPR multiple datasets can be fused, so to say, that they are not restricted to two data sources. Any set of time series collection can be entered into the MOGPR, that is, the use of variables from multiple optical and radar data sources, coming from multiple satellite missions, for example, Landsat, SPOT, the Sentinels, can be envisaged, as long as they all share a certain amount of information. This fusion idea was exemplified with the reconstruction of S2-like NDVI trends back in time thanks to the availability of historical Landsat data. These historical – as well latest – satellite data are nowadays easily accessible at the global scale on cloud-based platforms such as the Google Earth Engine. Accordingly, in the present era of freely available continuous and multi-source satellite data streams, there is no doubt that fused time series processing will become indispensable in producing accurate cloud-free data and subsequent vegetation phenology monitoring.

However, just like any method, the MOGPR fusion method is not without limitations, especially in view of operational processing over large areas. The main weakness of MOGPR is its high computational cost, which makes it inefficient if the per-pixel training has to be applied over many pixels. To mitigate this drawback, a more efficient usage of MOGPR should be developed to enable the exploitation of spatial homogeneities. For instance, polygons from land cover maps, if available, enable the reduction of the computational time by moving from a pixel-wise to an object-oriented approach, if we accept losing details about the spatial heterogeneity of each polygon. Yet, it also implies that the homogeneity of the same polygon must be revised when dealing with multiple-year time series to avoid averaging different crop types in space or time. Another remark is that MOGPR does not provide added value when the gaps in time are small; in that case, little changes occurred and a single-source interpolation, for example, by means of GPR, is good enough (Pipia et al. 2019).

As a final remark, the MOGPR algorithm has been implemented as a GUI module into the latest version of the DATimeS toolbox (v. 1.10). As such, the user has only to insert two time series datasets, and to make sure the data are prepared in the same grid. The module does all further processing and provides as output the multi-output time series maps. Moreover, the algorithm can process either pixel based or polygon based.

The latter enables processing of time series predictions on a per-field scale, which can be beneficial for monitoring agricultural practices at a larger scale.

5.7. Conclusions

Consistent time series of remote sensing imagery at high spatiotemporal resolution is critical to identify the precise date when vegetation seasonal event occurs. In this chapter, we reviewed the recent advances available in image time series processing for the extraction of information about phenology trends. We first reviewed the standard gap-filling methods used to reach consistent and cloud-free datasets and introduced machine learning (ML) alternatives. As a demonstration case, we tested common gap-filling methods as well as ML methods for the reconstruction of a cloud-free Sentinel-2 LAI image based on a multi-year time series over an agricultural region in Spain. When comparing against a reference image, it appeared that particularly the fitting algorithms, Whittaker and Gaussian processes regression (GPR), were proven to be accurate. Based on the best-performing gap-filling method, cloud-free reconstructed time series images were generated and subsequently used for the extraction of phenology indicators, such as start- and end-of-season, maximum value and day of maximum value. These indicators quantify the vegetation dynamics and can eventually be used to estimate valuable and quantitative agricultural information such as day of harvesting and crop yield.

In the second part of the chapter, we introduced a novel ML fusion method to further improve the reconstruction of time series data. The fusion method is based on multi-output GPR (MOGPR) where the complementary information of multiple data sources is exploited. We demonstrated the fusion of both optical and all-weather radar data. This approach was proven to be particularly advantageous for long gapped time series, such as long-lasting cloud cover, where fitting methods relying on optical data alone notoriously fail. It allows us to conclude that the MOGPR method shows great potential in cloud data recovery, time series data reconstruction and thus phenology monitoring. Further studies are now required to advance toward a more efficient usage of the MOGPR fusion algorithm. For instance, spatial homogeneities can be exploited to move from a pixel-wise to an object-oriented approach, accepting the loss of the possible spatial heterogeneity within the polygon. Other two aspects to be examined include multiple radar descriptors to MOGPR model over specific crop types as well as combining multiple optical data sources, for example, fusing time series of Sentinel-2 with historical Landsat data. Finally, each of the methods explored in this chapter have been implemented into a single toolbox, called DATimeS. DATimeS is a stand-alone toolbox for time series analysis that brings together conventional interpolation techniques and versatile ML algorithms (e.g. GPR and MOGPR) and can then extract a suite of phenology indicators. DATimeS aspires to be a generic and innovative time series toolbox that provides sufficient flexibility to capture the principal features without imposing too heavy computational or inferential burdens for gap-filling and subsequent phenological analysis. The toolbox can be freely downloaded at: http://artmotoolbox.com.

5.8. Acknowledgments

This chapter was funded by the European Research Council (ERC) under the ERC-2017-STG SENTIFLEX project (grant agreement 755617). It was also partially supported by Generalitat Valenciana (SEJIGENT/2021/001) and the European Union – NextGenerationEU (ZAMBRANO 21-04). It was also supported by the Action CA17134 SENSECO (Optical synergies for spatiotemporal sensing of scalable ecophysiological traits) funded by COST (European Cooperation in Science and Technology: www.cost.eu).

5.9. References

Abburu, S. and Golla, S. (2015), Satellite image classification methods and techniques: A review. *International Journal of Computer Applications*, 119(8), 20–25.

Akaike, H. (1998). *Information Theory and an Extension of the Maximum Likelihood Principle*. Springer, New York.

Álvarez, M., Rosasco, L., Lawrence, N. (2012). Kernels for vector-valued functions: A review. *Foundations and Trends in Machine Learning*, 4(3), 195–255.

Amin, E., Verrelst, J., Rivera-Caicedo, J., Pasqualotto, N., Delegido, J., Verdú, A.R., Moreno, J. (2018). The Sensagri Sentinel-2 LAI green and brown product: From algorithm development towards operational mapping. In *IGARSS 2018 IEEE International Geoscience and Remote Sensing Symposium*. IEEE, Valencia.

Araya, S., Ostendorf, B., Lyle, G., Lewis, M. (2018). Cropphenology: An R package for extracting crop phenology from time series remotely sensed vegetation index imagery. *Ecological Informatics*, 46, 45–56.

Atzberger, C. and Eilers, P. (2011). A time series for monitoring vegetation activity and phenology at 10-daily time steps covering large parts of South America. *International Journal of Digital Earth*, 4(5), 365–386.

Bai, T., Zhang, N., Mercatoris, B., Chen, Y. (2019). Jujube yield prediction method combining Landsat 8 vegetation index and the phenological length. *Computers and Electronics in Agriculture*, 162, 1011–1027.

Beck, P., Atzberger, C., Høgda, K., Johansen, B., Skidmore, A. (2006). Improved monitoring of vegetation dynamics at very high latitudes: A new method using MODIS NDVI. *Remote Sensing of Environment*, 100(3), 321–334.

Belda, S., Pipia, L., Rivera-Caicedo, J., Amin, E., Verrels, J. (2020). Datimes: A machine learning time series GUI toolbox for gap-filling and vegetation phenology trends detection. *Environmental Modelling & Software Journal*, 127, 104666.

Boles, S., Xiao, X., Liu, J., Zhang, Q., Munkhtuya, S., Chen, S., Ojima, D. (2004). Land cover characterization of temperate East Asia using multi-temporal VEGETATION sensor data. *Remote Sensing of Environment*, 90(4), 477–489.

Bolton, D. and Friedl, M. (2013). Forecasting crop yield using remotely sensed vegetation indices and crop phenology metrics. *Agricultural and Forest Meteorology*, 173, 74–84.

Bréda, N. (2003). Ground-based measurements of leaf area index: A review of methods, instruments and current controversies. *Journal of Experimental Botany*, 54(392), 2403–2417.

Camps-Valls, G., Bruzzone, L., Rojo-Alvarez, J.L., Melgani, F. (2006). Robust support vector regression for biophysical variable estimation from remotely sensed images. *IEEE Geoscience and Remote Sensing Letters*, 3(3), 339–343.

Cao, R., Chen, J., Shen, M., Tang, Y. (2015). An improved logistic method for detecting spring vegetation phenology in grasslands from MODIS EVI time-series data. *Agricultural and Forest Meteorology*, 200, 9–20.

Chelton, D. and Wentz, F. (2005). Global microwave satellite observations of sea surface temperature for numerical weather prediction and climate research. *Bulletin of the American Meteorological Society*, 86(8), 1097–1116.

Chen, J., Jönsson, P., Tamura, M., Gu, Z., Matsushita, B., Eklundh, L. (2004). A simple method for reconstructing a high-quality NDVI time-series data set based on the Savitzky-Golay filter. *Remote Sensing of Environment*, 91(3–4), 332–344.

Chevrel, M., Courtois, M., Weill, G. (1981). The SPOT satellite remote sensing mission. *Photogrammetric Engineering & Remote Sensing*, 47, 1163–1171.

Cloude, S. (2010). *Polarisation: Applications in Remote Sensing*. Oxford University Press, Oxford.

De Beurs, K. and Henebry, G. (2004). Land surface phenology, climatic variation, and institutional change: Analyzing agricultural land cover change in Kazakhstan. *Remote Sensing of Environment*, 89(4), 497–509.

De Beurs, K. and Henebry, G. (2005). A statistical framework for the analysis of long image time series. *International Journal of Remote Sensing*, 26(8), 1551–1573.

De Beurs, K. and Henebry, G. (2010). Spatio-temporal statistical methods for modelling land surface phenology. In *Phenological Research*, Hudson, I.L. and Keatley, M.R. (eds). Springer, Berlin.

DeFries, R. and Chan, J. (2000). Multiple criteria for evaluating machine learning algorithms for land cover classification from satellite data. *Remote Sensing of Environment*, 74(3), 503–515.

Delbart, N., Toan, T., Kergoat, L., Fedotova, V. (2006). Remote sensing of spring phenology in boreal regions: A free of snow-effect method using NOAA-AVHRR and SPOT-VGT data (1982–2004). *Remote Sensing of Environment*, 101(1), 52–62.

Dierckx, W., Sterckx, S., Benhadj, I., Livens, S., Duhoux, G., Van Achteren, T., Francois, M., Mellab, K., Saint, G. (2014). PROBA-V mission for global vegetation monitoring: Standard products and image quality. *International Journal of Remote Sensing*, 35(7), 2589–2614.

Dong, Y. and Peng, C. (2013). Principled missing data methods for researchers. *Springerplus*, 2(1), 222.

Donlon, C., Berruti, B., Buongiorno, A., Ferreira, M.-H., Féménias, P., Frerick, J., Goryl, P., Klein, U., Laur, H., Mavrocordatos, C. et al. (2012). The Global Monitoring for Environment and Security (GMES) Sentinel-3 mission. *Remote Sensing of Environment*, 120, 37–57.

Drusch, M., Del Bello, U., Carlier, S., Colin, O., Fernandez, V., Gascon, F., Hoersch, B., Isola, C., Laberinti, P., Martimort, P. et al. (2012). Sentinel-2: ESA's optical high-resolution mission for GMES operational services. *Remote Sensing of Environment*, 120, 25–36.

Dusseux, P., Corpetti, T., Hubert-Moy, L., Corgne, S. (2014). Combined use of multi-temporal optical and radar satellite images for grassland monitoring. *Remote Sensing*, 6(7), 6163–6182.

Eerens, H., Haesen, D., Rembold, F., Urbano, F., Tote, C., Bydekerke, L. (2014). Image time series processing for agriculture monitoring. *Environmental Modelling & Software*, 53, 154–162.

Elmore, A., Guinn, S.M., Minsley, B., Richardson, A. (2012). Landscape controls on the timing of spring, autumn, and growing season length in mid-Atlantic forests. *Global Change Biology*, 18(2), 656–674.

Ewert, F. (2004). Modelling plant responses to elevated CO2: How important is leaf area index? *Annals of Botany*, 93(6), 619.

Ferrazzoli, P., Paloscia, S., Pampaloni, P., Schiavon, G., Solimini, D., Coppo, P. (1992). Sensitivity to microwave measurements to vegetation biomass and soil moisture content: A case study. *IEEE Transactions on Geoscience and Remote Sensing*, 30(4), 750–756.

Fisher, J., Mustard, J., Vadeboncoeur, M. (2006). Green leaf phenology at Landsat resolution: Scaling from the field to the satellite. *Remote Sensing of Environment*, 100(2), 265–279.

Ganguly, S., Friedl, M., Tan, B., Zhang, X., Verma, M. (2010). Land surface phenology from MODIS: Characterization of the collection 5 global land cover dynamics product. *Remote Sensing of Environment*, 114(8), 1805–1816.

Gómez, V., Medina, V., Bengoa, J., García, D. (2018). Accuracy assessment of a 122 classes land cover map based on Sentinel-2, Landsat 8 and Deimos-1 images and ancillary data. *IGARSS 2018–2018 IEEE International Geoscience and Remote Sensing Symposium*, 5453–5456.

Gorelick, N., Hancher, M., Dixon, M., Ilyushchenko, S., Thau, D., Moore, R. (2017). Google Earth Engine: Planetary-scale geospatial analysis for everyone. *Remote Sensing of Environment*, 202, 18–27.

Guanter, L., Kaufmann, H., Segl, K., Foerster, S., Rogass, C., Chabrillat, S., Sang, B. (2015). The EnMAP spaceborne imaging spectroscopy mission for earth observation. *Remote Sensing*, 7(7), 8830–8857.

Hermance, J. (2007). Stabilizing high-order, non-classical harmonic analysis of NDVI data for average annual models by damping model roughness. *International Journal of Remote Sensing*, 28(12), 2801–2819.

Hill, M. and Donald, G. (2003). Estimating spatio-temporal patterns of agricultural productivity in fragmented landscapes using AVHRR NDVI time series. *Remote Sensing of Environment*, 84(3), 367–384.

Hird, J. and McDermid, G. (2009). Noise reduction of NDVI time series: An empirical comparison of selected techniques. *Remote Sensing of Environment*, 113(1), 248–258.

Huete, A., Didan, K., Miura, T., Rodriguez, E., Gao, X., Ferreira, L. (2002). Overview of the radiometric and biophysical performance of the MODIS vegetation indices. *Remote Sensing of Environment*, 83(1–2), 195–213.

Jakimow, B., van der Linden, S., Thiel, F., Frantz, D., Hostert, P. (2020). Visualizing and labeling dense multi-sensor earth observation time series: The EO time series viewer. *Environmental Modelling & Software*, 125, 104631.

Jönsson, J. and Eklundh, L. (2004). TIMESAT – A program for analysing time-series of satellite sensor data. *Computers and Geosciences*, 30, 833–845.

Journel, A. and Huijbregts, C. (1978). *Mining Geostatistics*. Academic Press, London and New York.

Kraft, S., Del Bello, U., Bouvet, M., Drusch, M., Moreno, J. (2012). Flex: ESA's Earth explorer 8 candidate mission. *IEEE International Geoscience and Remote Sensing Symposium*, 7125–7128.

Kulkarni, S. and Rege, P. (2020). Pixel level fusion techniques for SAR and optical images: A review. *Information Fusion*, 59, 13–29.

Kumar, L. and Mutanga, O. (2018). Google Earth Engine applications since inception: Usage, trends, and potential. *Remote Sensing*, 10(10), 1509.

Labate, D., Ceccherini, M., Cisbani, A., De Cosmo, V., Galeazzi, C., Giunti, L., Melozzi, M., Pieraccini, S., Stagi, M. (2009). The PRISMA payload optomechanical design, a high performance instrument for a new hyperspectral mission. *Acta Astronautica*, 65(9–10), 1429–1436.

Lázaro-Gredilla, M., Titsias, M.K., Verrelst, J., Camps-Valls, G. (2014). Retrieval of biophysical parameters with heteroscedastic Gaussian processes. *IEEE Geoscience & Remote Sensing Letters*, 11(4), 838–842.

Lloyd, D. (1990). A phenological classification of terrestrial vegetation cover using shortwave vegetation index imagery. *International Journal of Remote Sensing*, 11(12), 2269–2279.

Mateo, F., Muñoz-Marí, J., Laparra, V., Verrelst, J., Camps-Valls, G. (2015). Learning structures in earth observation data with Gaussian processes. In *International Workshop on Advanced Analysis and Learning on Temporal Data*. Springer.

Mattia, F., Le Toan, T., Picard, G., Posa, F., D'Alessio, A., Notarnicola, C., Gatti, A., Rinaldi, M., Satalino, G., Pasquariello, G. (2003). Multitemporal C-band radar measurements on wheat fields. *IEEE Transactions on Geoscience and Remote Sensing*, 41(7), 1551–1560.

Melaas, E., Friedl, M., Zhu, Z. (2013). Detecting interannual variation in deciduous broadleaf forest phenology using Landsat TM/ETM+ data. *Remote Sensing of Environment*, 132, 176–185.

Moody, A. and Johnson, D. (2001). Land-surface phenologies from AVHRR using the discrete Fourier transform. *Remote Sensing of Environment*, 75(3), 305–323.

Pipia, L., Muñoz-Marí, J., Amin, E., Belda, S., Camps-Valls, G., Verrelst, J. (2019). Fusing optical and SAR time series for LAI gap-filling with multi-output Gaussian processes. *Remote Sensing of Environment*, 235, 111452.

Ramachandran, B., Justice, C., Abrams, M. (2011). *Land Remote Sensing and Global Environmental Change*, volume 11. Springer, New York.

Rasmussen, C.E. (2004). Gaussian processes in machine learning. In *Advanced Lectures on Machine Learning*, Bousquet, O., Luxburg, U., Rätsch, G. (eds). Springer, Berlin, Heidelberg.

Rasmussen, C.E. and Williams, C.K.I. (2006). *Gaussian Processes for Machine Learning*. MIT Press, Cambridge, MA.

Reed, B., Schwartz, M., Xiao, X. (2009). Remote sensing phenology: Status and the way forward. In *Phenology of Ecosystems Processes*, Noormets, A. (ed.). Springer, New York.

Ren, S., Chen, X., An, S. (2017). Assessing plant senescence reflectance index-retrieved vegetation phenology and its spatiotemporal response to climate change in the inner Mongolian grassland. *International Journal of Biometeorology*, 61(4), 601–612.

Roberts, D., Quattrochi, D., Hulley, G., Hook, S., Green, R. (2012). Synergies between VSWIR and TIR data for the urban environment: An evaluation of the potential for the Hyperspectral Infrared Imager (HyspIRI) Decadal Survey mission. *Remote Sensing of Environment*, 117, 83–101.

Sakamoto, T. (2018). Refined shape model fitting methods for detecting various types of phenological information on major US crops. *ISPRS Journal of Photogrammetry and Remote Sensing*, 138, 176–192.

Sakamoto, T., Yokozawa, M., Toritani, H., Shibayama, M., Ishitsuka, N., Ohno, H. (2005). A crop phenology detection method using time-series MODIS data. *Remote Sensing of Environment*, 96(3–4), 366–374.

Schmitt, M., Tupin, F., Zhu, X. (2017). Fusion of SAR and optical remote sensing data – Challenges and recent trends. In *2017 IEEE International Geoscience and Remote Sensing Symposium (IGARSS)*. IEEE, Fort Worth, 5458–5461.

Schwartz, M. (2013). *Phenology: An Integrative Environmental Science*, volume 2, Springer, Dordrecht.

Schwarz, G. (1978). Estimating the dimension of a model. *Annals of Statistics*, 6, 461–464.

Smola, A.J. and Schölkopf, B. (2004). A tutorial on support vector regression. *Statistics and Computing*, 14, 199–222.

Stuffler, T., Kaufmann, C., Hofer, S., Farster, K., Schreier, G., Mueller, A., Eckardt, A., Bach, H., Penné, B., Benz, U. (2007). The EnMAP hyperspectral imager – An advanced optical payload for future applications in Earth observation programmes. *Acta Astronautica*, 61(1–6), 115–120.

Tang, H., Li, Z., Zhu, Z., Chen, B., Zhang, B., Xin, X. (2015). Variability and climate change trend in vegetation phenology of recent decades in the greater Khingan mountain area, northeastern China. *Remote Sensing*, 7(9), 11914–11932.

Thayananthan, A., Navaratnam, R., Stenger, B., Torr, P., Cipolla, R. (2006). Multivariate relevance vector machines for tracking. In *Computer Vision – ECCV 2006*, Leonardis, A., Bischof, H., Pinz, A. (eds). Springer, Berlin, Heidelberg.

Townshend, J. (1994). Global data sets for land applications from the advanced very high-resolution radiometer: An introduction. *International Journal of Remote Sensing*, 15, 3319–3332.

Udelhoven, T. (2011). Timestats: A software tool for the retrieval of temporal patterns from global satellite archives. *IEEE Journal of Selected Topics in Applied Earth Observations and Remote Sensing*, 4(2), 310–317.

Ulaby, F., Moore, R., Fung, A. (1981). *Microwave Remote Sensing: Active and Passive, Volume I: Microwave Remote Sensing Fundamentals and Radiometry*. Artech House, Norwood.

Veloso, A., Mermoz, S., Bouvet, A., Le Toan, T., Planells, M., Dejoux, J., Ceschia, E. (2017). Understanding the temporal behavior of crops using Sentinel-1 and Sentinel-2-like data for agricultural applications. *Remote Sensing of Environment*, 199, 415–426.

Verbesselt, J., Hyndman, R., Newnham, G., Culvenor, D. (2010). Detecting trend and seasonal changes in satellite image time series. *Remote Sensing of Environment*, 114(1), 106–115.

Verrelst, J., Romijn, E., Kooistra, L. (2012). Mapping vegetation density in a heterogeneous river floodplain ecosystem using pointable CHRIS/PROBA data. *Remote Sensing*, 4(9), 2866–2889.

Verrelst, J., Rivera-Caceido, J., Moreno, J., Camps-Valls, G. (2013). Gaussian processes uncertainty estimates in experimental Sentinel-2 LAI and leaf chlorophyll content retrieval. *ISPRS Journal of Photogrammetry and Remote Sensing*, 86, 157–167.

Verrelst, J., Camps-Valls, G., Munoz-Marí, J., Rivera, J., Veroustraete, F., Clevers, J., Moreno, J. (2015). Optical remote sensing and the retrieval of terrestrial vegetation bio-geophysical properties – A review. *ISPRS Journal of Photogrammetry and Remote Sensing*, 108, 273–290.

Verrelst, J., Malenovský, Z., Van der Tol, C., Camps-Valls, G., Gastellu-Etchegorry, J., Lewis, P., North, P., Moreno, J. (2019). Quantifying vegetation biophysical variables from imaging spectroscopy data: A review on retrieval methods. *Surveys in Geophysics*, 40(3), 589–629.

Wang, C., Hunt, E., Zhang, L., Guo, H. (2013). Phenology-assisted classification of C3 and C4 grasses in the US Great Plains and their climate dependency with MODIS time series. *Remote Sensing of Environment*, 138, 90–101.

Wardlow, B. and Egbert, S. (2008). Large-area crop mapping using time-series MODIS 250 m NDVI data: An assessment for the US Central Great Plains. *Remote Sensing of Environment*, 112(3), 1096–1116.

Watson, D.J. (1947). Comparative physiological studies on the growth of field crops I: Variation in net assimilation rate and leaf area between species and varieties, and within and between years. *Annals of Botany*, 11(1), 41–76.

White, M. and Nemani, R. (2006). Real-time monitoring and short-term forecasting of land surface phenology. *Remote Sensing of Environment*, 104(1), 43–49.

White, M., Hoffman, F., Hargrove, W., Nemani, R. (2005). A global framework for monitoring phenological responses to climate change. *Geophysical Research Letters*, 32(4).

Wiseman, G., McNairn, H., Homayouni, S., Shang, J. (2014). RADARSAT-2 polarimetric SAR response to crop biomass for agricultural production monitoring. *IEEE Journal of Selected Topics in Applied Earth Observations and Remote Sensing*, 7(11), 4461–4471.

Wu, W., Yang, P., Tang, H., Zhou, Q., Chen, Z., Shibasaki, R. (2010). Characterizing spatial patterns of phenology in cropland of China based on remotely sensed data. *Agricultural Sciences in China*, 9(1), 101–112.

Xue, Z., Du, P., Feng, L. (2014). Phenology-driven land cover classification and trend analysis based on long-term remote sensing image series. *IEEE Journal of Selected Topics in Applied Earth Observations and Remote Sensing*, 7(4), 1142–1156.

Yu, L., Liu, T., Bu, K., Yan, F. (2017). Monitoring the long term vegetation phenology change in Northeast China from 1982 to 2015. *Scientific Reports*, 7(1), 14770.

Zeng, L., Wardlow, B., Xiang, D., Hu, S., Li, D. (2020). A review of vegetation phenological metrics extraction using time-series, multispectral satellite data. *Remote Sensing of Environment*, 237, 111511.

Zhang, X., Friedl, M., Schaaf, C., Strahler, A., Hodges, J., Gao, F., Reed, B., Huete, A. (2003). Monitoring vegetation phenology using modis. *Remote Sensing of Environment*, 84(3), 471–475.

Zhang, J., Rivard, B., Sánchez-Azofeifa, A. (2005). Spectral unmixing of normalized reflectance data for the deconvolution of lichen and rock mixtures. *Remote Sensing of Environment*, 95(1), 57–66.

Zhou, G., Liu, X., Liu, M. (2019). Assimilating remote sensing phenological information into the WOFOST model for rice growth simulation. *Remote Sensing*, 11(3), 268.

6

Data-Driven Spatio-Temporal Interpolation for Satellite-Derived Geophysical Tracers

Maxime BEAUCHAMP and Ronan FABLET

IMT Atlantique Bretagne Pays-de-la-Loire, Technopôle Brest-Iroise, France

6.1. Notations

$\mathbf{x} = \{\mathbf{x}_k\} \in \mathbb{R}^{T \times m}$	State vector, in our case corresponding to m gridcells discretizing domain $\mathcal{D}$
t_k	Time for index $k = 1, \cdots, T$
$\mathbf{y}_k \in \mathbb{R}^p$, Ω_k	Observations at time t_k on subdomain $\Omega_k \subset \mathcal{D}$
$\boldsymbol{\alpha}_k = \{\alpha_k^l\}$, $l = 1, \cdots, L$	Temporal coefficients of EOF decomposition at time t_k
$\overline{\Omega}_k$	Subdomain of $\mathcal{D}$ with missing data at time t_k
$\Phi = \{\phi^l\}$, $l = 1, \cdots, L$	EOF orthogonal basis
$\mathcal{M}_k$	Dynamical model from time t_{k-1} to t_k
$\mathbf{M}_k$	Linear/linearized model from time t_{k-1} to t_k
$\mathbf{A}_k$, $\mathbf{F}_k$	Surrogate operators of a dynamical model from time t_{k-1} to t_k
$\mathbf{K}_k$	Kalman gain at time t_k
$\mathbf{P}^{f/a}$	Error covariance matrix (forecast/analysis)

Multitemporal Earth Observation Image Analysis,
coordinated by Clément MALLET and Nesrine CHEHATA.

$\mathbf{x}_i^{f/a}, i = 1, \cdots, N$	Member or particle (forecast/analysis)
$\mathbf{E}^f \in \mathbb{R}^{m \times N}$	Ensemble forecast $\mathbf{E} = [\mathbf{x}_1^f, \cdots, \mathbf{x}_N^f]$
$\mathbf{E}^a \in \mathbb{R}^{m \times N}$	Ensemble analysis $\mathbf{E} = [\mathbf{x}_1^a, \cdots, \mathbf{x}_N^a]$
$\bar{\mathbf{x}}^f$	Ensemble mean (forecast) $\bar{\mathbf{x}}^f = \frac{1}{N} \sum_{i=1}^N \mathbf{x}_i^f$
$\bar{\mathbf{x}}^a$	Ensemble mean (analysis) $\bar{\mathbf{x}}^a = \frac{1}{N} \sum_{i=1}^N \mathbf{x}_i^a$
$\mathbf{X}^f \in \mathbb{R}^{m \times N}$	Normalized anomalies (forecast) $\mathbf{X}^f = \frac{1}{\sqrt{N-1}}[\mathbf{x}_1^f - \bar{\mathbf{x}}^f, \cdots, \mathbf{x}_N^f - \bar{\mathbf{x}}^f]$
$\mathbf{X}^a \in \mathbb{R}^{m \times N}$	Normalized anomalies (analysis) $\mathbf{X}^f = \frac{1}{\sqrt{N-1}}[\mathbf{x}_1^a - \bar{\mathbf{x}}^f, \cdots, \mathbf{x}_N^a - \bar{\mathbf{x}}^a]$
$\mathbf{u}_i \in \mathbb{R}^p$	Perturbed observations of member i $\mathbf{u}_i \sim \mathcal{N}(\mathbf{0}, \mathbf{R})$
$\mathbf{Y}^f \in \mathbb{R}^{p \times N}$	Innovation matrix (forecast) $\mathbf{Y}^f = \frac{1}{\sqrt{N-1}}[\mathbf{H}\mathbf{x}_i^f - \mathbf{u}_i - \mathbf{H}\bar{\mathbf{x}}^f - \bar{\mathbf{u}}]_{i=1,\cdots,N}$

6.2. Introduction

Over the last 20 years, new large-scale spatiotemporal observation datasets have been collected from remote sensing satellites, providing additional, and potentially more realistic, information to already huge datasets provided by the numerical simulation of physical models. Due to the irregular space-time sampling of these observations, the reconstruction of gap-free regularly gridded products is a key challenge. Although model-driven and optimal interpolation (OI) approaches are the state-of-the-art solutions for operational products, data-driven strategies have recently emerged as promising alternatives to improve the space-time resolution of interpolated products.

The general objective of this chapter is to provide an introduction to new data-driven and learning-based methodologies emerging from the machine learning (ML) and deep learning (DL) literature to solve for the interpolation problem of images with high missing data rates. It is currently an important research topic with an important literature exploring synergies between data assimilation and ML. We may cite among the recent applications the inference of unresolved scale parameterizations (Brajard et al. 2021; Bocquet et al. 2019) or the use of data assimilation (DA)-based parameterizations inference in numerical models involving ML (O'Gorman and Dwyer 2018; Rasp et al. 2018). We believe these developments to be representative of a new trend in earth observation data analytics, where operational tools will explore physics-informed learning paradigms more and more to best exploit the computational efficiency and flexibility of data-driven schemes with the genericity of physical principles.

It should be noted that other communities address the problem of filling the gaps in image time series without explicitly referring to DA. For instance, Ebrahimi and Lunasin (2012) and Bertalmio et al. (2001) use Partial Differential Equations (PDEs) to establish an analogy between the intensity of pixels and two-dimensional Navier–Stokes equations for incompressible fluids (see also Barbu (2018) and Schönlieb (2015) for more details and additional references). More recently, advances in DL opened up the development of new paradigms in the so-called *inpainting* techniques for video sequences whose content is spatially and temporally coherent, using supervised training (Huang et al. 2016; Kim et al. 2019; Le Thuc et al. 2017; Xu et al. 2019) for image time series with large missing data rates. Unsupervised configurations have also been proposed (Newson et al. 2014) to fill in the missing parts of images, extracting the information provided by its direct neighborhood, which is quite natural and directly relates to Markovian properties. In the same vein, Chang et al. (2019) presented a three-dimensional convolutional model to address the uncertainty of free-form masks and deal with the temporal coherence of the image sequence. In remote sensing (clouds and shadows), deep-based techniques have recently been used in spatial gap filling methods from mono- and multi-temporal imagery, somehow related to inpainting.

Besides the methods discussed above, this chapter will present a generic framework drawing from a DA formulation to explore the link between data-driven algorithms and both variational or statistical formulation of the interpolation problem. The notations are taken from the DA theoretical framework, which is historically used by researchers and operational systems to merge models and for geoscience applications. Far from opposing the two types of information, DA offers an appropriate formulation to deal with the intrinsic dynamics of natural systems, while taking both model and observational errors into account in order to provide an optimal estimation of the geophysical field. Because we draw from model-based DA to come up with a guide through new purely data-driven approaches, we remain interested in the dynamical nature of the spatiotemporal datasets. This is the reason why we refer to technical terms such as state space, (geophysical) fields or dynamical processes. For application-oriented issues, the readers more familiar with remote sensing may be reminded that both the presented methods and applications can easily be transposed to any spatiotemporal sequence of images with missing data. In the end, the difference is merely semantic.

This chapter is organized as follows. In section 6.2, we will start by providing a reminder of state-space formulation in DA, with a quick overview of both variational and statistical approaches used for estimation and prediction. In section 6.3, we review data-driven methods and focus on those which relate to DA formulations. In section 6.4, we introduce Observation System Simulation Experiments (OSSE), which is a classic framework used in geoscience for benchmarking purposes. Within an OSSE set-up, we illustrate an application over a selected panel of methods to satellite altimetry. In section 6.5, we further discuss future developments and new avenues in data-driven methods for the reconstruction of geophysical fields based on geophysical tracers, with an overview regarding alternate formulations of the reconstruction/prediction problem from a DL consideration.

6.3. Data assimilation

For interpolation problems, DA aims to provide efficient algorithms to couple/fuse physical models and observation data through some model-based information propagation from observational data to areas with missing data, and also deal with the computational big data challenge often encountered in geosciences due to very large datasets and high-dimensional numerical model outputs. That is why DA offers a useful framework and provides the appropriate formulation to deal with both model-data-coupled state estimation, but also purely data-driven learning-based methods, in which the dynamical prior is no longer model based, but driven by ML algorithms.

6.3.1. *Introduction*

Let us start by introducing some formal notations used in DA. We aim to estimate the state $\mathbf{x} = \{\mathbf{x}_k\}$ of a dynamical system where index $k = 1, \cdots, T$ refers to time t_k. Let us denote by:

$$\begin{cases} \mathbf{x}_{k+1} &= \mathcal{M}_{k+1}(\mathbf{x}_k) + \boldsymbol{\eta}_k \\ \mathbf{y}_k &= \mathcal{H}_k(\mathbf{x}_k) + \boldsymbol{\varepsilon}_k \end{cases}. \qquad [6.1]$$

The first equation is the *forecast step*, which describes the evolution of the system from time t_k to t_{k+1} according to the potentially nonlinear model $\mathcal{M}_{k+1}$. $\boldsymbol{\eta}_k$ is the model error $\mathbf{x}_{k+1} - \mathcal{M}_{k+1}(\mathbf{x}_k)$. For the sake of simplicity, this error may be assumed to be Gaussian, unbiased and uncorrelated over time. Its covariance matrix is denoted as $\mathbf{Q}_k \in \mathbb{R}^{m \times m}$:

$$\mathrm{E}[\boldsymbol{\eta}_k] = \mathbf{0}, \mathrm{E}[\boldsymbol{\eta}_k \boldsymbol{\eta}_l^{\mathrm{T}}] = \mathbf{Q}_k \delta_{kl}.$$

The second equation is the *observation equation*, with $\mathcal{H}_k$ as the observation operator, potentially time dependent, at time t_k. The observation error $\boldsymbol{\varepsilon}_k$ is also assumed to be Gaussian, unbiased and uncorrelated over time. Its covariance matrix is denoted $\mathbf{R}_k \in \mathbb{R}^{p_k \times p_k}$:

$$\mathrm{E}[\boldsymbol{\varepsilon}_k] = \mathbf{0}, \mathrm{E}[\boldsymbol{\varepsilon}_k \boldsymbol{\varepsilon}_l^{\mathrm{T}}] = \mathbf{R}_k \delta_{kl}.$$

The two types of error are also considered to be uncorrelated:

$$E[\eta_k \varepsilon_l^{\mathsf{T}}] = \mathbf{0}.$$

Let us point out that in geosciences, the dimension p_k of the observation vector is generally much smaller than the state space dimension, that is, $p_k << m$. In what follows, index k is omitted for the observation vector size for the sake of clarity. In addition, state $\mathbf{x}$ is also associated with an error covariance matrix $\mathbf{P}$ because the true state will differ from the estimated state.

6.3.2. *Variational versus statistical DA*

Broadly speaking, a vast family of DA methods stems from the minimization of some energy or function which involves two terms, a dynamical prior and an observation term. We may distinguish two main categories of DA approaches (Evensen 2009): variational DA and statistical DA.

6.3.2.1. *Variational data assimilation*

Variational DA relies on a continuous state-space formulation. The state analysis $\mathbf{x}^a$ results in a gradient-based minimization of the defined variational cost:

$$\mathbf{x}^a = \arg\min_{\mathbf{x}} \mathcal{J}(\mathbf{x}),$$

where the variational cost function $\mathcal{J}(\mathbf{x}) = \mathcal{J}_\Phi(\mathbf{x}, \mathbf{y}, \Omega)$ is generally the sum of an observation term and a regularization term involving an operator Φ, which is typically a dynamical prior based on a physical model:

$$\begin{aligned}\mathcal{J}_\Phi(\mathbf{x}, \mathbf{y}, \Omega) &= \mathcal{J}^o(\mathbf{x}, \mathbf{y}, \Omega) + \mathcal{J}_\Phi^b(\mathbf{x}) \\ &= \lambda_1 ||\mathbf{y} - \mathcal{H}(\mathbf{x})||_\Omega^2 + \lambda_2 ||\mathbf{x} - \Phi(\mathbf{x})||^2,\end{aligned}$$

where $\mathcal{H}$ is the observation operator, $\Omega = \{\Omega_k\}$ is the set of subdomains of $\mathcal{D}$ with observations at time t_k, $k = 1, \cdots, T$ and $\lambda_{1,2}$ are predefined or tunable scalar weights. This formulation of functional $\mathcal{J}_\Phi(\mathbf{x}, \mathbf{y}, \Omega)$ directly relates to weak constraint 4D-Var (Carrassi et al. 2018) (see Algorithm 6.1), which aims to estimate the optimal trajectory of a system in a predefined data assimilation window (DAW), given the additional constraint of model uncertainty in the objective function $\mathcal{J}$.

Algorithm 6.1: 4D-Var algorithm

Data:
- $\mathbf{y} = \{\mathbf{y}_k\}$: observations at time t_k, $k = 1, \cdots, T$
- $\mathcal{M} : \mathbf{x}_{k+1} = \mathcal{M}(\mathbf{x}_k)$: a dynamical model

Preprocessing:
- Specify stopping criteria ε and/or i_{MAX}
- Specify α the gradient-descent step

$i \leftarrow 0$
while $\mathbf{x}^{(i+1)} - x^{(i)} > \varepsilon$ *and* $i < i_{\text{MAX}}$ **do**
  Compute $\mathcal{J}_{\mathcal{M}}(\mathbf{x}^{(i)}, \mathbf{y}, \Omega)$
  Compute $\nabla_{\mathbf{x}} \mathcal{J}_{\Phi}(\mathbf{x}^{(i)}, \mathbf{y}, \Omega)$
  Update $\mathbf{x}^{(i+1)} = x^{(i)} - \alpha \nabla_{\mathbf{x}} \mathcal{J}_{\Phi}(\mathbf{x}^{(i)}, \mathbf{y}, \Omega)$
end

Result: $\mathbf{x}^a$: the analyzed state sequence

We briefly review key theoretical aspects of variational DA. We may first remind the reader that for inverse problems with time-related processes, the minimization of functional $\mathcal{J}_{\Phi}$ usually involves iterative gradient-based algorithms, and in particular, a request to consider the adjoint method in classic model-based variational DA schemes (Asch et al. 2016), where operator Φ identifies to a deterministic model $\mathbf{x}_{k+1} = \mathcal{M}(\mathbf{x}_k)$:

$$\mathbf{x}^{(i+1)} = \mathbf{x}^{(i)} - \alpha \nabla_{\mathbf{x}} \mathcal{J}_{\Phi}(\mathbf{x}^{(i)}, \mathbf{y}, \Omega)$$

This is a key point for the end-to-end learning method presented in section 6.4.4, where meta-learning schemes[1] may be used to avoid the complex analytical derivation of adjoint models by rather exploiting the numerical implementation of the adjoint operator through a neural network-based operator Φ and the associated automatic differentiation for cost function $\mathcal{J}_{\Phi}$ embedded in state-of-the-art DL frameworks.

The readers interested in more details concerning adjoint methods in DA and the different variational frameworks, especially 3D-Var (stationary case) and 4D-Var (non-stationary case) schemes, should refer to Asch et al. (2016) for an exhaustive presentation of DA theory, or to Carrassi et al. (2018) for a complete review of DA methods in the geosciences.

1 Meta-learning refers to a specific area in machine learning which addresses the design of ML algorithms to assist in both the training and optimization of ML models using automatic differentiation tools (Andrychowicz et al. 2016).

6.3.2.2. *Statistical data assimilation*

Statistical DA schemes generally rely on iterative formulations of stochastic filtering techniques such as Kalman and particle filters.

At the core of all statistical DA techniques, OI is a linear least-square interpolator. As suggested by its name, it is optimal, meaning it refers to minimum variance linear estimation. It is also known as (simple) kriging in spatial statistics (Chilès and Delfiner 2012; Chilès and Desassis 2018). Let us denote $\mathbf{x}_k \in \mathbb{R}^m$ the state space to estimate at time t_k and the space-time observation dataset along the DAW $\mathbf{y} = \{\mathbf{y}_{k-L}, \cdots, \mathbf{y}_{k+L}\} \in \mathbb{R}^p$. The optimally interpolated state $\mathbf{x}_k^\star$ given the observations $\mathbf{y}$ is:

$$\mathbf{x}_k^\star = \mathbf{\Sigma}_{\mathbf{xy}}\mathbf{\Sigma}_{\mathbf{yy}}^{-1}\mathbf{y}, \qquad [6.2]$$

with the associated error covariance matrix $\mathbf{P}_k$:

$$\mathbf{P}_k = \mathbf{\Sigma}_{\mathbf{xx}} - \mathbf{\Sigma}_{\mathbf{xy}}\mathbf{\Sigma}_{\mathbf{yy}}^{-1}\mathbf{\Sigma}_{\mathbf{yx}}, \qquad [6.3]$$

where $\mathbf{\Sigma}_{\mathbf{xy}} = \mathrm{Cov}\{\mathbf{x}_k, \mathbf{y}\} \in \mathbb{R}^{m \times \sum_{k-L}^{k+L} p_k}$ denotes the (cross-)covariance between $\mathbf{x}_k$ and $\mathbf{y}$, and $\mathbf{\Sigma}_{\mathbf{yy}} = \mathrm{Cov}\{\mathbf{y}, \mathbf{y}\} \in \mathbb{R}^{\sum_{k-L}^{k+L} p_k \times \sum_{k-L}^{k+L} p_k}$ denotes the covariance between the observations and $\mathbf{\Sigma}_{\mathbf{xx}} = \mathrm{Cov}\{\mathbf{x}_k, \mathbf{x}_k\} \in \mathbb{R}^{m \times m}$.

This family of algorithms has known many operational applications in spaceborne earth observation for the release of satellite-derived gap-free gridded products. In section 6.5 we provide a typical example for the interpolation of sea surface height (SSH) altimetric datasets from nadir satellite altimeters. The operational approach initially proposed in Traon et al. (1998), referred to as the DUACS (Data Unification and Altimeter Combination System) operational system (Ballarotta et al. 2019), relies on OI.

Algorithm 6.2: Optimal interpolation algorithm

Data:

- $\mathbf{y}_{k-L:k+L}$: observations along the DAW t_k

Preprocessing:

- Specify a spatiotemporal covariance model $\mathrm{Cov}(\cdot, \cdot)$

1 $\mathbf{x}_k^\star = \mathbf{\Sigma}_{\mathbf{xy}}\mathbf{\Sigma}_{\mathbf{yy}}^{-1}\mathbf{y}_k$

2 $\mathbf{P}_k = \mathbf{\Sigma}_{\mathbf{xx}} - \mathbf{\Sigma}_{\mathbf{xy}}\mathbf{\Sigma}_{\mathbf{yy}}^{-1}\mathbf{\Sigma}_{\mathbf{yx}}$

Result: $\mathbf{x}_k^\star$: interpolated current field at time t_k

The OI (see Algorithm 6.2) involves the modeling of the spatiotemporal covariance structure of the dynamical process $\mathbf{x}$, generally under simplified assumptions such as the constant covariance structure in space and in time, which is often not realistic. An alternative option is to estimate spatiotemporal covariances based on a historical run of a deterministic numerical model, that is, a long-enough simulation to estimate the covariances based on empirical statistics. Over the last few decades, a new path has been opened up by the seminal paper by Lindgren et al. (2011), which provides new solutions for dealing with interpolation of Gaussian processes (GP) with large datasets under physical constraints. New families of spatial (and spatiotemporal) covariances can be explored by making an explicit link between the discretized version of GP, namely, Gaussian Markov random fields (GMRF) and the stochastic partial differential equations (SPDE) (see, for example, Carrizo Vergara (2018)). These stochastic equations (Lindgren et al. 2011; Cameletti et al. 2011) can explicitly integrate differential operators, such as advection and diffusion (Fuglstad et al. 2015a,b; Hildeman et al. 2021), to interpolate observational datasets with complex local anisotropic covariances, rather than using isotropic space-time covariances or simplified model-based anisotropies.

OI can also be expressed to combine both observational and numerical model-based datasets. In DA, we refer to the BLUE (Best Linear Unbiased Estimator), whose equations are provided in equations [6.5a] and [6.5b]. They constitute the basis of the so-called statistical, and fundamentally Bayesian, DA (Asch et al. 2016).

The *prior* knowledge for state $\mathbf{x}_k$ is typically provided by an evolution model $\mathcal{M}$. The *posterior* distribution of $\mathbf{x}_k$ given the data $\mathbf{y}_k$ is given by Bayes' rule:

$$p(\mathbf{x}_k|\mathbf{y}_k) \propto p(\mathbf{y}_k|\mathbf{x}_k)p(\mathbf{x}_k).$$

In a Gaussian context with a linear observation operator $\mathbf{H}$, we have:

$$\begin{cases} \mathbf{x}_k & \sim \mathcal{N}(\boldsymbol{\mu}_k, \mathbf{B}) \\ \mathbf{y}_k|\mathbf{x}_k & \sim \mathcal{N}(\mathbf{H}_k\mathbf{x}_k, \mathbf{R}_k) \end{cases}, \qquad [6.4]$$

with $\boldsymbol{\mu}_k$ as the expectation of $\mathbf{x}_k$, $\mathbf{B} = \mathrm{E}[\mathbf{u}_k\mathbf{u}_k^\mathsf{T}]$ as the error covariance matrix of the model and $\mathbf{R}_k = \mathrm{E}[\mathbf{v}_k\mathbf{v}_k^\mathsf{T}]$ as the error covariance matrix of the observations. Let us note that $\mathbf{B}$ is not indexed by k. In a classic OI scheme, the error covariance matrix of the model is considered constant (and learned on a historical run, for instance).

Then, by Bayes again, the posterior distribution of $\mathbf{x}_k$ knowing $\mathbf{y}_k$ is Gaussian:

$$\mathbf{x}_k|\mathbf{y}_k \sim \mathcal{N}(\mathbf{x}_k^\star, \mathbf{P}_k),$$

with:

$$\begin{aligned} \mathbf{x}_k^\star &= \left(\mathbf{H}_k^\mathsf{T}\mathbf{R}_k^{-1}\mathbf{H}_k + \mathbf{B}^{-1}\right)^{-1}\left(\mathbf{H}_k^\mathsf{T}\mathbf{R}_k^{-1}\mathbf{y}_k + \mathbf{B}^{-1}\boldsymbol{\mu}_k\right) \\ &= \boldsymbol{\mu}_k + \mathbf{K}_k\left(\mathbf{y}_k - \mathbf{H}_k\boldsymbol{\mu}_\mathbf{k}\right), \end{aligned} \qquad [6.5a]$$

$$\mathbf{P}_k = \left(\mathbf{H}_k^\mathsf{T}\mathbf{R}_k^{-1}\mathbf{H}_k + \mathbf{B}^{-1}\right)^{-1} = \boldsymbol{\Sigma}_{\mathbf{x}|\mathbf{y}} = \left(\mathbf{I} - \mathbf{K}_k\mathbf{H}_k\right)\mathbf{B}, \qquad [6.5b]$$

where the optimal gain matrix $\mathbf{K}_k$ (at time t_k) is written as:

$$\mathbf{K}_k = \mathbf{B}_k\mathbf{H}_k^{\mathsf{T}}\left(\mathbf{R}_k + \mathbf{H}_k\mathbf{B}\mathbf{H}_k^{\mathsf{T}}\right)^{-1}.$$

Interestingly, this optimal formulation of the analyzed state exactly corresponds to the state minimizing the 3DVar cost function, which relates to the stationary case of 4DVar formulation:

$$\mathcal{J}_{\mathbf{x}} = \frac{1}{2}\left(\mathbf{x}_k - \boldsymbol{\mu}_k\right)^{\mathsf{T}}\mathbf{B}^{-1}\left(\mathbf{x}_k - \boldsymbol{\mu}_k\right) + \frac{1}{2}\left(\mathbf{H}_k\mathbf{x}_k - \mathbf{y}_k\right)^{\mathsf{T}}\mathbf{R}^{-1}\left(\mathbf{H}_k\mathbf{x}_k - \mathbf{y}_k\right),$$

which makes the formal link between the statistical DA frameworks and the optimal control theory used in the variational formulation.

Looking at this BLUE expression (see 6.3), it is also nothing else than OI of the innovations $\mathbf{d}_k = \mathbf{y}_k - \mathbf{H}_k\boldsymbol{\mu}_{\mathbf{k}}$, where the optimal weights matrix of the (time-fixed) innovations is:

$$\mathbf{K} = \boldsymbol{\Sigma}_{\mathbf{xy}}\boldsymbol{\Sigma}_{\mathbf{yy}}^{-1},$$

where:

$$\boldsymbol{\Sigma}_{\mathbf{yy}} = \mathrm{Cov}\left[\mathbf{y}_k, \mathbf{y}_k\right] = \mathbf{H}_k\mathbf{B}\mathbf{H}_k^{\mathsf{T}} + \mathbf{R}$$

$$\boldsymbol{\Sigma}_{\mathbf{xy}} = \mathrm{Cov}\left[\mathbf{x}_k, \mathbf{y}_k\right] = \mathbf{B}\mathbf{H}_k^{\mathsf{T}}$$

The notation $\mathbf{K}$ defines the so-called *Kalman gain* in the Kalman filter recursion (Asch et al. 2016), as we will discuss below.

Algorithm 6.3: BLUE algorithm

Data:
- $\mathbf{x}_k$: current state at time t_k
- $\mathbf{y}_k$: current observation at time t_k

Preprocessing:
- Specify a covariance matrix for the (background) error $\mathbf{B}$
- Specify a covariance matrix for the observational error $\mathbf{R}_k$
- Specify an observation operator $\mathbf{H}_k$

$\mathbf{K}_k = \mathbf{B}\mathbf{H}_k^{\mathsf{T}}\left(\mathbf{R}_k + \mathbf{H}_k\mathbf{B}\mathbf{H}_k^{\mathsf{T}}\right)^{-1}$

$\mathbf{x}_k^{\star} = \mathbf{x}_k + \mathbf{K}_k\left(\mathbf{y}_k - \mathbf{H}_k\mu\right)$

$\boldsymbol{\Sigma}(\mathbf{x}_k^{\star} - \mathbf{x}_k) = \left(\mathbf{I} - \mathbf{K}_k\mathbf{H}_k\right)\mathbf{B}$

Result: $\mathbf{x}_k^{\star}$: interpolated current field at time t_k

From a probabilistic point of view, we can see the spatiotemporal interpolation problem as a Bayesian filtering problem, where the main objective is to evaluate the posterior probabilities $p(\mathbf{x}_{k+1}|\mathbf{y}_{1:k})$ (*posterior* pdf state $\mathbf{x}_{k+1}$ given the observations up to time t_k) and $p(\mathbf{x}_{k+1}|\mathbf{y}_{1:k+1})$ (*posterior* pdf state $\mathbf{x}_{k+1}$ given the observations up to time t_{k+1}). Under given assumptions, in particular if the dynamical and observational models are linear, these distributions are also Gaussian. Their means and covariances are obtained by sequential recursion of the Kalman filter (see Algorithm 6.4), with the BLUE-based analysis step:

$$\begin{cases} \mathbf{x}^f_{k+1} = \mathcal{M}(\mathbf{x}^a_k), & \text{[6.6a]} \\ \mathbf{P}^f_{k+1} = \mathbf{M}\mathbf{P}^a_k\mathbf{M}^\mathsf{T} + \mathbf{Q}_k, & \text{[6.6b]} \\ \mathbf{x}^a_{k+1} = \mathbf{x}^f_{k+1} + \mathbf{K}_{k+1}\left(\mathbf{y}_{k+1} - \mathbf{H}_{k+1}\mathbf{x}^f_{k+1}\right), & \text{[6.6c]} \\ \mathbf{P}^a_{k+1} = \left(\mathbf{I} - \mathbf{K}_{k+1}\mathbf{H}_{k+1}\right)\mathbf{P}^f_{k+1}, & \text{[6.6d]} \end{cases}$$

where the Kalman gain matrix at time t_{k+1}, denoted $\mathbf{K}_{k+1}$ is defined by:

$$\mathbf{K}_{k+1} = \mathbf{P}^f_{k+1}\mathbf{H}^\mathsf{T}_{k+1}\left(\mathbf{R}_k + \mathbf{H}_{k+1}\mathbf{P}_{k+1}\mathbf{H}^\mathsf{T}_{k+1}\right)^{-1}.$$

Algorithm 6.4: Kalman Filter algorithm

Data:
- $\mathbf{x} = \{\mathbf{x}_k\}$: model field at time t_k
- $\mathbf{y} = \{\mathbf{y}_k\}$: current observation at time t_k

Preprocessing:
- Initialize a covariance matrix for the (background) error $\mathbf{P}_0$
- Specify a covariance matrix for the observational error $\mathbf{R}_k$
- Specify an observation operator $\mathbf{H}_k$

$\mathbf{x}^f_1 = \mathcal{M}(\mathbf{x}_0)$

$\mathbf{P}^f_1 = \mathbf{M}\mathbf{P}_0\mathbf{M}^\mathsf{T} + \mathbf{Q}_1$ **for** $k = 1, \cdots, T$ **do**

  $\mathbf{x}^a_{k+1} = \mathbf{x}^f_{k+1} + \mathbf{K}_{k+1}\left(\mathbf{y}_{k+1} - \mathbf{H}_{k+1}\mathbf{x}^f_{k+1}\right)$

  $\mathbf{P}^a_{k+1} = \left(\mathbf{I} - \mathbf{K}_{k+1}\mathbf{H}_{k+1}\right)\mathbf{P}^f_{k+1}$

  $\mathbf{x}^f_{k+1} = \mathcal{M}(\mathbf{x}^a_k)$

  $\mathbf{P}^f_{k+1} = \mathbf{M}\mathbf{P}^a_k\mathbf{M}^\mathsf{T} + \mathbf{Q}_k$

end

Result: $\mathbf{x}^a$: analyzed sequence of state $\mathbf{x}$

In current applications and in particular for geoscience applications, the model is strongly nonlinear and the distributions are no longer Gaussian. Different variants have been proposed to overcome this problem by introducing the tangent linear operator in a so-called extended Kalman filter, or by using "square-root" approaches (see, for example, Asch et al. (2016)). Due to the imperfection of the tangent linear model for strongly nonlinear processes, ensemble-based DA methods have become particularly popular (Evensen 2009). The key idea is to approximate the *posterior* mean $\mathbf{x}^f_{k+1}$ and its covariance $\mathbf{P}^f_{k+1}$ by the two first moments of the N ensemble members, or particles in case of particle filters (see, for example, Van Leeuwen et al. (2019)). The recursion of the ensemble Kalman filter (EnKF) is then defined by:

$$
\begin{cases}
\mathbf{x}^f_{i,k+1} = \mathcal{M}(\mathbf{x}^a_{i,k}), \text{ß} = 1, \cdots, N, & [6.7a] \\
\mathbf{P}^f_{k+1} = \mathbf{X}^f (\mathbf{X}^f)^\mathsf{T}, & [6.7b] \\
\mathbf{x}^a_{i,k+1} = \mathbf{x}^f_{i,k+1} + \mathbf{K}_{k+1} \left(\mathbf{y}_{k+1} - \mathbf{H}_{k+1} \mathbf{x}^f_{i,k+1} \right), & [6.7c] \\
\mathbf{P}^a_{k+1} = \left(\mathbf{I} - \mathbf{K}_{k+1} \mathbf{H}_{k+1} \right) \mathbf{P}^f_{k+1}, & [6.7d]
\end{cases}
$$

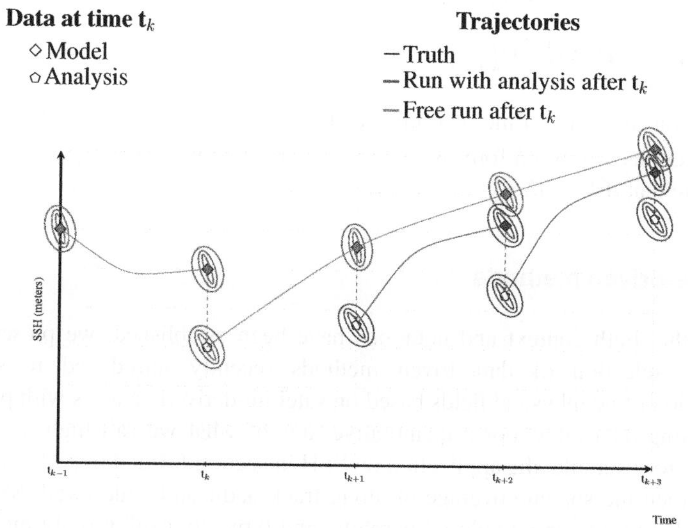

Figure 6.1. *Sketch of the sequential ensemble-based Kalman filter. For a color version of this figure, see www.iste.co.uk/mallet/multitemporal.zip*

Figure 6.1 shows the differences in behavior between the free evolution of the model and its "assimilated" version, where the different observations $\mathbf{y}_{1:k}$ are used to correct/analyze the numerical outputs of the model at each observation time.

Many variants of the EnKF exist, which are divided into two large families: stochastic approaches, which require the introduction of a random noise $\mathbf{u}_{i,k} \sim \mathcal{N}(\mathbf{0}, \mathbf{R}_k) \in \mathbb{R}^p$ for tracking each of the members, and deterministic approaches, involving a "square-root" formalism, in which both the state and square root of the error covariance matrix are propagated by the model. It is beyond the scope of this chapter to introduce all these variants. The reader may refer to Asch et al. (2016) for an exhaustive presentation of the many EnKF flavors. An important point for the next section introducing the analog data assimilation (AnDA) is to remember that both families can be used in this data-driven assimilation scheme.

If sequential DA enables us to assimilate the observations forward in time through filtering algorithms, smoothing versions are also available to assimilate the observations backward in time along the DA window. We then speak of *reanalysis*. Regarding the Kalman smoother formulation (see, for example, Tandeo et al. (2020)), it refers to the filtering algorithm formulation:

$$\begin{cases} \mathbf{x}_k^s = \mathbf{x}_k^a + \mathbf{K}_k^s(\mathbf{x}_{k+1}^s - \mathbf{x}_{k+1}^f), & \text{[6.8a]} \\ \mathbf{K}_k^s = \mathbf{P}_k^a \mathbf{M}_k^{\mathrm{T}} \left(\mathbf{P}_{k+1}^f\right)^{-1}, & \text{[6.8b]} \end{cases}$$

where $\mathbf{x}_k^a$, the analysis at time t_k, is updated by a smoother Kalman gain $\mathbf{K}_k^s$ weighting of the differences between forecast state $\mathbf{x}_{k+1}^f$ and smoothed state $\mathbf{x}_{k+1}^s$ at time t_{k+1}. Let us note that if $k = T$, $\mathbf{x}_{k+1}^s = \mathbf{x}_{k+1}^a$.

6.4. Data-driven methods

Now that both context and notations have been established, we present in this section a selection of data-driven methods recently introduced to solve for interpolation of geophysical fields based on satellite-derived datasets with potentially high missing data rates. For a quantitative idea of what we call high missing data rates, see, for example, the application on SSH in section 6.5 and in particular, Figure 6.8, in which the spatial coverage of along-track nadir and wide-swath SWOT data are provided by the green-colored barplots, and typically exhibit daily missing data rates between 75% and 95%.

6.4.1. *DINEOF schemes*

DINEOF (data interpolating empirical orthogonal function) is a methodology for the reconstruction of missing data using an empirical orthogonal function (EOF) decomposition of the whole dataset. First proposed by Beckers and Rixen (2003), it was used in many applications in oceanography (see, for example, Alvera-Azcárate et al. (2005, 2009)). We start by giving EOF decomposition fundamentals, then a list of methods using it, before presenting, in detail, the DINEOF algorithm and its VE-DINEOF evolution (in which the optimal EOF coefficients are variable), as proposed by Ping et al. (2016).

The EOF method applies to a given space-time field $\mathbf{x} = \{\mathbf{x}_k\}$, where $\mathbf{x}_k \in \mathbb{R}^m$ is the spatial field at time t_k, $k = 1, \cdots, T$. It consists of decomposing the information contained in the dataset into different orthogonal spatial modes. The decomposition involves a constant spatial structure with a time-dependent coefficient. Let us note

$$\mathbf{x}_k = \sum_{l=1}^{L} \alpha_k^l().\phi_l(\mathbf{x}) = \boldsymbol{\alpha}_k.\Phi, \qquad [6.9]$$

where $\boldsymbol{\alpha}_k = \{\alpha_k^l\}, l = 1, \cdots, L$ is the temporal coefficient vector $\in \mathbb{R}^L$ related to each spatial mode l in the EOF decomposition and $\Phi = \{\phi_l\} \in \mathbb{R}^{L\times m}$ denotes the basis of l spatial modes.

Algorithm 6.5: DINEOF algorithm (Beckers and Rixen 2003)

Data:
- $\mathbf{x} \in \mathbb{R}^{T\times m}$: vectors $\mathbf{x}_k$ for $k = 1, \cdots, T$
- $L_{\max}$: the maximum number of EOF coefficients

List of procedures:
- EOF ($\mathbf{x}$): EOF decomposition of $\mathbf{x}$
- recEOF ($\boldsymbol{\alpha}$, Φ): Reconstruct the field $\mathbf{x}$ based on EOF coefficients $\boldsymbol{\alpha}$ and EOF basis Φ

Init:
- 1) $i = 1$, $\mathbf{x}^{(i),*}$ denotes reconstruction of $\mathbf{x}$ at ith iteration of the algorithm
- 2) $\mathbf{x}^{(1),*} = \mathbf{x}$ and add missing data $\overline{\Omega}^v$ in $\mathbf{x}^{(1),*}$ for validation. $\mathbf{x}^*_{\overline{\Omega}^v}$ denotes the reconstruction field restricted to $\overline{\Omega}^v$
- 3) Initialize new missing data to 0, that is, $\mathbf{x}^{(1),*}_{\overline{\Omega}^v} = \mathbf{0}$ (*non-bias* assumption)
- 4) Initialize rmse $^{(i)}$ = RMSE $(\mathbf{x}_{\overline{\Omega}^v}, \mathbf{x}^{(1),*}_{\overline{\Omega}^v})$

L $\leftarrow$ 1
rmse $^{(0)}$ $\leftarrow +\infty$
while *rmse* $^{(i)}$ - *rmse* $^{(i-1)} > \epsilon$ *and* $L < L_{max}$ **do**
  while *rmse* $^{(i)}$ - *rmse* $^{(i-1)}$ **do**
    $\{\boldsymbol{\alpha}, \Phi\} \leftarrow$ EOF $(\mathbf{x}^{(i),*})$
    $\mathbf{x}^{(i),*}_{\overline{\Omega}^v} \leftarrow$ recEOF ($\boldsymbol{\alpha}$, Φ) with L modes
    $i \leftarrow i + 1$
    $\mathbf{x}^{(i),*}_{\overline{\Omega}^v} \leftarrow \mathbf{x}^{(i-1),*}_{\overline{\Omega}^v}$
    rmse $^{(i)}$ = rmse $(\mathbf{x}_{\overline{\Omega}^v}, \mathbf{x}^{(i),*}_{\overline{\Omega}^v})$
  end
  L $\leftarrow$ L + 1
  $\{\boldsymbol{\alpha}, \Phi\} \leftarrow$ EOF $(\mathbf{x}^{(i),*})$
  $\mathbf{x}^{(i),*}_{\overline{\Omega}^v} \leftarrow$ recEOF ($\boldsymbol{\alpha}$, Φ) with L modes
  $i \leftarrow i + 1$
  $\mathbf{x}^{(i),*}_{\overline{\Omega}^v} \leftarrow \mathbf{x}^{(i-1),*}_{\overline{\Omega}^v}$
  rmse $^{(i)}$ = rmse $(\mathbf{x}_{\overline{\Omega}^v}, \mathbf{x}^{(i),*}_{\overline{\Omega}^v})$
end

Result: $\mathbf{x}^*$

The terms α_k^l and ϕ_l satisfy the following orthogonality conditions:

$$\sum_{i=1}^{m} \phi_l . \phi_{l'} = \delta_{ll'}, \text{i.e.,} \, \mathbf{\Phi}^\mathsf{T}\mathbf{\Phi} = \mathbf{I},$$

$$\frac{1}{T}\sum_{k=1}^{T} \alpha_k^l . \alpha_k^{l'} = \frac{1}{T}\sum_{k=1}^{T} (\alpha_k^l)^2 = \lambda_l \delta_{ll'}, \text{i.e.,} \, \boldsymbol{\alpha}^\mathsf{T}\boldsymbol{\alpha} = \boldsymbol{\lambda}^\mathsf{T}\mathbf{I},$$

with $\boldsymbol{\lambda} = \{\lambda_l\}$ as the L singular values of $\mathbf{x}_k$ singular value decomposition (SVD) (see, for example, Golub and van Loan (2013)). In other words:

– the spatial sum over $\mathcal{D}$ of the product between two modes ϕ_l and $\phi_{l'}$ is equal to 1 if $l = l'$, 0 otherwise;

– the temporal sum of the product between coefficients α_k^l and $\alpha_k^{l'}$ related to the two spatial modes ϕ_l and $\phi_{l'}$ is equal to the lth singular value λ_l of the EOF decomposition if $l = l'$, 0 otherwise.

The initial algorithm consists of performing the SVD decomposition of $\mathbf{x} = [\mathbf{x}_k]_{T \times m}$:

$$\mathbf{x} = \mathbf{U}_{T\times T}^{-1}\mathbf{\Sigma}_{T\times m}\mathbf{V}_{m\times m}^\mathsf{T}, \qquad [6.10]$$

when extracting the $L << m$ singular values and vectors after rearranging the singular values of $\mathbf{\Sigma}$ in decreasing order. The variability of the dataset is decomposed into a reduced number of modes (the ones that explain a satisfactory percentage of the total variance) into $\widetilde{\mathbf{x}} = \boldsymbol{\alpha}_{T\times L}\mathbf{\Phi}_{L\times m}$, where $\mathbf{\Phi}$ contains the L first columns of $\mathbf{V}$ and $\boldsymbol{\alpha}$ contains the temporal coefficients corresponding to the L modes retained in the final decomposition.

Note that:

$$\mathbf{C} = \mathbf{x}^\mathsf{T}\mathbf{x} = (\boldsymbol{\alpha}\mathbf{\Phi})^\mathsf{T}(\boldsymbol{\alpha}\mathbf{\Phi}) = \mathbf{\Phi}^\mathsf{T}(\boldsymbol{\alpha}^\mathsf{T}\boldsymbol{\alpha})\mathbf{\Phi},$$

$$\boldsymbol{\alpha} = \mathbf{x}\mathbf{\Phi}^{-1} = \mathbf{x}\mathbf{\Phi}^\mathsf{T}.$$

Looking at the formulation of matrix $\mathbf{C}$, this directly relates to ensemble-based error covariance formulation in DA:

$$\mathbf{P} = \mathbf{X}\mathbf{X}^\mathsf{T},$$

where columns of $\mathbf{X}$ can also be EOFs, scaled by their singular value, which is exactly the core of the EOF-based interpolation methods.

Many variations of the EOF-based iterative reconstruction methods have been proposed when dealing with datasets with potentially large missing data rates. Among all these methods, the traditional approach (Bjornsson and Venegas 1997; Storch and Zwiers 1999) (see Algorithm 6.5) modifies the covariance matrix used in the EOF decomposition using a weighting based on the number of paired observations, thus redefining the principal components according to the same weighting. The RSEOF approach (see Taylor et al. (2013) for more details), modifies the traditional approach by recursively solving the first mode of the decomposition, whose reconstruction field is then removed from the original field. The iterations are performed until a given number of modes L, providing a satisfactory percentage of total variance.

Algorithm 6.6: VE-DINEOF algorithm (Ping et al. 2016)

Data:
- $\mathbf{x} \in \mathbb{R}^{T \times m}$: vectors $\mathbf{x}_k$ for $k = 1, \cdots, T$
- L_{max}: the maximum number of EOF coefficients

List of procedures:
- EOF ($\mathbf{x}$): EOF decomposition of $\mathbf{x}$
- recEOF ($\boldsymbol{\alpha}$, Φ): Reconstruct the field $\mathbf{x}$ based on EOF coefficients $\boldsymbol{\alpha}$ and EOF basis Φ

Init:
- 1) $i = 1$, $\mathbf{x}^{(i),*}$ denotes the reconstruction of $\mathbf{x}$ at $i-$th iteration of the algorithm
- 2) $\mathbf{x}^{(1),*} = \mathbf{x}$ and add missing data $\overline{\Omega}^v$ in $\mathbf{x}^{(1),*}$ for validation. $\mathbf{x}^*_{\overline{\Omega}^v}$ denotes the reconstruction field restricted to $\overline{\Omega}^v$
- 3) Initialize new missing data to 0, that is, $\mathbf{x}^{(1),*}_{\overline{\Omega}^v} = \mathbf{0}$ (*non-bias* assumption)
- 4) Initialize rmse $^{(i)}$ = RMSE $(\mathbf{x}_{\overline{\Omega}^v}, \mathbf{x}^{(1),*}_{\overline{\Omega}^v})$

rmse $^{(0)} \leftarrow +\infty$
while *rmse* $^{(i)}$ - *rmse* $^{(i-1)} > \epsilon$ *and* $N < N_{max}$ **do**
    $\{\boldsymbol{\alpha}, \Phi\} \leftarrow$ EOF $(\mathbf{x}^{(i),*})$
    Init rmse2 $[L_{\text{max}}]$
    for l *in* $1, \cdots, L_{max}$ **do**
        $\mathbf{x}^{(i),*}_{\overline{\Omega}^v} \leftarrow$ recEOF ($\boldsymbol{\alpha}$, Φ) with l modes
        rmse2 $[l]$ = rmse $(\mathbf{x}_{\overline{\Omega}^v}, \mathbf{x}^{(i),*}_{\overline{\Omega}^v})$
    end
    $i \leftarrow i + 1$
    $L \leftarrow \underset{l}{\text{argmin}}$ (rmse2 $[l]$)
    rmse $^{(i)}$ = rmse2 $[L]$
end

Result: $\mathbf{x}^*$

The DINEOF approach (Beckers and Rixen 2003) fills in the missing parts by iteratively decomposing the dataset via a singular value decomposition until the best solution is found w.r.t. a validation training dataset. This is done by gradually including more EOF modes in the reconstruction field, until the error converges. An improved version of the DINEOF algorithm was proposed by Ping et al. (2016) (see Algorithm 6.6). Unlike DINEOF, the value of L is allowed to change during the iterations: it is determined by cross-validation according to the reconstruction performance at the ith iteration. The initial SVD decomposition is performed on the whole dataset and an RMSE score is computed for all possible reconstructions with $l = 1, \cdots, L_{\text{max}}$ modes. An optimal value of L is found based on the RMSE minimization. Until convergence, a new SVD decomposition is carried out and the optimal value of L is updated. Convergence is reached when the RMSE between two successive iterations is less than ϵ. To save computation time and avoid unnecessary iterations, a maximum number N_{max} of SVD decompositions is also predefined. These two parameters often request sensitivity studies to yield satisfactory results.

For the sake of comparison with AnDA (section 6.4.2) and Kalman-based neural networks (section 6.4.3) algorithms, VE-DINEOF is often applied after splitting the domain into small patches $\mathbf{x}_i, i = 1, \cdot, N_p$ (see section 6.4.2.2 for more details). State $\mathbf{x}$ is finally reconstructed by simply averaging the overlapping patches.

6.4.2. *Analog data assimilation*

Instead of using a deterministic model $\mathcal{M}$ for the dynamical evolution of state $\mathbf{x}$ from time t_k to t_{k+1}, the AnDA uses a surrogate operator $\mathcal{A}$, leading to a classic DA state-space formulation (see equation 6.1):

$$\begin{cases} \mathbf{x}_{k+1} & = \mathcal{A}_{k+1}(\mathbf{x}_k) + \boldsymbol{\mu}_k \\ \mathbf{y}_k & = \mathcal{H}_k(\mathbf{x}_k) + \boldsymbol{\varepsilon}_k \end{cases}. \quad [6.11]$$

This operator $\mathcal{A}$, referred to as the analog forecasting operator, is based on a catalog $\mathcal{C} = \{(\mathbf{x}_k, \mathbf{x}_{k+1})\}$ of observation data or numerical simulations, made up of two consecutive state vectors (the potential analogs and their successors) with time step $t_{k+1} - t_k$.

6.4.2.1. *Analog forecasting operator*

The general principle of the analog forecasting operator is to predict state $\mathbf{x}_{k+1}$ at time t_{k+1} through a linear combination of the K nearest neighbors $s_i\{\mathbf{x}_k\}$, $i = 1, \cdots, K$, the so-called analogs $a_i\{\mathbf{x}_k\}$ of $\mathbf{x}_k$ in catalog $\mathcal{C}$:

$$\mathbf{x}_{k+1}^f = \sum_{i=1}^{K} \omega_i s_i\{\mathbf{x}_k^a\}. \quad [6.12]$$

The nearest neighbors of state $\mathbf{x}_k$ are preliminary identified by a k-d tree-based search (Bentley 1975) and the weights $\omega_i, i = 1 \cdots, K$ are given by:

$$\omega_i = \frac{g\left(\mathbf{x}_k^a, a_i\{\mathbf{x}_k^a\}\right)}{\sum_{i=1}^K g\left(\mathbf{x}_k^a, a_i\{\mathbf{x}_k^a\}\right)}, \qquad [6.13]$$

where g is a Gaussian kernel, that is, $\forall(u,v)$ state variables $\mathbf{x}$, $g(u,v) = \exp(-\lambda||uv||^2)$, with $||.||$ as the Euclidean distance and λ as a scaling parameter (which corresponds to an exponential covariance with unit variance and range $1/\lambda$). Other kernels can be considered as proposed in Zhao and Giannakis (2014).

Three different specifications of the analog forecasting operator have been proposed in Lguensat et al. (2017):

– locally constant: $\mathbf{x}_{k+1}^f = \sum_{i=1}^K \omega_i s_i\{\mathbf{x}_k^a\}$. In a Gaussian framework, and by denoting $\mathbf{S}_f = \frac{1}{\sqrt{K-1}}[\mathbf{s}_1 - \bar{\mathbf{s}}, \cdots, \mathbf{s}_N - \bar{\mathbf{s}}]$ with N the number of members (at time t_k), the covariance $\mathbf{\Sigma}_{k+1}^f$ of $\mathbf{x}_{k+1}^f$ is given by $\mathbf{S}_f \mathbf{S}_f^\mathsf{T}$;

– locally incremental: $\mathbf{x}_{k+1}^f = x_k + \sum_{i=1}^K \omega_i \tau_i[\mathbf{x}_k]$, with $\tau_i[\mathbf{x}_k] = s_i\{\mathbf{x}_k^a\} - a_i\{\mathbf{x}_k^a\}$;

– locally linear: $\mathbf{x}_{k+1}^f = \beta_{0,k} + \beta_{1,k}(\mathbf{x}_k)$, with $\beta_{0,k}$ and $\beta_{1,k}$ as the regression parameters of the successors $s_i\{\mathbf{x}_k^a\}$ on the $a_i\{of\mathbf{x}_k^a\}$. Then, $\boldsymbol{\xi}_i(\mathbf{x}_k) = s_i\{\mathbf{x}_k^a\} - \beta_{0,k} - \beta_{1,k}(a_i\{\mathbf{x}_k^a\})$. Again, in a Gaussian framework, and by denoting $\Xi_f = \frac{1}{\sqrt{K-1}}[\boldsymbol{\xi}_1 - \bar{\boldsymbol{\xi}}, \cdots, \boldsymbol{\xi}_N - \bar{\boldsymbol{\xi}}]$ (at time t_k), the covariance $\mathbf{\Sigma}_{k+1}^f$ of $\mathbf{x}_{k+1}^f$ is given by $\Xi_f \Xi_f^\mathsf{T}$.

The regression-based locally linear operator generally leads to better forecasting and assimilation performance (Lguensat et al. 2019; Lopez-Radcenco et al. 2019). They also make the exploitation of additional covariates to state conditional forecasting operators simpler (see, for example, Lguensat et al. (2019)).

6.4.2.2. *Patch-based approaches*

Looking for nearest neighbors in the AnDA processing obviously faces the curse of dimensionality when it comes to handling three-dimensional satellite-derived datasets or numerical model outputs (Platzer et al. 2021). Even when considering only the two-dimensional surface state, the dimension m of state $\mathbf{x}$ is often at least of 10^4. To overcome this problem, the state $\mathbf{x}$ is decomposed as $\mathcal{P} = \{\mathbf{x}_i\}, i = 1, \cdots, N_p$, a set of N_p overlapping patches where $\mathbf{x}_i$ denotes the patch centered at the point $\mathbf{s}_i$. Along this line, an EOF-based decomposition of each patch is made to again reduce the dimensionality of the problem: in the application proposed by Lopez-Radcenco et al. (2019), 50 EOF coefficients are used in the patch-based AnDA version (see Algorithm 6.7).

Algorithm 6.7: Patch-based AnDA algorithm

Data: - $\mathbf{x}_{train} = \{\mathbf{x}_{k',train}\}, ' = 1, \cdots, T$: training dataset
- $\mathbf{x}_k$: state at time t_k
- $\mathbf{y}_k$: observations at time t_k
- $\{\mathbf{s}_i\}, i = 1, \cdots, N_p$, central positions of patches $\mathbf{x}_i$
- K: number of nearest neighbors

List of procedures:
- EOF ($\mathbf{x}_{train}$): Perform the EOF decomposition of $\mathbf{x}_{train}$
- findKnn ($\boldsymbol{\alpha}$, $\mathcal{C}$, K): Find the K nearest neighbors of the EOF coefficients $\boldsymbol{\alpha}$ in catalog $\mathcal{C}$
- LgmKnn ($\boldsymbol{\alpha}$, $\mathcal{C}$): Combine the successors of $\boldsymbol{\alpha}$ in catalog $\mathcal{C}$
- projEOF ($\mathbf{x}$, Φ): Project $\mathbf{x}$ on the EOF database Φ
- EnKF/EnKS ($\boldsymbol{\alpha}^f$, $\boldsymbol{\alpha}^o$): Ensemble-based Kalman Filter/Smoother analysis of the coefficients $\boldsymbol{\alpha}^f$ given the observations $\boldsymbol{\alpha}^o$
- recEOF ($\boldsymbol{\alpha}$, Φ): Reconstruction of $\mathbf{x}$ with EOF coefficients $\boldsymbol{\alpha}$ and basis Φ
- averagePatches ($\{\mathbf{x}_i\}$): Average overlapping patches areas of the $\mathbf{x}_i$'s

Preprocessing:
$\{\mathbf{x}_{i,train}\}$, ß $= 1, \cdots, N_p \leftarrow$ ExtractPatch ($\mathbf{x}_{train}$)
for i *in* $1, \cdots, N_p$ **do**
  $\{\boldsymbol{\alpha}_{i,k'}, \Phi_i\} \leftarrow$ EOF ($\mathbf{x}_{i,train}$)
  $\mathcal{C}_i = \{\}$
  for k' *in* $1, \cdots, K-1$ **do**
    $\mathcal{C}_i \leftarrow \{\mathcal{C}_i, (\boldsymbol{\alpha}_{i,k'-1}, \boldsymbol{\alpha}_{i,k'})\}$

$\{\mathbf{x}_{i,k}\}$, ß $= 1, \cdots, N_p \leftarrow$ ExtractPatch ($\mathbf{x}_k$)
$\{\mathbf{y}_{i,k}\}$, ß $= 1, \cdots, N_p \leftarrow$ ExtractPatch ($\mathbf{y}_k$)
for i *in* $1, \cdots, N_p$ **do**
  $\boldsymbol{\alpha}^a_{i,k} \leftarrow$ projEOF ($\mathbf{x}^a_{i,k}$, Φ_i)
  $\{\mathcal{C}_{i,n_j}\}, j = 1, \cdots, K \leftarrow$ findKnn ($\boldsymbol{\alpha}^a_{i,k}$, $\mathcal{C}_i$, K)
  $\mathcal{F}(\boldsymbol{\alpha}_{i,k}) = \boldsymbol{\alpha}^f_{i,k+1} \leftarrow$ LgmKnn ($\boldsymbol{\alpha}^a_{i,k}$, $\{\mathcal{C}_{i,n_j}\}$)
  $\boldsymbol{\alpha}^o_{i,k+1} \leftarrow$ projEOF ($\mathbf{y}_{i,k+1}$, Φ_i)
  $\boldsymbol{\alpha}^a_{i,k+1}$ EnKF / EnKS ($\boldsymbol{\alpha}^f_{i,k+1}$, $\boldsymbol{\alpha}^o_{i,k+1}$)
  $\mathbf{x}^a_{i,k+1} \leftarrow$ recEOF ($\boldsymbol{\alpha}^a_{i,k+1}$, Φ_i)

Postprocessing:
$\mathbf{x}^a_{k+1} \leftarrow$ averagePatches ($\{\mathbf{x}^a_{i,k+1}\}$)
Result: $\mathbf{x}^a_{k+1}$

In this configuration, a catalog $\mathcal{C}_i = \{(\mathbf{x}_{i,k}, \mathbf{x}_{i,k+1})\}$ is built for each patch $\mathbf{x}_i$:

$$\mathbf{x}_i(t_k) = \mathbf{x}_{i,k} = \sum_{l=1}^{L} \alpha^l_{i,k} \cdot \phi_{i,l} = \boldsymbol{\alpha}^{\mathrm{T}}_{i,k} \cdot \phi_i \qquad [6.14]$$

where:

– $\mathbf{x}_{i,k}$ denotes the patch i of state $\mathbf{x}$ at time t_k;

– $\boldsymbol{\alpha}_{i,k} = \{\alpha^l_{i,k}\}, l = 1, \cdots, L$, with $\alpha^l_{i,k}$ as the lth coefficient in the EOF decomposition of $\mathbf{x}_{i,k}$;

– $\Phi_i = \{\phi_{i,l}\}, l = 1, \cdots, L$ is the EOF set of basis functions for the patch $\mathbf{x}_i$.

Lastly, the reconstruction of $\mathbf{x}_{i,k}$ is obtained by sequential ensemble assimilation of the EOF decomposition coefficients $\boldsymbol{\alpha}_{i,k}$, which represent the projection of patch $\mathbf{x}_i$ on the lower dimensional EOF subspace. In other words, during the forecast step, the distribution of the coefficients $\boldsymbol{\alpha}_{i,k+1}$ at time t_{k+1} given the coefficients $\boldsymbol{\alpha}^a_{i,k}$ at time t_k is Gaussian:

$$\boldsymbol{\alpha}_{i,k+1} \sim \mathcal{N}\left(\boldsymbol{\alpha}^f_{i,k+1}, \boldsymbol{\Sigma}^f_{i,k+1}\right)$$

with mean and covariance estimated by one of the three analog forecasting schemes of the analog operator described above.

6.4.3. *Neural network–based Kalman filters*

Instead of using an analog forecasting operator, we may explore using a neural network (NN)-based operator for the dynamical evolution of state $\mathbf{x}$ (see, for example, Ouala et al. (2018) and Brajard et al. (2021)). The prediction step involves two NNs $\mathcal{F}$ and $\mathcal{G}$ to estimate, respectively, *prior* mean $\mathbf{x}^f_{k+1}$ and covariance $\mathbf{P}^f_{k+1}$ of state $\mathbf{x}_{k+1}$ at time t_{k+1} given previous *posterior* analysis $\mathbf{x}^a_k$ and $\mathbf{P}^a_k$ at time t_k.

$$\begin{cases} \mathbf{x}^f_{k+1} &= \mathcal{F}(\mathbf{x}^a_k) \\ \mathbf{P}^f_{k+1} &= \mathcal{G}(\mathbf{x}^a_k, \mathbf{P}^a_k) \end{cases}. \qquad [6.15]$$

Within such a formulation, and still assuming a linear observation operator, the analysis step is provided by a Kalman filter (see equations [6.6c] and [6.6d]), avoiding the use of ensemble-based statistical methods, thus considerably reducing the complexity of the approach (see Algorithm 6.8). In addition, Ouala et al. (2018) show that this parameterization of the covariance is more efficient than estimating it from ensemble-based propagation of NN $\mathcal{F}$.

The dimensionality problem is again solved by decomposing $\mathbf{x}_k$ into N_p patches of dimension $p \times p$. This decomposition is explicitly supported by the architecture of the NN.

Algorithm 6.8: Patch-based Kalman-based neural networks algorithm

Data:
- $\mathbf{x}_{train} = \{\mathbf{x}_{k',train}\}, k' = 1, \cdots, T$: training dataset
- $\mathbf{x}_k$: state at time t_k
- $d\mathbf{y}_k$: observations at time t_k
- $\{\mathbf{s}_i\}, i = 1, \cdots, N_p$, central positions of patches $\mathbf{x}_i$ $\mathbf{p}(\mathbf{s}_i, \cdot)$

List of procedures:
- EOF ($\mathbf{x}_{train}$): Perform the EOF decomposition of the data $\mathbf{x}_{train}$
- trainF1 ($\mathbf{x}$): Train the neural network $\mathcal{F}_1$ for the prediction of the mean at the patch level
- trainF2 ($\mathbf{x}$): Train the neural network $\mathcal{F}_2$ for the reconstruction of the global mean
- trainG ($\mathbf{x}$, $\mathbf{P}$): Train the neural network $\mathcal{G}$
- projEOF ($\mathbf{x}$, Φ): Project $\mathbf{x}$ on the EOF basis Φ
- KF / KS ($\boldsymbol{\alpha}^f$, $\boldsymbol{\alpha}^o$): Kalman filter/smoother analysis of the coefficients $\boldsymbol{\alpha}^f$ given the observations $\boldsymbol{\alpha}^o$

Preprocessing:

$\{\mathbf{x}_{i,train}\}, \text{ß} = 1, \cdots, N_p \leftarrow$ ExtractPatch ($\mathbf{x}_{train}$)

for i *in* $1, \cdots, N_p$ **do**

  $\{\boldsymbol{\alpha}_{i,k'}, \Phi_i\} \leftarrow$ EOF ($\mathbf{x}_{i,train}$)

  $\mathcal{F}_{1,i} \leftarrow$ trainF1 ($\{\boldsymbol{\alpha}_{i,k'}\}$)

$\mathcal{F}_2 \leftarrow$ trainF2 ($\{\boldsymbol{\alpha}_i\}$)

for i *in* $1, \cdots, N_p$ **do**

  $\mathcal{G}_i \leftarrow$ trainG ($\{\boldsymbol{\alpha}_{i,k'}, \mathbf{P}_{i,k'}\}$)

$\mathbf{x}^f_{k+1} \leftarrow \mathcal{F}(\mathbf{x}^a_k)$

$\{\mathbf{P}^f_{i,k+1}\} \leftarrow \mathcal{G}(\mathbf{x}^a_k, \mathbf{P}^a_{i,k})$ $\{\mathbf{x}^f_{i,k+1}\}, \text{ß} = 1, \cdots, N_p \leftarrow$ ExtractPatch ($\mathbf{x}^f_k$)

$\{\mathbf{y}_{i,k}\}, \text{ß} = 1, \cdots, N_p \leftarrow$ ExtractPatch ($\mathbf{y}_k$)

for i *in* $1, \cdots, N_p$ **do**

  $\mathbf{x}^a_{i,k+1}, \mathbf{P}_{i,k+1} \leftarrow$ KF/KS ($\mathbf{x}^f_{i,k}, \mathbf{y}_{i,k}$)

Postprocessing:

$\mathbf{x}^a_{k+1} \leftarrow \mathcal{F}_2(\{\mathbf{x}^a_{i,k+1}\})$

$\mathbf{P}^a_{k+1} \leftarrow \{\mathbf{P}^a_{i,k+1}\}, \text{ß} = 1, \cdots, N_p$

Result: $\mathbf{x}^a_{k+1}, \mathbf{P}^a_{k+1}$

6.4.3.1. *Neural-network propagation for the mean state*

Regarding the first NN-based model $\mathcal{F}$ for the prediction of the mean $\mathbf{x}^f_{k+1}$, the network architecture is composed of three main elements. Using the same notation already introduced in equation [6.14], the prediction of the coefficients $\boldsymbol{\alpha}^f_{i,k+1}$ is ensured by a network $\mathcal{F}_{1,i}$, $i = 1, \cdots, N_p$ with residual architecture and bilinear parameterization, which emulates a fourth order Runge-Kutta (RK4) scheme. Indeed, RK4 may be restated using a graphical model (see Figure 6.2), involving a four-layer residual net (LeCun et al. 2015), each layer sharing the same operator, which is a NN implementation B of an ODE-based dynamical operator.

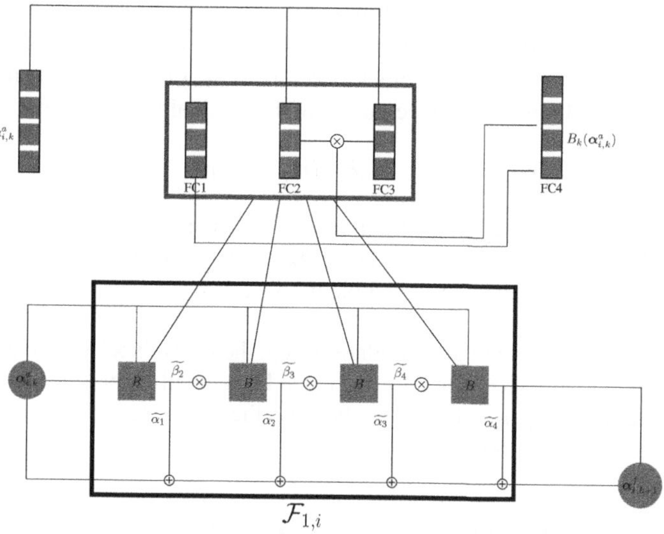

Figure 6.2. *ResNet formulation of $\mathcal{F}_1$ with bilinear parameterization emulating a RK4 scheme. For a color version of this figure, see www.iste.co.uk/mallet/multitemporal.zip*

The block B is a combination of fully connected (FC) layers. Specifically, a final layer FC4 combines linear terms, resulting from the linear combination of the first layer FC1 input variables, and bilinear terms, that is, products between two linear combinations of layers FC2 and FC3 input variables. In such a configuration, the parameterization of the architecture is based on the number of linear and nonlinear terms, defined by the number of hidden nodes in the FC layers (FC1, FC2 and FC3). The emulation of the dynamic model then amounts to learning the weights of these FC layers. In Fablet et al. (2017), three linear neurons and three bilinear neurons are used to emulate the chaotic Lorentz-63 model (Lorenz 1963). In Ouala et al. (2018), 60 linear neurons, 100 bilinear neurons and 10 FC layers with ReLU activation are used in the spatiotemporal interpolation of sea surface temperature (SST) fields.

Last, the reconstruction of the initial state $\mathbf{x}^f_{k+1}$ is obtained by combining all the patches $\mathbf{x}^f_{i,k+1} = \boldsymbol{\alpha}^f_{i,k+1}\Phi_i$, whose weighting is learnt by $\mathcal{F}_2$, an additional convolutional network (CNN).

6.4.3.2. *Neural-network propagation for the state covariance*

Regarding the second NN-based model $\mathcal{G}$ for the prediction of the error covariance matrix, we also get back to a patch-based decomposition of the overall covariance matrix. More precisely, $\mathcal{G}_i$, $i = 1, \cdots, N_p$ is parameterized by a covariance matrix $\widetilde{\mathbf{P}}_{i,k} = \text{Cov}(\boldsymbol{\alpha}_{i,k})$ with diagonal block structure in the EOF lower dimensional space, which leads to full rank covariance matrix in the original patch space:

$$\mathbf{P}^f_{i,k+1} = \mathcal{G}_i(\mathbf{x}^a_{i,k}, \mathbf{P}^a_{i,k}) = \Phi_i^{\mathsf{T}} \mathcal{F}_i^2, (\boldsymbol{\alpha}^a_{i,k}, \boldsymbol{\Sigma}_0)\Phi_i \qquad [6.16]$$

where the dataset used for training is collected from $\mathcal{N}(\boldsymbol{\alpha}_{i,k'}, \boldsymbol{\Sigma}_0)$ with $\boldsymbol{\alpha}_{i,k'} = \mathcal{F}_i(\boldsymbol{\alpha}_{i,k'-1})$ and the covariance $\boldsymbol{\Sigma}_0 = \text{Cov}(\boldsymbol{\alpha})$ is given by the empirical EOF coefficient covariance for the whole training dataset.

Let us note that during the assimilation step, the analysis $\mathbf{x}^a_k$ is reconstructed from all the patches analysis $\mathbf{x}^a_{i,k}$ but the covariance $\mathbf{P}^a_k$ is finally represented as a collection of covariances $\{\mathbf{P}^a_{i,k}\}$, $i = 1, \cdots, N_p$ because the complete covariance matrix is not necessary to run the algorithm.

6.4.4. *Variational end-to-end learning-based methods*

Recently, new NN-based architectures have emerged to solve inverse problems based on variational formulation. More precisely, restarting from the variational DA scheme (Asch et al. 2016), the state analysis $\mathbf{x}^a$ is obtained by solving the minimization problem:

$$\mathbf{x}^a = \arg\min_{\mathbf{x}} \mathcal{J}(\mathbf{x})$$

where the variational cost function $\mathcal{J}(\mathbf{x}) = \mathcal{J}_\Phi(\mathbf{x}, \mathbf{y}, \Omega)$ is the sum of an observation term and a regularization term involving a dynamical prior operator Φ:

$$\begin{aligned} \mathcal{J}_\Phi(\mathbf{x}, \mathbf{y}, \Omega) &= \mathcal{J}^o(\mathbf{x}, \mathbf{y}, \Omega) + \mathcal{J}^b_\Phi(\mathbf{x}) \\ &= \lambda_1 ||\mathbf{y} - \mathcal{H}(\mathbf{x})||^2_\Omega + \lambda_2 ||\mathbf{x} - \Phi(\mathbf{x})||^2 \end{aligned} \qquad [6.17]$$

The parameterization of operator Φ may exploit some physical priors, for instance, through some ODE-based parameterization as mentioned in the previous section. We refer the reader to Fablet et al. (2020) for details on such strategies. Here, we rather focus on another class of parameterization for Φ, which relates to Gibbs models considered in statistical physics. The resulting neural architectures are referred to as GENN architectures and can be seen as a way of embedding Markovian priors in CNN (Fablet et al. 2019). This solution provides promising results on satellite-derived sea surface fields (Fablet et al. 2021; Beauchamp et al. 2020). In addition, DL automatic differentiation tools are involved for the computation of gradient operator $\nabla_{\mathbf{x}} \mathcal{J}_\Phi$, given the architecture of operator Φ: it can be seen as a composition of operators such as tensors, convolutions and activation functions.

The proposed end-to-end architecture consists of embedding an iterative gradient-based solver based on a specific variational representation (Fablet et al. 2020). We consider observations $\mathbf{y}$ on the associated observation domain Ω as inputs. At the initialization of the state sequence $\mathbf{x}$ at the first iteration $\mathbf{x}^{(0)}$, the missing areas $\overline{\Omega}$ are filled with a large-scale component given, for instance, by an OI.

$$\begin{cases} \mathbf{x}^{(i+1)} & = \mathbf{x}^{(i)} - \alpha \nabla_{\mathbf{x}} \mathcal{J}_{\Phi}(\mathbf{x}^{(i)}, \mathbf{y}, \Omega) \\ \mathbf{x}^{(k+1)}(\Omega) & = \mathbf{y}(\Omega) \\ \mathbf{x}^{(k+1)}(\overline{\Omega}) & = \mathbf{x}^{(k+1)}(\overline{\Omega}) \end{cases} \quad [6.18]$$

where superscript $(k) = 0, \cdots, N_{\text{GB}}$ denotes the kth iteration of the gradient-based descent, see the corresponding sketch in Figure 6.3.

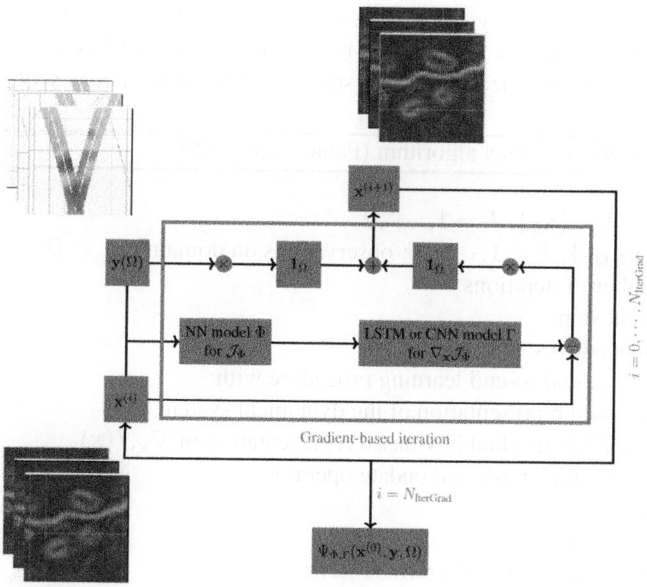

Figure 6.3. *Sketch of the gradient-based algorithm: the upper left stack of images corresponds to an example of a temporal sequence of SSH observations with missing data used as inputs. The upper right stack of images is an example of the intermediate reconstruction of the SSH gradient at iteration i, while the bottom left stack of images identifies the updated reconstruction fields used as new inputs after each iteration of the algorithm. For a color version of this figure, see www.iste.co.uk/mallet/multitemporal.zip*

Let us denote by Γ this iterative update operator. Following meta-learning schemes (Andrychowicz et al. 2016), an LSTM-based operator is considered. Let us note that we may consider other iterative solvers, including fixed-point procedures and other recurrent cells such as CNN-based operators.

Regarding the residual LSTM-based representation (Fablet et al. 2020) of operator Γ, the ith iterative update of the solver is given by:

$$\begin{cases} g^{(i+1)} = LSTM\left[\alpha \cdot \nabla_{\mathbf{x}} \mathcal{J}_{\Phi}(\mathbf{x}^{(i)}, \mathbf{y}, \Omega), h(i), c(i)\right] \\ x^{(i+1)} = x^{(i)} - \mathcal{T}\left(g^{(i+1)}\right) \end{cases}, \quad [6.19]$$

where $g^{(i+1)}$ is the LSTM output using as input gradient $\nabla_{\mathbf{x}} \mathcal{J}_{\Phi}(\mathbf{x}^{(i)}, \mathbf{y}, \Omega)$, while $h(i)$ and $c(i)$ denote the internal states of the LSTM (Arras et al. 2019), α is a normalization scalar and $\mathcal{T}$ can be a linear or convolutional mapping to be chosen according to validation performance.

Overall, let us denote by $\Psi_{\Phi,\Gamma}(\mathbf{x}^{(0)}, \mathbf{y}, \Omega)$ the output of the end-to-end learning scheme given architectures for both NN-based operators Φ and Γ (see Figure 6.3 and Algorithm 6.9), the initialization $\mathbf{x}^{(0)}$ of state $\mathbf{x}$ and the observations $\mathbf{y}$ on domain Ω.

Algorithm 6.9: 4DVarNN algorithm (Fablet et al. 2020)

Data:
- $\mathbf{x} \in \mathbb{R}^{T \times m} = \{\mathbf{x}_k\}$, $k = 1, \cdots, T$
- $\mathbf{y}_\Omega = \{\mathbf{y}_{k,\Omega_k}\}$, $k = 1, \cdots, T$: observations on domains $\Omega_k \subset \mathcal{D}$
- N_I: number of iterations
- η: gradient step

List of procedures:
- Train_$\Psi_{\Phi,\Gamma}$: end-to-end learning procedure with:
  * Φ: NN-based representation of the dynamical system
  * $GradLSTM$: residual NN-based representation of $\nabla_{\mathbf{x}} \mathcal{J}(\mathbf{x})$
  * Γ: iterative gradient-based update operator

$\mathbf{x} = \Phi(\mathbf{y})$
while $i < N_I$ **do**
  $\mathbf{x}^{(i+1)} \leftarrow \mathbf{x}^{(i)} - \eta \times GradLSTM(\mathbf{x}^{(i)})$
  $N_I \nearrow; \eta \searrow; i \leftarrow i + 1$
end
Train $\Psi_{\Phi,\Gamma}$
Result: $\mathbf{x}^{\star} \leftarrow \Psi_{\Phi,\Gamma}(\mathbf{x}^{(0)}, \mathbf{y}, \Omega)$

Then, the joint learning of operators $\{\Phi, \Gamma\}$ is stated as the minimization of a reconstruction cost:

$$\arg\min_{\Phi,\Gamma} \mathcal{L}(\mathbf{x}, \mathbf{x}^\star) \text{ s.t. } \mathbf{x}^\star = \Psi_{\Phi,\Gamma}(\mathbf{x}^{(0)}, \mathbf{y}, \Omega). \qquad [6.20]$$

In case of a supervised learning, where targets are gap free, $\mathcal{L}$ is defined by:

$$\mathcal{L}(\mathbf{x}, \mathbf{x}^\star) = \lambda_1 ||\mathbf{x} - \mathbf{x}^\star||^2 + \lambda_2 ||\nabla_{\mathbf{x}} - \nabla_{\mathbf{x}^\star}||^2 + \lambda_3 ||\mathbf{x} - \Phi(\mathbf{x}^\star)||^2 + \lambda_4 ||\mathbf{x} - \Phi(\mathbf{x})||^2,$$

that is, the L2-norm of the difference between state $\mathbf{x}$ and reconstruction $\mathbf{x}^\star$ with additional terms related to the gradient of state $\mathbf{x}$, and constraints on operator Φ to ensure realistic reconstruction of the dynamics when applying operator Φ on both $\mathbf{x}^\star$ and $\mathbf{x}$.

In case of an unsupervised learning, given the observations $\mathbf{y}$ on domain Ω and hidden state $\mathbf{x}$, the 4DVar cost function may be used:

$$\mathcal{L}(\mathbf{x}, \mathbf{x}^\star) = \lambda_1 ||\mathbf{y} - \mathcal{H}(\mathbf{x})||_\Omega^2 + \lambda_2 ||\mathbf{x} - \Phi(\mathbf{x})||^2,$$

with weights λ_1 and λ_2 indicating how reliable the observations are.

Let note that when replacing both the CNN/LSTM cells by the identity operator and the minimization function $\mathcal{J}_\Phi(\mathbf{x}, \mathbf{y}, \Omega)$ by its single regularization term $\mathcal{J}_\Phi^b(\mathbf{x})$, the gradient-based solver simply leads to a parameter-free fixed-point version of the algorithm, the same used in Beauchamp et al. (2020) and Fablet et al. (2019), which is similar to the DINEOF approach (see Figure 6.4).

$$\begin{cases} \mathbf{x}^{(k+1)} & = \psi\left(\mathbf{x}^{(k)}\right) \\ \mathbf{x}^{(k+1)}(\Omega) & = \mathbf{y}(\Omega) \\ \mathbf{x}^{(k+1)}(\overline{\Omega}) & = \mathbf{x}^{(k+1)}(\overline{\Omega}) \end{cases} \qquad [6.21]$$

This fixed-point solver is parameter-free and easily implemented as an NN in a joint solution with the NN-parameterization of $\mathcal{J}_\Phi$ for the interpolation problem.

The state sequences $\mathbf{x}_{k\pm l} = \mathbf{x}_{k-l:k+l}$ of length $2l+1$ are used in the training for the interpolation problem. The idea is to optimize the results for the time t_k at the center of the window $[t_{k-l}; t_{k+l}]$. The value of l has to be chosen according to the dynamics of the geophysical field considered.

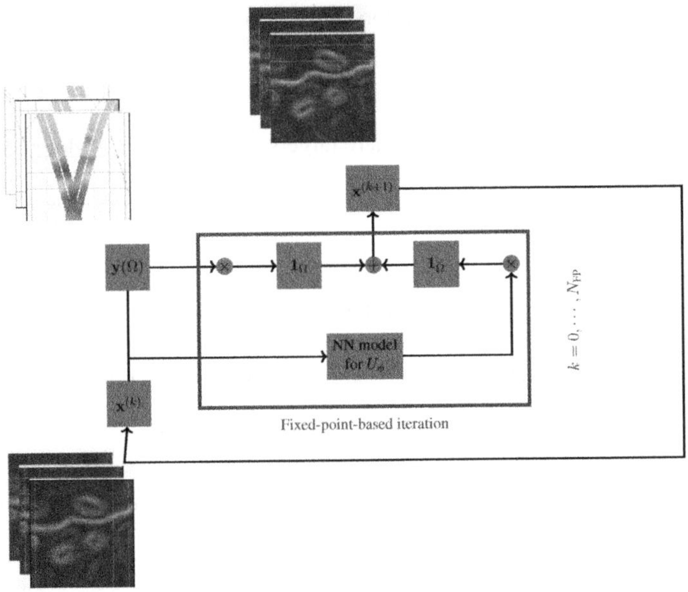

Figure 6.4. *Sketch of the iterative fixed-point algorithm: the upper left stack of images corresponds to an example of SSH observations temporal sequence with missing data used as inputs. The upper right stack of images is an example of intermediate reconstruction of the SSH gradient at iteration i, while the bottom left stack of images identifies the updated reconstruction fields used as new inputs after each iteration of the algorithm. For a color version of this figure, see www.iste.co.uk/mallet/multitemporal.zip*

6.5. Application to satellite-derived ocean surface topography datasets

As illustration for the application of the data-driven and learning-based approaches presented in previous section, we consider satellite altimeter data for the reconstruction of ocean surface topography, referred to as the space-time interpolation of the SSH. We consider a $10° \times 10°$ region along the Gulfstream as the case-study area (see Figure 6.5). The selected region is mainly driven by energetic mesoscale dynamics[2], which makes the reconstruction challenging because current state-of-the-art interpolation techniques only recover the spatial horizontal scales up to 100 km. To illustrate how data-driven and learning-based strategies may help to

2 Ocean mesoscale dynamics have typical horizontal scales of less than 100 km and timescales on the order of a month.

solve for the oversmoothing of the state-of-the-art OI, we use an OSSE with the NEMO (Nucleus for European Modeling of the Ocean) model NATL60 high-resolution basin-scale configuration (Molines 2018). Doing so, it enables the comparison between the different reconstruction methods and the NATL60 numerical outputs used as ground truth (GT) here.

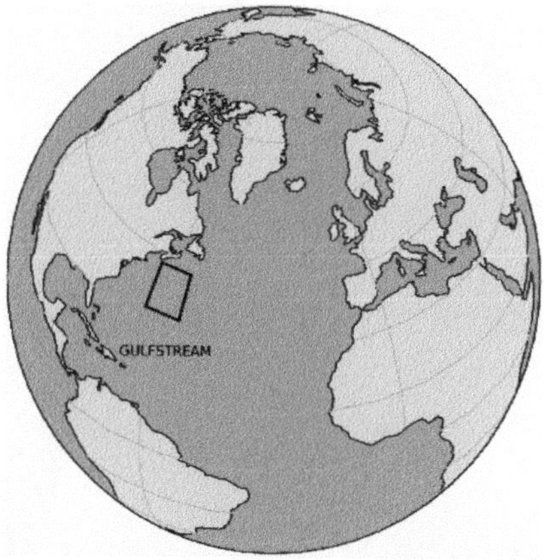

Figure 6.5. *Considered case-study area along the Gulfstream in the northwest atlantic ocean*

Based on this one-year long simulation, we generate pseudo along-track nadir data for the current capabilities of the observation system (Ballarotta et al. 2019) and pseudo wide-swath SWOT data in the context of the upcoming SWOT mission (Metref et al. 2020). Realistic noise patterns are added to these noise-free pseudo observations in order to simulate observational nadir/SWOT errors (Dufau et al. 2016; Esteban-Fernandez 2014; Gaultier and Ubelmann 2010). The two types of observations are merged (see Figure 6.6) to produce one-year long daily datasets of partial and noisy observations of the idealized GT. Last, the DUACS operational system (CMEMS/C3S Copernicus program) provides the OI baseline (Taburet et al. 2019) as daily gridded ($0.25^\circ \times 0.25^\circ$) product.

As a selection of data-driven methods, our experiments involve the AnDA (section 6.4.2), VE-DINEOF (section 6.4.1) and the variational end-to-end learning framework (section 6.4.4) to cover both EOF-based, analog and neural approaches. We aim to solve smaller scales than the 100 km spatial resolution currently handled by operational OI products, more adapted to estimate large-scale dynamics. Along this line, we rely on the following multiscale decomposition:

$$\mathbf{x} = \overline{\mathbf{x}} + d\mathbf{x} + \epsilon \qquad [6.22]$$

and all the interpolations methods used here will apply to the anomaly field $d\mathbf{x}$, seen as the difference between the original field $\mathbf{x}$ and the large-scale components provided by the OI. In the end, we hope that the effective resolution estimated for the anomaly field $d\mathbf{x}$ will be better than the OI-based representation of the dynamics.

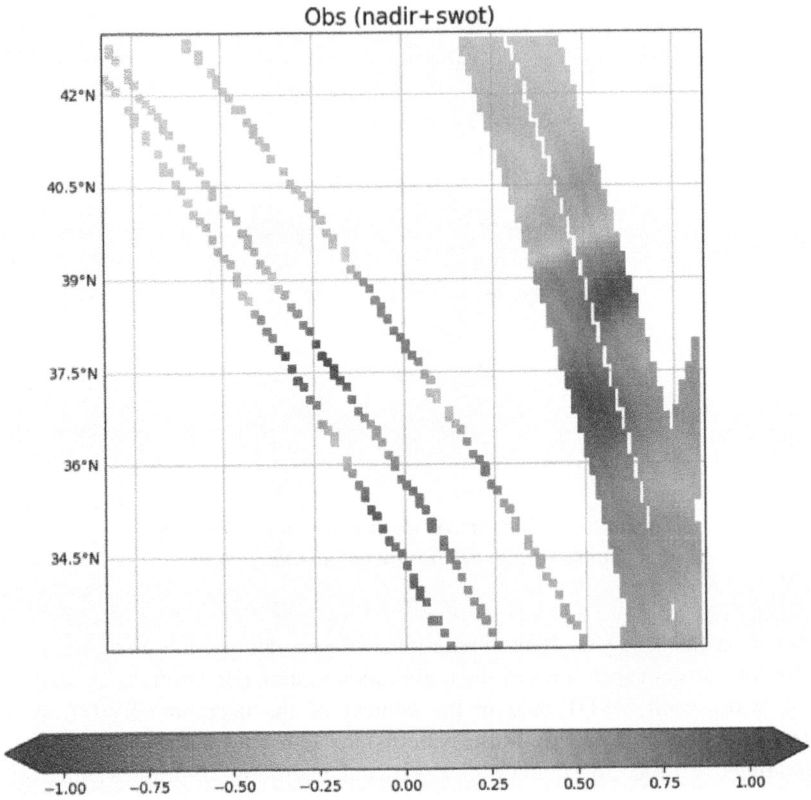

Figure 6.6. *One ($d = 0$) day accumulated along-track nadir and wide-swath SSH pseudo-observations (meters) on August 4, 2013. For a color version of this figure, see www.iste.co.uk/mallet/multitemporal.zip*

Regarding the setup of each data-driven method, we consider the following experimental settings:

1) The observation error covariance matrix $\mathbf{R}_k = Cov(\varepsilon_k)$ gets values increasing according to a parametric function of the hourly time lag between the observations $\mathbf{y}_k$ and the day to estimation time t_k.

As in Lopez-Radcenco et al. (2019), a patch-based version of AnDA coupled with an EOF-based representation of the individual patches is used. The anomaly field $d\mathbf{x}$

is split into 169 vectorized patches $\mathbf{p}(\mathbf{s}, t)$ of sizes 1°× 1°, corresponding to 20 pixels × 20 pixels, with overlapping areas of 5 pixels.

2) VE-DINEOF is also applied in its patch-based version.

3) Regarding the variational NN-based methods with Gibbs energy–based representation of the dynamics and either a fixed point (FP) or gradient-based (Grad) solver, the architecture of operator ϕ comprises: an initial 4 × 4 average pooling, a Conv2D layer with 40 filters, 11 × 11 kernel, ReLu activation and a zero-weight constraint on the center of the convolution window, a 1 × 1 Conv2D layer with 40 filters, a 10-unit ResNet composed of an initial mapping to an initial 200 × 200 × (5 × 40) space with a Conv2D+ReLu layer and a linear 1x1 Conv2D+ReLu layer with 40 filters. The final layer applies a linear upsampling to the input shape using 4×4 Conv2DTranspose layer. Note that ϕ involves a total of ≈450,000 parameters.

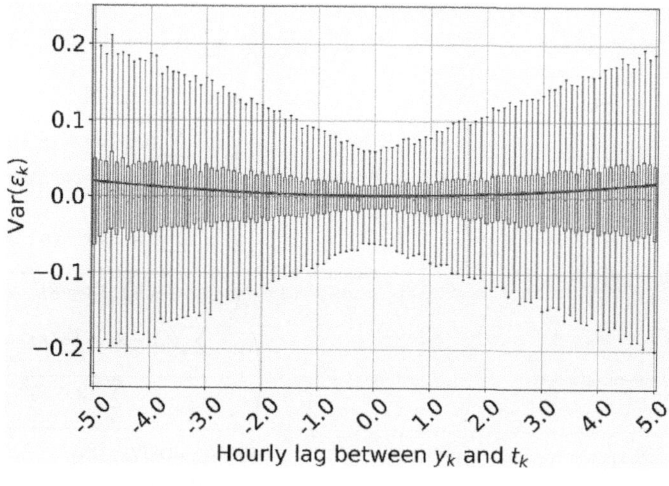

Figure 6.7. *Variance of the observation error ε_k as a function of the hourly lag between the observations $\mathbf{y}_k$ and the day to estimation time t_k. The blue barplots represent the conditional distributions of ε_k according to hourly time lags, the dotted red line represents their variances and the solid black line represents the corresponding parametric fit. For a color version of this figure, see www.iste.co.uk/mallet/multitemporal.zip*

Here, $\widetilde{\mathcal{D}}$ denotes the gridded version of domain $\mathcal{D}$ and $|\widetilde{\mathcal{D}}|$ is then the number of grid nodes of $\widetilde{\mathcal{D}}$. DSP denotes the density power spectrum, as introduced by Welch (1967).

For the evaluation, we use four 20-day long validation periods distributed along the one-year long dataset (see the starting dates reported on the X-axis in Figure 6.8) to account for seasonal variabilities. Regarding the metrics detailed in Table 6.1, daily normalized RMSE (nRMSE) time series are provided to give an overview of the performance of the data-driven interpolators. Additional correlation and variance scores are displayed together with RMSE as Taylor diagrams. We also give global reconstruction scores (R-score) for the known SSH field areas (Ω), the interpolation performance (I-score) for the missing data areas ($\overline{\Omega}$) and the reconstruction performance of the trained NN-based representation of the SSH dynamics for FP/Grad-GENN when applied to gap-free SSH fields (AE-score). In the spectral domain, signal-to-noise ratios are computed to assess up to which spatial scale the different interpolators are able to reproduce the GT.

Name	Formula
nRMSE	$\text{nRMSE}(t_k)=\sqrt{\frac{1}{\lvert\widetilde{\mathcal{D}}\rvert}\sum_{\widetilde{\mathcal{D}}}(\mathbf{x}_k-\hat{\mathbf{x}}_k)^2}\Big/\sigma_{\mathbf{x}}$
Error variance	$\sigma^2_{\mathbf{x}\_\hat{\mathbf{x}}}(t_k)=\frac{1}{\lvert\widetilde{\mathcal{D}}\rvert}\sum_{\widetilde{\mathcal{D}}}\left[(\mathbf{x}_k-\hat{\mathbf{x}}_k)-\overline{(\mathbf{x}_k-\hat{\mathbf{x}}_k)}\right]^2$
Correlation	$\text{COR}(t_k)=\frac{\text{Cov}(\mathbf{x}_k,\hat{\mathbf{x}}_k)}{\sigma(\mathbf{x}_k)\sigma(\hat{\mathbf{x}}_k)}$
Reconstruction score	$\text{R-score}=100\times\left(1-\frac{\sum_{\Omega}((\mathbf{x}-\overline{\mathbf{x}})-(\hat{\mathbf{x}}-\overline{\hat{\mathbf{x}}}))^2}{\sum_{\Omega}(\mathbf{x}-\overline{\mathbf{x}})^2}\right)$
Interpolation score	$\text{I-score}=100\times\left(1-\frac{\sum_{\overline{\Omega}}((\mathbf{x}-\overline{\mathbf{x}})-(\hat{\mathbf{x}}-\overline{\hat{\mathbf{x}}}))^2}{\sum_{\overline{\Omega}}(\mathbf{x}-\overline{\mathbf{x}})^2}\right)$
Auto-encoder score	$\text{AE-score}=100\times\left(1-\frac{\sum_{\widetilde{\mathcal{D}}}((\mathbf{x}-\overline{\mathbf{x}})-(\psi(\mathbf{x})-\overline{\psi(\mathbf{x})}))^2}{\sum_{\widetilde{\mathcal{D}}}(\mathbf{x}-\overline{\mathbf{x}})^2}\right)$
Radially averaged power spectrum (RAPS)	$\text{RAPS}(\lambda)=\text{DSP}(\hat{\mathbf{x}}_k)(\lambda)$
Signal-to-Noise Ratio	$\text{SNR}(\lambda)=\frac{\text{DSP}(\mathbf{x}_k-\hat{\mathbf{x}}_k)(\lambda)}{\text{DSP}(\mathbf{x}_k)(\lambda)}$

Table 6.1. *Temporal and spectral statistics used to assess the performance of the interpolators in the Observation System Simulation Experiment*

The daily nRMSE of AnDA, VE-DINEOF, FP-GENN and Grad-GENN are displayed in Figure 6.8, where it is quite clear that Grad-GENN with iterative gradient-based solver outperforms the other methods. Let us note that AnDA exhibits better scores at the limits of the validation periods, which is explained by strong persistence of the SSH mesoscale dynamics, thus providing good short-term forecasts of analog operators.

Table 6.2 displays SSH and ∇_{SSH} R/I/AE-scores after applying a 150 km high-pass filter to enhance small spatial scales structures. The 10-day-long center window of each validation period is used to compute these scores to exclude artificial performance of AnDA at both the start and beginning of the validation periods. Once again, Grad-GENN outperforms the other methods. The reported performance along with its plug-and-play implementation make Grad-GENN particularly appealing for the reported case study.

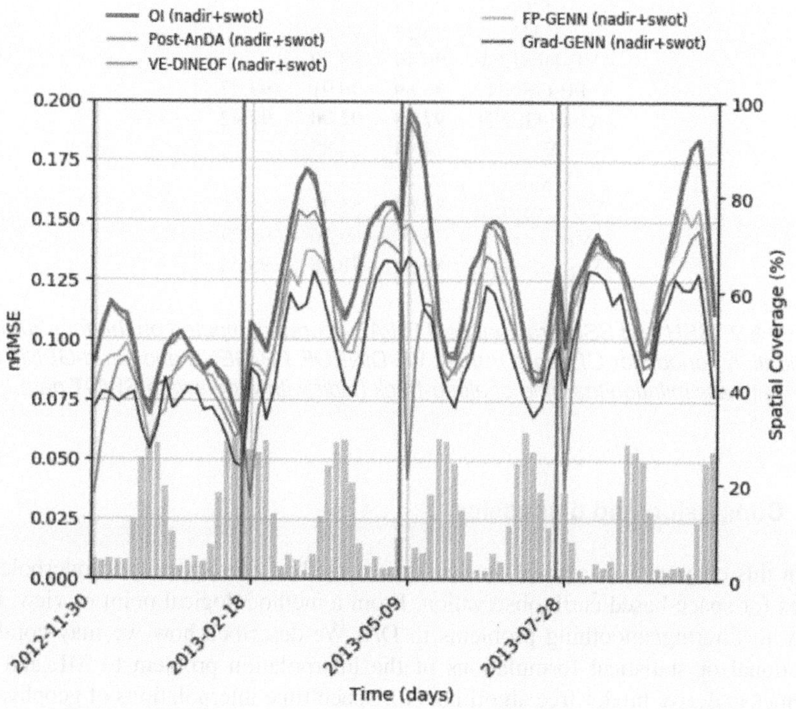

Figure 6.8. *Daily spatial nRMSE time series computed on the four 20-day validation periods for OI, (post-)AnDA, VE-DINEOF, FP-ConvAE and FP-GENN. The spatial coverage of along-track nadir and wide-swath SWOT data are provided by the green-colored barplots. For a color version of this figure, see www.iste.co.uk/mallet/multitemporal.zip*

Grad-GENN and AnDA have a closer spectrum (Figure 6.9) w.r.t. the GT and succeed in catching up the submesoscale range up to 70 km, if applying a threshold of 0.5 on signal-to-noise ratio. In addition, Grad-GENN seems to maintain a plateau up to 55 km, which demonstrates the importance of the solver in our architecture, where the simplified fixed-point version does not enable us to improve the OI spectrum on this particular region. To further enhance the vizualization of the improvements brought by the different interpolators, Figure 6.10 depicts the true spatial SSH gradient, as well as its global reconstruction based on OI, (post-)AnDA, VE-DINEOF, FP-GENN and Grad-GENN with joint use of along-track nadir and wide-swath pseudo-observations on August 4, 2013 characterized by significant patterns of the Gulfstreeam current.

Model type	R-score	I-score	AE-score
OI	96.39	88.72	–
AnDA	96.23	90.85	–
VE-DINEOF	96.49	88.79	–
FP-GENN	96.69	90.91	99.27
Grad-GENN	**97.68**	**92.90**	99.42
∇_{OI}	74.94	57.85	–
∇_{AnDA}	80.25	64.46	–
$\nabla_{\text{VE-DINEOF}}$	81.62	55.21	–
$\nabla_{\text{FP-GENN}}$	78.90	63.56	91.27
$\nabla_{\text{Grad-GENN}}$	**86.51**	**70.18**	93.54

Table 6.2. *SSH and SSH gradient field R/I/AE-scores computed on the four 20-day validation periods for OI, (post-)AnDA, VE-DINEOF, FP-GENN and Grad-GENN for joint assimilation/learning of along-track nadir with wide-swath SWOT data*

6.6. Conclusion and discussion

In this chapter, we focused on data-driven methods for space-time interpolation issues for space-based earth observation. From a methodological point of view, they relate to filtering/smoothing problems in DA. We described how we may combine variational or statistical formulations of the interpolation problem to ML and DL schemes to derive model-free algorithms for space-time interpolations of geophysical fields based on satellite-derived measurements. Through an example on satellite-derived SSH dynamics, we illustrated that data-driven schemes, especially NN-based schemes can result in a significant gain compared to the operational OI baseline: up to 30% relative gain on the SSH daily root mean squared error and a coherence of the spatial scales up to 60 km. As it mentions, the new avenues provided by NN-based methods are very encouraging: in particular, end-to-end learning illustrated here with the FP/Grad-GENN solution is a "plug-and-play" approach whose conceptual use easily enables its implementation on new datasets. Being backed on a variational formulation, these neural approaches go beyond black-box schemes as they may embed physical-sound priors in the specification of the observation models and the space-time priors. They open up a variety of methodological and operational research avenues. First, NN-based algorithms enable us to learn from the single observations without requiring any numerical simulations. As they provide trainable end-to-end architectures, we may envision the training of criterion-specific methods that may target features of interest, for instance, regarding

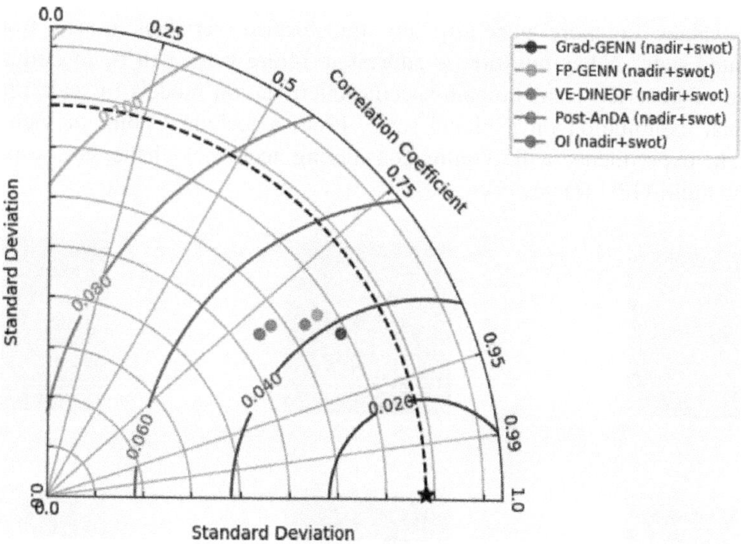

(a) Taylor diagram

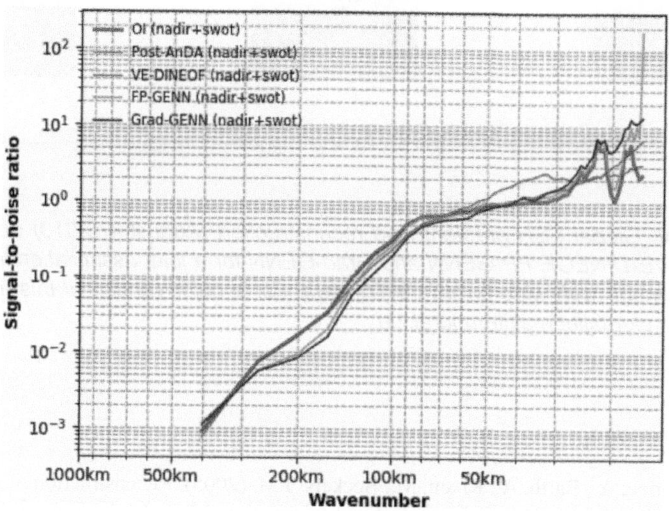

(b) Signal-to-Noise Ratio

Figure 6.9. *Taylor diagram and signal-to-noise ratio computed on the four 20-day validation periods for OI, (post-)AnDA, VE-DINEOF, FP-ConvAE and FP-GENN computed for joint assimilation/learning with along-track nadir and wide-swath SWOT data. For a color version of this figure, see www.iste.co.uk/mallet/ multitemporal.zip*

extreme values or more generally any transformed variables issued from the interpolated state. A key question to address in future work will be to evaluate the existence of generic versus domain-specific interpolation models to scale up to an operational exploitation on a global scale. From a technical point of view, such large-scale experiments will require considering technical challenges, especially regarding multi-GPU frameworks.

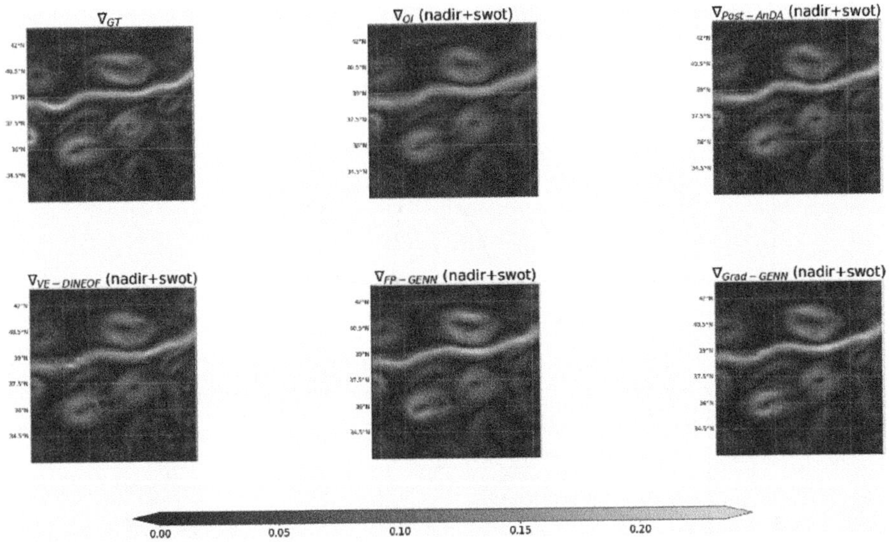

Figure 6.10. *Global SSH gradient field reconstruction (August 4, 2013) obtained by OI, AnDA, VE-DINEOF, FP-GENN and Grad-GENN for a joint assimilation/learning of along-track nadir with wide-swath SWOT data. For a color version of this figure, see www.iste.co.uk/mallet/multitemporal.zip*

6.7. References

Alvera-Azcárate, A., Barth, A., Rixen, M., Beckers, J.M. (2005). Reconstruction of incomplete oceanographic data sets using empirical orthogonal functions: Application to the Adriatic Sea surface temperature. *Ocean Modelling*, 9(4), 325–346.

Alvera-Azcárate, A., Barth, A., Sirjacobs, D., Beckers, J.-M. (2009). Enhancing temporal correlations in EOF expansions for the reconstruction of missing data using DINEOF. *Ocean Science*, 5(4), 475–485.

Andrychowicz, M., Denil, M., Gomez, S., Hoffman, M.W., Pfau, D., Schaul, T., Shillingford, B., De Freitas, N. (2016). Learning to learn by gradient descent. In *Advances in Neural Information Processing Systems 29*, Lee, D., Sugiyama, M., Luxburg, U., Guyon, I., Garnett, R. (eds). Curran Associates, Inc., Red Hook [Online]. Available at: https://proceedings.neurips.cc/paper_files/paper/2016/file/fb87582825f9d28a8d42c5e5e5e8b23d-Paper.pdf.

Arras, L., Arjona-Medina, J., Widrich, M., Montavon, G., Gillhofer, M., Müller, K.-R., Hochreiter, S., Samek, W. (2019). Explaining and interpreting LSTMs. In *Explainable AI: Interpreting, Explaining and Visualizing Deep Learning*, Samek, W., Montavon, G., Vedaldi, A., Hansen, L.K., Müller, K.-R. (eds). Springer International Publishing, Cham.

Asch, M., Bocquet, M., Nodet, M. (2016). *Data Assimilation: Methods, Algorithms, and Applications*. Society for Industrial and Applied Mathematics, Philadelphia.

Ballarotta, M., Ubelmann, C., Pujol, M.-I., Taburet, G., Fournier, F., Legeais, J.-F., Faugère, Y., Delepoulle, A., Chelton, D., Dibarboure, G. et al. (2019). On the resolutions of ocean altimetry maps. *Ocean Science*, 15(4), 1091–1109.

Barbu, T. (2018). *Variational and PDE Models for Image Interpolation*. Springer, Cham.

Beauchamp, M., Fablet, R., Ubelmann, C., Ballarotta, M., Chapron, B. (2020). Intercomparison of data-driven and learning-based interpolations of along-track nadir and wide-swath SWOT altimetry observations. *Remote Sensing*, 12(22) [Online]. Available at: https://www.mdpi.com/2072-4292/12/22/3806.

Beckers, J.M. and Rixen, M. (2003). EOF calculations and data filling from incomplete oceanographic datasets. *Journal of Atmospheric and Oceanic Technology*, 20(12), 1839–1856.

Bentley, J.L. (1975). Multidimensional binary search trees used for associative searching. *Communications of the ACM*, 18(9), 509–517.

Bertalmio, M., Bertozzi, A.L., Sapiro, G. (2001). Navier-Stokes, fluid dynamics, and image and video inpainting. *Proceedings of the 2001 IEEE Computer Society Conference on Computer Vision and Pattern Recognition. CVPR 2001*, 1, I–I.

Bjornsson, H. and Venegas, S. (1997). A manual for EOF and SVD analyses of climate data. Technical report, CCGCR report no. 97-1, McGill University, Montreal.

Bocquet, M., Brajard, J., Carrassi, A., Bertino, L. (2019). Data assimilation as a learning tool to infer ordinary differential equation representations of dynamical models. *Nonlinear Processes in Geophysics*, 26(3), 143–162.

Brajard, J., Carrassi, A., Bocquet, M., Bertino, L. (2021). Combining data assimilation and machine learning to infer unresolved scale parametrization, 379(2194), 20200086.

Cameletti, M., Lindgren, F., Simpson, D., Rue, H. (2011). Spatio-temporal modeling of particulate matter concentration through the SPDE approach. *AStA Advances in Statistical Analysis*, 97(2), 109–131.

Carrassi, A., Bocquet, M., Bertino, L., Evensen, G. (2018). Data assimilation in the geosciences: An overview of methods, issues, and perspectives. *WIREs Climate Change*, 9(5), e535.

Carrizo Vergara, R. (2018). Development of geostatistical models using stochastic partial differential equations. Thesis, Mines ParisTech – PSL University.

Chang, Y.-L., Liu, Z.Y., Lee, K.-Y., Hsu, W. (2019). Free-form video inpainting with 3D gated convolution and temporal patchgan. *arXiv:1904.10247*.

Chilès, J.-P. and Delfiner, P. (2012). *Geostatistics: Modeling Spatial Uncertainty*, 2nd edition. Wiley, New York.

Chilès, J.-P. and Desassis, N. (2018). *Fifty Years of Kriging*. Springer International Publishing, Cham.

Dufau, C., Orsztynowicz, M., Dibarboure, G., Morrow, R., Le Traon, P.-Y. (2016). Mesoscale resolution capability of altimetry: Present and future. *Journal of Geophysical Research: Oceans*, 121(7), 4910–4927.

Ebrahimi, M.A. and Lunasin, E. (2012). The Navier-Stokes-Voight model for image inpainting. *IMA Journal of Applied Mathematics*, 78(5), 869–894.

Esteban-Fernandez, D. (2014). SWOT project mission performance and error budget document. Technical report, JPL, NASA.

Evensen, G. (2009). *Data Assimilation*. Springer, Berlin, Heidelberg.

Fablet, R., Ouala, S., Herzet, C. (2017). Bilinear residual neural network for the identification and forecasting of dynamical systems. *arXiv: 1712.07003*.

Fablet, R., Drumetz, L., Rousseau, F. (2019). End-to-end learning of energy-based representations for irregularly-sampled signals and images. *arXiv: 1910.00556*.

Fablet, R., Drumetz, L., Rousseau, F. (2020). Joint learning of variational representations and solvers for inverse problems with partially-observed data. *arXiv: 2006.03653*.

Fablet, R., Beauchamp, M., Drumetz, L., Rousseau, F. (2021). Joint interpolation and representation learning for irregularly sampled satellite-derived geophysical fields. *Frontiers in Applied Mathematics and Statistics*, 7, 25.

Fuglstad, G.-A., Lindgren, F., Simpson, D., Rue, H. (2015a). Exploring a new class of non-stationary spatial Gaussian random fields with varying local anisotropy. *Statistica Sinica*, 25(1), 115–133.

Fuglstad, G.-A., Simpson, D., Lindgren, F., Rue, H. (2015b). Does non-stationary spatial data always require non-stationary random fields? *Spatial Statistics*, 14, 505–531.

Gaultier, L. and Ubelmann, C. (2010). SWOT simulator documentation. Technical report, JPL, NASA.

Golub, G.H. and van Loan, C.F. (2013). *Matrix Computations*, 4th edition. JHU Press, Baltimore [Online]. Available at: http://www.cs.cornell.edu/cv/GVL4/golubandvanloan.htm.

Hildeman, A., Bolin, D., Rychlik, I. (2021). Deformed SPDE models with an application to spatial modeling of significant wave height. *Spatial Statistics*, 42, 100449.

Huang, J.-B., Kang, S.B., Ahuja, N., Kopf, J. (2016). Temporally coherent completion of dynamic video. *ACM Transactions on Graphics*, 35(6), 196:1–196:11.

Kim, D., Woo, S., Lee, J., Kweon, I.S. (2019). Deep video inpainting. *CoRR*, abs/1905.01639.

Le T.T., Almansa, A., Gousseau, Y., Masnou, S. (2017). Motion-consistent video inpainting. *ICIP 2017: IEEE International Conference on Image Processing*, Beijing.

LeCun, Y., Bengio, Y., Hinton, G. (2015). Deep learning. *Nature*, 521(7553), 436–444.

Lguensat, R., Tandeo, P., Ailliot, P., Pulido, M., Fablet, R. (2017). The analog data assimilation. *Monthly Weather Review*, 145(10), 4093–4107.

Lguensat, R., Viet, P.H., Sun, M., Chen, G., Fenglin, T., Chapron, B., Fablet, R. (2019). Data-driven interpolation of sea level anomalies using analog data assimilation. *Remote Sensing*, 11(7) [Online]. Available at: https://www.mdpi.com/ 2072-4292/11/7/858.

Lindgren, F., Rue, H., Lindström, J. (2011). An explicit link between Gaussian fields and Gaussian Markov random fields: The stochastic partial differential equation approach. *Journal of the Royal Statistical Society: Series B (Statistical Methodology)*, 73(4), 423–498.

Lopez-Radcenco, M., Pascual, A., Gomez-Navarro, L., Aissa-El-Bey, A., Chapron, B., Fablet, R. (2019). Analog data assimilation of along-track nadir and wide-swath SWOT altimetry observations in the Western Mediterranean sea. *IEEE Journal of Selected Topics in Applied Earth Observations and Remote Sensing*, 12(7), 2530–2540.

Lorenz, E.N. (1963). Deterministic nonperiodic flow. *Journal of Atmospheric Sciences*, 20(2), 130–141.

Metref, S., Cosme, E., Le Guillou, F., Le Sommer, J., Brankart, J.-M., Verron, J. (2020). Wide-swath altimetric satellite data assimilation with correlated-error reduction. *Frontiers in Marine Science*, 6, 822.

Molines, J.-M. (2018). MEOM-configurations NATL60-CJM165: NATL60 code used for CJM165 experiment [Online]. Avalable at: https://zenodo.org/record/1210116#.XDpmc89Khp8.

Newson, A., Almansa, A., Fradet, M., Gousseau, Y., Pérez, P. (2014). Video inpainting of complex scenes. *SIAM Journal on Imaging Sciences*, 7, 1993–2019.

O'Gorman, P.A. and Dwyer, J.G. (2018). Using machine learning to parameterize moist convection: Potential for modeling of climate, climate change, and extreme events. *Journal of Advances in Modeling Earth Systems*, 10(10), 2548–2563.

Ouala, S., Fablet, R., Herzet, C., Chapron, B., Pascual, A., Collard, F., Gaultier, L. (2018). Neural network based Kalman filters for the spatio-temporal interpolation of satellite-derived sea surface temperature. *Remote Sensing*, 10(12), 1864.

Ping, B., Su, F., Meng, Y. (2016). An improved DINEOF algorithm for filling missing values in spatio-temporal sea surface temperature data. *PLOS ONE*, 11(5), e0155928.

Platzer, P., Yiou, P., Naveau, P., Filipot, J.-F., Thiebaut, M., Tandeo, P. (2021). Probability distributions for analog-to-target distances. *Journal of the Atmospheric Sciences*, American Meteorological Society, Boston, 78(10), 3317–3335.

Rasp, S., Pritchard, M.S., Gentine, P. (2018). Deep learning to represent subgrid processes in climate models. *Proceedings of the National Academy of Sciences*, 115(39), 9684–9689.

Schönlieb, C.-B. (2015). *Partial Differential Equation Methods for Image Inpainting*. Cambridge University Press, Cambridge.

Storch, H.V. and Zwiers, F.W. (1999). *Statistical Analysis in Climate Research*. Cambridge University Press, Cambridge.

Taburet, G., Sanchez-Roman, A., Ballarotta, M., Pujol, M.-I., Legeais, J.-F., Fournier, F., Faugere, Y., Dibarboure, G. (2019). DUACS DT2018: 25 years of reprocessed sea level altimetry products. *Ocean Science*, 15(5), 1207–1224.

Tandeo, P., Ailliot, P., Bocquet, M., Carrassi, A., Miyoshi, T., Pulido, M., Zhen, Y. (2020). A review of innovation-based methods to jointly estimate model and observation error covariance matrices in ensemble data assimilation. *Monthly Weather Review*, 148(10), 3973–3994.

Taylor, M., Losch, M., Wenzel, M., Schröter, J. (2013). On the sensitivity of field reconstruction and prediction using empirical orthogonal functions derived from gappy data. *Journal of Climate*, 26(22), 9194–9205.

Traon, P.-Y., Nadal, F., Ducet, N. (1998). An improved mapping method of multisatellite altimeter data. *Journal of Atmospheric and Oceanic Technology*, 15(2), 522–534.

Van Leeuwen, P.J., Künsch, H.R., Nerger, L., Potthast, R., Reich, S. (2019). Particle filters for high-dimensional geoscience applications: A review. *Quarterly Journal of the Royal Meteorological Society*, 145(723), 2335–2365.

Welch, P. (1967). The use of fast Fourier transform for the estimation of power spectra: A method based on time averaging over short, modified periodograms. *IEEE Transactions on Audio and Electroacoustics*, 15(2), 70–73.

Xu, R., Li, X., Zhou, B., Loy, C.C. (2019). Deep flow-guided video inpainting. In *Proceedings of the IEEE/CVF Conference on Computer Vision and Pattern Recognition (CVPR)*, Long Beach, 3718–3727.

Zhao, Z. and Giannakis, D. (2014). Analog forecasting with dynamics-adapted kernels. *arXiv: 1412.3831*.

7

Recent Advances in Tropical Cyclone Forecasting Using Machine Learning on Reanalysis and Remote Sensing

Sophie GIFFARD-ROISIN[1,2]
[1] *Université Grenoble Alpes, France*
[2] *Université Savoie Mont Blanc, CNRS, IRD, IFSTTAR, ISTerre, Chambéry, France*

In this chapter, we will show how the recent advances of machine learning (ML) and in particular deep learning (DL) can be applied to one major natural hazard forecasting: tropical cyclones (TC). Cyclones, hurricanes and typhoons refer to the same rare and complex phenomena characterized by strong winds surrounding a low-pressure area. The precision of the forecasting trajectory and intensity is crucial for the protection of people and property. However, their evolution depends on many factors at different scales, altitudes and times, which leads to modeling difficulties. As the dynamical models evolve, their forecast accuracy improves; however, historical TC databases have scarcely been used to further improve the precision. In recent years, ML methods have gained accuracy and are now competing with the dynamical models on some forecasting tasks. To do so, they have incorporated historical spatial information (remote sensing or reanalysis) as well as temporal information (past time series). Moreover, they often needed to combine different input data types, from scalar values to 3-D images. We will show how these difficulties have been successively tackled and the remaining open problems.

Multitemporal Earth Observation Image Analysis,
coordinated by Clément MALLET and Nesrine CHEHATA.

7.1. Background

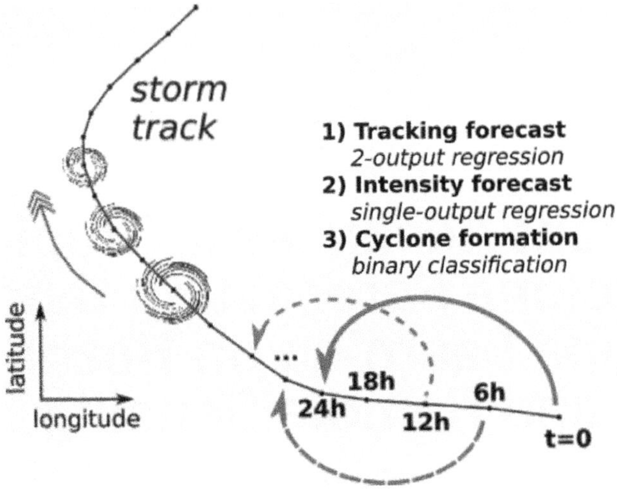

Figure 7.1. The three TC forecasting problems tackled in this chapter. Example of a 24-h forecast. The dots indicate the time at which the data and ground truth are available (here, every 6 h). For a color version of this figure, see www.iste.co.uk/mallet/multitemporal.zip

7.1.1. *Problem statement*

TC forecasting aims to solve the following questions:

1) **Tracking:** where will the TC be located in x hours or days?

2) **Intensity:** how strong will the TC be in x hours or days?

3) **Formation:** will the current tropical depression or storm evolve in a tropical cyclone?

We can notice that question (3) can be contained in (2) because (i) the force of a TC is defined by its maximal sustained wind speed[1] and (ii) a TC is defined as a tropical storm having its maximal sustained wind speed higher than 64 knots (118 km/h).

We can formulate the three questions in terms of supervised ML. The *tracking* problem is a multi-output regression (estimate both x and y coordinates), the *intensity*

1 Maximal recorded wind speed at 10 m high, sustained for a minimum period of one minute. It is calculated in knots (1 knot = 1.85 km/h).

problem is a single-output regression (estimating the wind speed value) and the *formation* problem can be seen as a binary classification. The goal of this chapter is to show how ML can be efficient on these tasks, and how we can overcome the three following challenges: processing temporal data, processing spatial data and combining different data sources.

Note that there exists a second branch of ML TC forecasting, one that aims to improve the current numerical models by developing physics-based ML (Loridan et al. 2017; Jiang et al. 2018; Kim et al. 2019), but we will not develop it here.

7.1.2. *Existing TC forecasts methods*

Today, tropical cyclone forecasts (tracking and intensity) are provided by numerous numerical guidance models[2] that can be separated into two groups: dynamical models and statistical models. Dynamical models solve the physical equations governing motions in the atmosphere and they are influenced by physical models – convective schemes (such as Kain-Fritsch or simplified Arakawa Schubert), cloud microphysics, land surface model, ocean model, sea/land ice model, planetary boundary layer scheme, surface layer scheme, longwave and shortwave radiation schemes and subgrid-scale diffusion – and by their data assimilation methods (such as 4D-VAR, for example, in Leidner et al. (2003)). They are computationally demanding and long to process: modern numerical weather prediction models typically need several hours to complete a forecast cycle[3]. Current forecasts produced by regional specialized meteorological centers, like the American Official NHC Forecast (OFCL), are driven by consensus or ensemble methods able to combine different dynamical models[1] (up to 20 models for the Global Ensemble Forecast System[4]). Statistical models, in contrast, are based on historical relationships between storm behavior and various other parameters (DeMaria et al. 2005). However, they are based on simple regressions on few statistical features (such as 12-h wind changes, day of the year, 200 hPa estimated temperature, current and maximum intensity, etc.). By incorporating large spatial atmospheric data in a statistical model using state-of-the-art ML methods, we can improve the accuracy while reducing the calculation time.

2 NHC track and intensity models: https://www.nhc.noaa.gov/modelsummary.shtml [Accessed 2 March 2021].

3 Available at: https://web.uwm.edu/hurricane-models/models/models.html [Accessed 2 March 2021].

4 GEFS: https://www.ncdc.noaa.gov/data-access/model-data/model-datasets/global-ensemble-forecast-system-gefs [Accessed 12 December 2018].

7.1.3. *Available data*

7.1.3.1. *Best tracks*

The TC historical data consists mainly of "best track" information: the a posteriori recording of the TC locations every time step, a time step ranging from 10 min to 6 h depending on the database and the ocean basin. Together with its locations, the main databases usually add some other information such as the maximal sustained wind speed, from which the category of the TC is derived. The database that is mainly used in this chapter groups most of the historical "best tracks" provided by different international governmental agencies. It is maintained by NOAA and is called IBTrACS (International Best Track Archive for Climate Stewardship (Knapp et al. 2010)).

By removing the tracks containing missing data, the IBTrACS database is still composed of more than 3,000 tropical and extra-tropical storm tracks since 1979, as shown in Figure 7.2. They are defined by the 6-h – or less – center locations, latitude and longitude. The database includes both hemispheres and the number of records per TC varies from 2 to 120 time steps. In total, the database counts more than 90,000 time steps between 1979 and 2018.

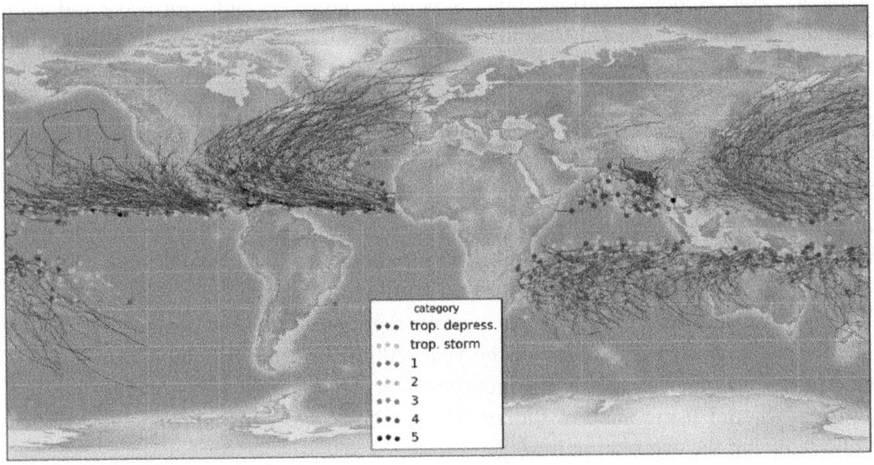

Figure 7.2. *IBTrACS database since 1979: 3,000 tropical/extra-tropical storm tracks. Dots = initial position, color = maximal storm strength (Saffir–Simpson scale, ranging the TCs from category 1 to category 5 based on their strength). For a color version of this figure, see www.iste.co.uk/mallet/multitemporal.zip*

7.1.3.2. *Reanalysis data*

The trajectory of a TC depends on large-scale atmospheric flows, which can be extracted from *reanalysis*. Reanalysis is a systematic approach to produce datasets for climate monitoring and research (Dee et al. 2011; Hersbach et al. 2020). They cover the entire globe from the Earth's surface to well above the stratosphere and estimate hundreds of available variables. The raw reanalysis sensor data can include, but are not limited to radiosonde, satellite, buoy, aircraft and ship reports. By combining these observed data with an unchanging data assimilation system and an unchanging atmospheric model, the reanalysis generates dynamically consistent and better estimates of the climate of the atmospheric fields or states at discrete points in space and time than by using observations or models alone.

Extracting analyzed atmospheric fields from reanalysis data in order to augment the information at every time step can be easily performed. Moreover, many reanalyses also provide forecast fields (predictions of future atmospheric states), which can be used as TC forecast inputs too. One of the reanalysis used in this chapter is the ERA-Interim database (Dee et al. 2011), which covers the data-rich period since 1979. ERA-Interim is produced by the European Centre for Medium-Range Weather Forecasts (ECMWF) in near-to-real time. The spectral resolution is T255 (around 80 km), and the time resolution is 6 h. There are 60 vertical pressure levels until 0.1 hPa (altitude around 64 km). Since 2018, the ERA-5 reanalysis (Hersbach et al. 2020), which is more accurate and provides a finer resolution in time and space than the ERA-Interim, is available and should be preferred.

7.1.3.3. *Remote sensing data*

In the scope of having near-to-real-time TC forecasts, reanalysis computation increase the forecast time by at least few hours. That is why treating directly "raw" satellite data can be very useful. Specifically, the long-wave infrared satellite imaging provides an estimate of the cloud-top temperature, which is a proxy for regions of deeper convections (Piñeros et al. 2008). Moreover, microwave satellite imaging has been used to observe the TC intensity variations as it is based on the emission and scattering of cloud and precipitation particles (Kerns and Chen 2013). Both can be available at a good resolution in time and space by polar-orbiting meteorological satellite (such as the radiometer WindSat on Coriolis) and by geostationary operational environmental satellites (GEOS, such as the China FengYun-2C). However, the information contained in the remote sensing images is not as rich as the reanalysis. Moreover, they are also noisier, making the forecast more difficult.

7.1.4. *Machine learning background*

7.1.4.1. *Deep learning*

Artificial intelligence (AI) is a general concept that refers to machines being able to carry out specific tasks in a way that would be considered "smart". ML is a branch of AI, which provides the system with the tools to automatically learn and improve from experience without being explicitly programmed. ML is thus the development of computer programs that can access data and use it to learn for themselves. Among ML, the family of methods that have shown many outbreaks in the recent years is an extension of artificial neural networks (ANNs) called DL. The reason for this outbreak is that DL methods are designed to treat abstract concepts, with multiple linear and nonlinear transformations in an architecture sharing many similarities with brain neurons.

7.1.4.2. *Convolutional neural networks*

A convolutional neural network (CNN) is a DL architecture widely adopted as a very effective model for analyzing images or image-like data for pattern recognition (Krizhevsky et al. 2012; Milletari et al. 2016). A CNN is structured in layers: an input layer connected to the data, an output layer connected to the quantities to estimate and multiple hidden layers in between. The hidden layers of a CNN typically consist of convolutional layers, pooling layers, fully connected (FC) layers and normalization layers. The convolutional operations are inspired by the cortex visual system, where each neuron only processes data for its receptive field. FC layers, usually at the end of the network, connect every neuron in one layer to every neuron in another layer. The advantage of CNN is that it can learn to recognize spatial patterns by exploiting translation invariance (i.e. all parts of the image are processed in a similar way), and thus can extract features automatically while considerably reducing the number of parameters.

7.1.4.3. *Recurrent neural networks and Transformer*

Recurrent neural networks (RNNs) are a class of ANNs that are suited to model temporal dynamic behavior for a time sequence. Unlike feedforward neural networks (NNs, like CNNs), RNNs can use their internal state (memory) to process sequences of inputs. An LSTM (long short term memory) network is a particular RNN used in different time-series applications. Even though the LSTM networks are among the most successful methods for predicting time series events, they are still difficult to train, and simpler CNNs can outperform LSTMs on some specific spatiotemporal tasks (Bai et al. 2018).

Some more recent recurrent architectures such as the gated recurrent unit (GRU) (Chung et al. 2014) and Transformer (Vaswani et al. 2017) have shown to be particularly efficient, for example, in the natural language processing (NLP) domain. GRU models are able to handle long-term dependencies in sequential problems thanks to their gating units that solve the vanishing gradient problem of RNNs. Transformer is an end-to-end memory network based on a recurrent attention mechanism in an encoder-decoder structure, instead of sequence-aligned recurrence. The idea is that the Transformer model is paying attention to all of the states of the encoder, instead of paying attention only to the last state of the encoder as is usually done with RNN. The attention mechanism thus extracts information from the whole sequence by using a weighted sum of all of the past encoder states. In the remote sensing field, GRU and Transformer are starting to be used in time series applications, such as multi-image super-resolution (Arefin et al. 2020) or crop classification (Rußwurm and Körner 2020). For the scope of TC forecasting, this capability of learning how to aggregate a time sequence can be very interesting as the atmospheric flows are defined by dynamical phenomena.

7.2. Handling spatiotemporal data for TC forecasting

Handling both spatial and temporal varying input data is fundamental for TC forecasting. We will discuss how to deal with such data structures in this section.

7.2.1. *Temporal feature structure: time series versus fixed time steps inputs*

The past states of a TC is fundamental for its forecasting, and the TC forecasting literature using historical storm tracks is rich. Specifically, many studies have attempted to forecast tracks or intensities by using the "best tracks" databases adding only few easily available features (such as the intensity of the current storm, the ocean basin, the season day, etc.). We invite the reader to look at a review on this topic by Chen et al. (2020) for a complete list. The advantage of such methods is that the input data at every time step can be organized in a single vector, as it consists only of a short list of features. Yet, in order to take benefit of the TC time series, more than one input vector is needed: at least two time steps are useful to capture the dynamics of the TC. Two approaches can be used to deal with time series inputs.

The first one is to use a fixed number of past time step inputs, for example two, by staking both feature vectors. This enables the use of standard ML methods as well as feed-forward DL methods, such as ANNs or CNNs. Yet, it does not allow us to use longer time series, and may be limited to provide longer term forecasts, as stacking time steps does not account for the temporal aspect. We will show the results of such approaches in sections 7.3 and 7.4.

The second option is to use some of the methods that are specifically designed to time series inputs, such as RNNs. They have been experimented in recent works (Moradi Kordmahalleh et al. 2016; Gao et al. 2018). Yet, these studies are only using a list of precomputed features (without 2D fields), so they cannot capture the complex dynamical atmospheric and oceanic flows that governs the TC evolution. Using both recurrent and convolutional architectures require another level of complexity that can be difficult to develop.

We will first see how the first approach, using a fixed number of past time steps, reveals satisfactory results for TC forecasting. We will then see how the second approach has recently shown some nice progress on spatiotemporal data by using state-of-the-art RNNs such as GRU or Transformers, coupled with CNNs.

7.2.2. *Spatial feature structure: fixed versus moving frame of reference*

In order to take into account spatial inputs such as reanalysis or remote sensing images, there exists different strategies to represent image-like data as inputs for training a DL network. The challenge is to retrieve interesting higher order features from 2-D or 3-D grids of data, such as the pressure, the wind speeds, the sea temperature or some remote sensing signal: how can we train the network in order to leverage from the historical storm track neighborhoods?

The first ML forecast models considered a fixed regional map for tracking storms, of a size of 160 × 80° (longitude/latitude) for Mudigonda et al. (2017) and Kim et al. (2019) and of the size of the Korean peninsula area (around 30 × 30°) for Rüttgers et al. (2018). However, a fixed region for tropical cyclone forecast has three major limitations. First, the tracked storm must stay in the region even though tracks often cross oceans (see Figure 7.2), forcing the input size to be large in order to cover a complete ocean basin, even if it leads to memory issues (Mudigonda et al. 2017). Moreover, learning local phenomena on a large and non-centered image can be difficult. Finally, it prevents information transfer between storms coming from different basins or regions, where ground truth data is scarce.

In more recent works (Giffard-Roisin et al. 2018a, 2018b, 2020; Boussioux et al. 2022), we can see the advantage of using a moving reference frame instead of a fixed map that follows the TC center. The spatial data is cropped images (e.g. 25 × 25°) centered on the TC location. In this way, the computation time is reduced (input images up to 20 times smaller) and we can infer information from storms coming from a large number of TC basins from both hemispheres. The size of the grid is usually motivated by the fact that meaningful information can be extracted from the movement of air masses around the TC, while focusing on the storm only (10 × 10°) bounds the most historic extreme TC.

7.2.3. *Combining different input data types*

As different input data types coming from different sources are carrying complementary information, many studies have investigated how to combine them in a unified framework. While spatial data can be structured on 2-D or 3-D grids, the tracking information are usually treated as a simple 1D vector. Moreover, handling time as a dimension can also add a level of complexity. First, one solution is to create separate learning pipelines for each data type and combine their results a posteriori. Another solution is to directly build a model that is able to take as input a variety of data types. We will see how the different options are investigated in some recent studies.

7.3. Application 1: intensity forecasting from spatiotemporal reanalysis, a hackathon experiment

In this section, we summarized the findings of the challenge made on September 2018 at the Climate Informatics hachathon[5]. The hackathon focused on forecasting the hurricane intensities, and there were 35 participants. Specifically, the goal was to predict the TC intensity (24-h forecast) using information from past storms since 1979. The contest, dataset, methods, participants, successes and challenges are discussed in the following section.

Goal: this challenge proposed to design the best algorithm to predict for a large number of TC the 24-h forecast intensity every 6 h, so four time steps in the future (see Figure 7.1). The database is composed of the IBTrACS database (3,000 TC, see section 7.1.3.1). Moreover, we also provide some 700 hPa and 1000 hPa feature maps of the neighborhood of the storm (from ERA-interim reanalysis database) that can be viewed as images centered on the current storm location (see Figure 7.3).

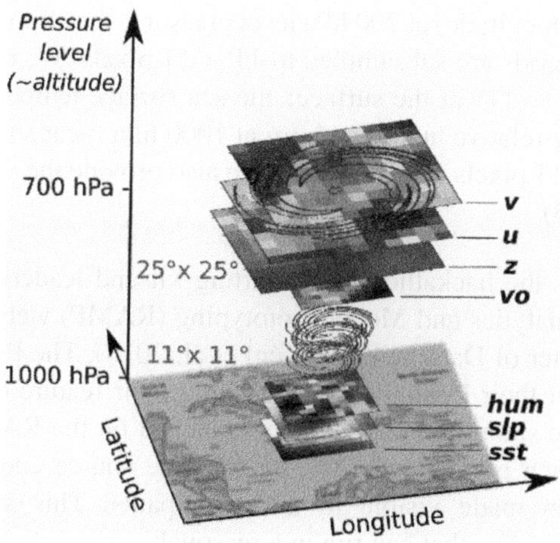

Figure 7.3. *Application 1, feature data: centered reanalysis maps of wind fields (u and v), altitude (z), sea surface temperature (sst), sea level pressure (slp), humidity (hum) and vorticity (vo). The color represents here the amplitude of every field in this data example. For a color version of this figure, see www.iste.co.uk/mallet/multitemporal.zip*

5 Climate Informatics is an annual international workshop (since 2011) and now a conference (since 2020). 2018 edition: https://www2.cisl.ucar.edu/events/workshops/climate-informatics/2018/home. 2020 edition: https://ci2020.web.ox.ac.uk/. The first day of the conference is dedicated to a hackathon on a new subject each year.

7.3.1. *Data and pipeline*

Tracks and intensity: the raw storm track data used in this study is composed of more than 3,000 TC tracks since 1979 (see Figure 7.2 and section 7.1.3.1). The intensity can be measured as the maximum sustained wind over a period of 1 min at 10-m height. This speed, calculated every 6 h, is usually explained in knots $(1\text{kt} = 0.514\text{m/s})$ and is used to define the hurricane category from the Saffir–Simpson scale.

Features: the data consist of a set of features relative to one time step of every TC. Firstly, the 0D features are a set of simple features: latitude, longitude, current windspeed, hemisphere, Jday predictor (Gaussian function of Julian day of storm init: peak day of the hurricane season), initial windspeed of the storm, last 12 h windspeed change, storm basin and current distance to the land. Secondly, the reanalysis data: at each time step, we extracted seven grids of meteorological parameters centered on the current storm location. Their choice is based on the forecast literature, on personal experience and on known hypotheses of storm strengthening. We provided three maps of 25 × 25° (latitude/longitude) at 700 hPa level pressure: the altitude z, and the u and v wind fields. These grids are subsampled to 11 × 11 pixels. We provide some more localized maps of 11 × 11° at the surface: the sea surface temperature sst, surface level pressure slp, the relative humidity hum at 1000 hPa (near surface). These grids are sampled to 11 × 11 pixels (1 pixel = 1°). We also provide the vorticity at 700 hPa $vo700$ (see Figure 7.3).

RAMP platform: the hackathon data, starting kit and leaderboard were hosted through the Rapid Analytics and Model Prototyping (RAMP) website, developed by the Paris-Saclay Center of Data Science (Kégl et al. 2018). The RAMP site requires participants to submit their Python code describing their feature extraction and ML model processes. The submitted code is then launched on the RAMP web server to train and evaluate each participant's ML model. The source code for each team's submissions are then made visible to all participants. This system encourages participants to build models that can run in a reasonable amount of time, discourages cheating and increases the amount of cooperation among different teams.

Pipeline: specifically, participants were asked to write two parts: a *feature extractor* and a *regressor* (see Figure 7.4). The 3,000 storms have been randomly separated in a training set (half of the storms), a test set and a local starting kit. The local starting kit was available for download and includes only one-fourth storms of the total database. Once a submission was submitted to the platform, a cross-validation was performed on the train set and the result was shown on the *public leaderboard*. The test set's (one-fourth of the whole dataset) results were hidden, and the *private leaderboard* was revealed only at the end of the hackathon in order to prevent overfitting. The metric used is the RMSE (root mean square error) across all TC time instants.

Participants: there were 38 participants from which 18 were students. In terms of expertise, less than 40% answered *climate/weather*, while more than 70% answered *computer science/python*. Based on their self-reported expertise, the participants were divided into 12 teams of three to four people so that each area of expertise would be covered by at least one team member.

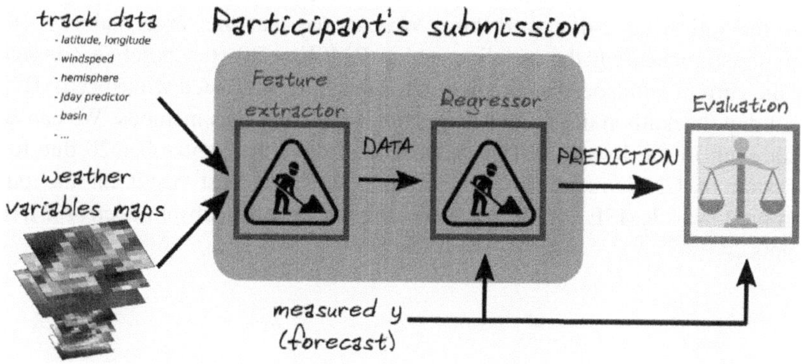

Figure 7.4. *Application 1: Pipeline of the TC forecast hackathon. For a color version of this figure, see www.iste.co.uk/mallet/multitemporal.zip*

7.3.2. Methods used by the participants

Feature extraction: many ideas were explored by the participants. A large majority performed a feature selection or feature analysis (using handcrafted significant variables or methods such as principal component analysis (PCA)). Many teams also used historical features (from previous time steps of the same TC). While the 0D data were easy to process, the reanalysis data (small images) were harder and thus subject to a variety of processing. Some methods used all of the $11 \times 11 \times 7$ reanalysis features independently, but the majority developed a pre-processing according to climate science knowledge: for each map, they selected a subset of the mean, the maximum, the minimum, the central value and the min-max distance. Finally, few methods treated them as images (for CNNs).

Regression: a large majority trained a random forest (RF) regression with different hyperparameter sets. RF performed well in case interesting features were selected and in case historical features were used too. Some teams tried also gradient boosting or multi-layer perceptron regressors. Finally, few teams took the time to write and train a CNN for the reanalysis maps, yet often without incorporating the 0D data in the features.

7.3.3. Results

Public leaderboard results: we show in Figure 7.5 the scores of the *public leaderboard* (cross-validation score using the training set, score visible during the event) with respect to time. There were more than 50 online submissions, mostly concentrated during the hackathon day, even if some participants continued to submit during the following 2-day workshop. We can see that the large majority of the submissions are beating the baseline which RMSE = 20 (the baseline consists in using the current windspeed as prediction for the 24-h forecasted windspeed). During the first day (hackathon day), the best results were using RFs methods. We can see a breakthrough in the lowest RMSE (yellow line) after the September 20 due to the submission of CNN models by few teams. The final best result on the public leaderboard was RMSE = 9.9 corresponding to a mean absolute error MAE = 6.3 knots.

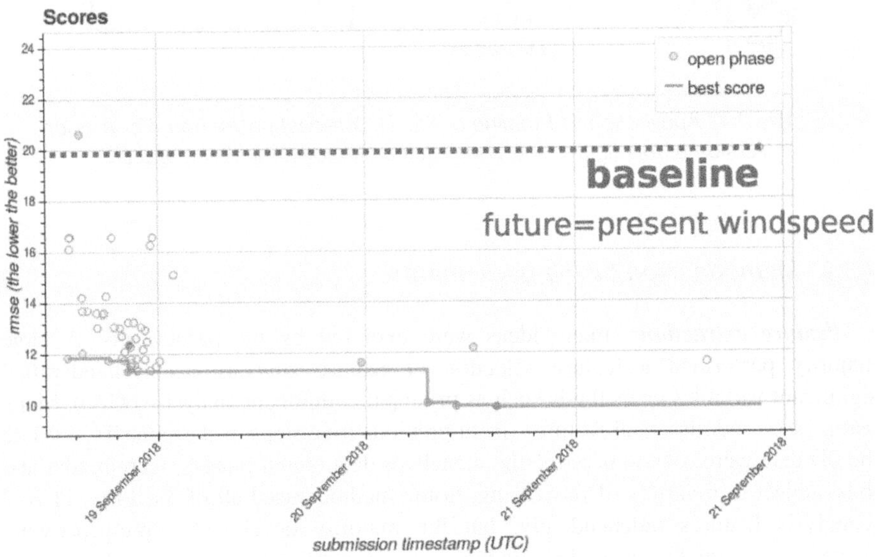

Figure 7.5. *Scores (RMSE) of the submissions wrt. time on the* public leaderboard *(cross-validation score using the online training set, score visible during the event). For a color version of this figure, see www.iste.co.uk/mallet/multitemporal.zip*

Private leaderboard results: at the end of the open phase, the best submission of every participant on the *public leaderboard* is used to predict the results on the *hidden* test set, leading to the *private competition leaderboard* shown in Table 7.1. The best RMSE (13.6) is obtained with a gradient boosting method, and the feature extractor uses all 0D features together with the mean and the center of every reanalysis map. The second best uses RF using historical 0D data. The best team reaches MAE = 9.7 knots. For comparison, let us recall that a storm is considered to be a TC if its maximum windspeed exceeds 65 knots. We can see that the baseline on this new test set has a comparable score (RMSE = 20.9) to the one on the public leaderboard (RMSE = 19.9): both sets have the same distributions. Finally, the column *Rank move* tells us the rank difference of the submission with respect to the *public leaderboard*, accessible during the hackathon.

Rank	Team	Rank move	RMSE	MAE (knots)	Train time (s)	Test time (s)	Reg. method
1	yeliz	+5	13.6	9.7	31	5	Grad. boosting
2	pjn	+2	13.6	9.8	3,099	20	Random forest
3	changgt	+2	13.8	10	39	13	Random forest
4	ftonini	+6	13.9	10	93	9	Random forest
5	djgagne	+8	13.9	10.1	164	11	Conv. nets
21	**baseline**	-	20.9	14.3	1	4	*Baseline* *

Table 7.1. Private competition leaderboard*: winning first five teams on the hidden test set.* Rank move = *rank difference of the submission with respect to the* public leaderboard. **The baseline consists of using the current windspeed as prediction for the 24-h forecasted windspeed*

Overfitting: we can see from the *rank move* in Table 7.1 that the winning teams were not the ones with the best results on the *public leaderboard*. Moreover, the best score of the test set (RMSE = 13.6) is very different from the one on the online training set (RMSE = 9.9, see Figure 7.5). Even if the online training set was not directly accessible (only the submissions scores and the codes were), it seems to indicate an overfitting on the training set by many teams. Each team was able to submit every 15 min. This type of task and dataset, having a limited number of examples but with a large variability, is known to be more subject to overfitting. We can also note that very different methods are in the top 5 (gradient boosting, RF and CNN), which seems to indicate that overfitting is not restricted to one method. We can clearly see the importance of having a hidden test set for such applied research problems.

Dissemination: the codes made to use this RAMP are freely available on github (www.github.com/ramp-kits/storm_forecast/). The processed data can also be shared to anyone by sending an email to the corresponding author.

Limitations of a hackathon: in this hackathon, the input data format and the output format were settled in advance by the organizers. It did not allow the participants to use methods such as RNNs. Moreover, the time duration of the challenge was 2 days, so more complex methods, requiring parameter tuning, were not possible. We will now show a second example on a more elaborate study.

7.4. Application 2: trajectory forecasting using fused deep learning

This section summarizes the work of Giffard-Roisin et al. (2020), in which a fusion of CNNs was developed in order to forecast 24-h TC trajectories (see Figure 7.6). Here, the output consists of two values $(\delta long_{24h}, \delta lat_{24h})$ corresponding to the longitudinal and latitudinal displacement of the TC during 24 h.

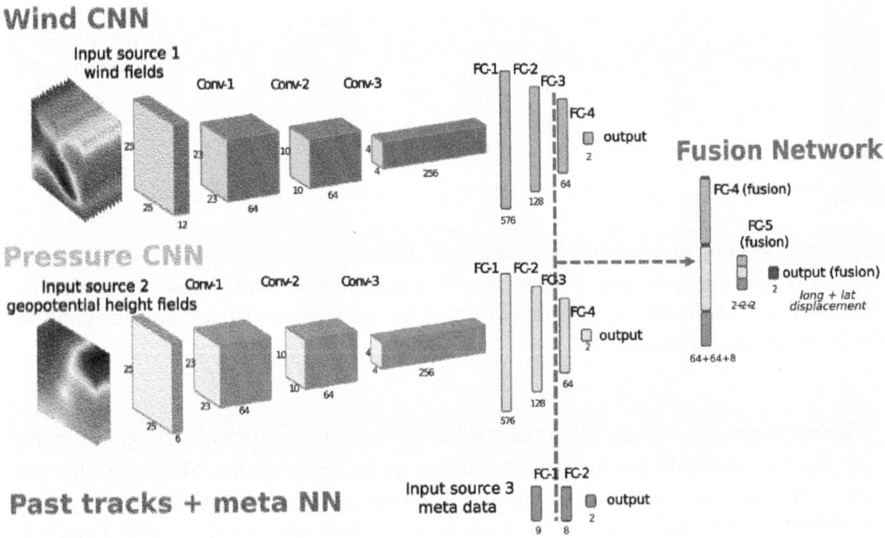

Figure 7.6. *General architecture: the three types of data feed three neural networks trained separately. The final fused network is retrained before predicting the 24-h forecast displacement. For a color version of this figure, see www.iste.co.uk/mallet/multitemporal.zip*

7.4.1. *Data processing*

7.4.1.1. *Feature selection*

In this section, the storm track data and reanalysis outputs are used to forecast tropical cyclone tracks. We can classify them into three types of information:

– **Past displacements (0D)**: we define a displacement as the values $(\delta long_{\Delta t}, \delta lat_{\Delta t})$ between the TC center locations at different times. The time difference, Δt, being in a multiple of 6 h. We used here the current displacement (i.e. between times $t - 6h$ and t) and the past displacement (between $t - 12h$ and $t - 6h$). We can refer to such features as 0D because they are not defined on a spatial grid.

– **Meta data (0D)**: all of the features extracted from the IBTrACS database are used, as they contain crucial information related to the TCs: the current center-point latitude and longitude, the current windspeed at the center of the storm, the current distance to land and the Jday predictor (Gaussian function of *Julian day of storm init – peak day of the tropical cyclone season in the hemisphere*, see DeMaria et al. (2005)).

– **Wind fields u and v and geopotential height fields z (spatial fields, 2-D)**: a sparse feature selection technique (automatic relevance determination, based on linear regression to the target displacement shift) over the 10 available reanalysis fields on pressure levels highlighted the usefulness of two reanalysis fields in particular: wind fields and the geopotential height. Wind fields are the direct observations of the atmospheric flows, so their importance is clear. In order to have a moving frame of reference, the wind fields of the neighborhood of the storm at every time step from the ERA-interim reanalysis database was extracted (see Figure 7.7). They correspond to the u-wind and v-wind fields on a $25 \times 25°$ grid centered on the current TC location, at three atmospheric pressure levels (700, 500 and 225 hPa). The idea is to capture the air movements at the different levels of the troposphere: low, mid and high clouds. Similar to wind fields, the geopotential height (or iso-pressure latitude) fields were extracted from the neighborhood of the storm at every time step on a $25 \times 25°$ grid centered on the current storm location at three atmospheric pressure levels. In order to capture the dynamics, the wind fields and the geopotential height measured at times t and $t - 6h$ were extracted at the same location.

7.4.1.2. *Dataset separation*

The TC were randomly separated into three datasets as following: training (60%)/validation (20%)/testing (20%). Thus, the TCs in the test set have never been seen before by the learning algorithm. Then, within each set, all time instants were treated independently. The training set was used for optimizing the parameters of the NNs (back-propagation). The validation set was used to select the architecture of the network (section 7.4.2). Finally, the test set was kept hidden and was only used to show the final prediction accuracy at test-time (section 7.4.3).

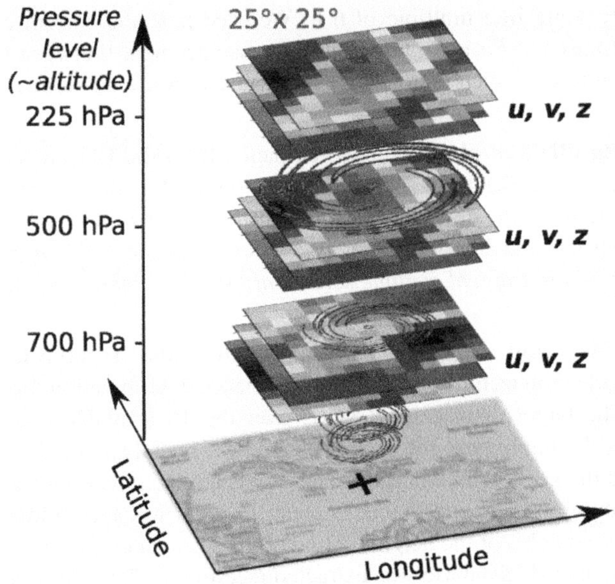

Figure 7.7. *Application 2: Global atmospheric grids centered on the storm location: wind fields (u and v) and geopotential height (z). The color represents the amplitude of each field in this data example. For a color version of this figure, see www.iste.co.uk/mallet/multitemporal.zip*

7.4.2. Methodology: a Deep Fusion Model

7.4.2.1. *Overview*

In order to jointly use the three different data sources, fusion neural network (NN) architecture was developed. An overview of the architecture is shown in Figure 7.6. The fusion architecture is divided into three branches: a Wind CNN, a Pressure CNN and a Past tracks + metadata NN. The Wind CNN and Pressure CNN are 2-D CNNs that take atmospheric fields (longitude, latitude, stacked over height and time) as input, while the Past tracks + meta NN is a small NN which takes 0D features as input (stacked over time). Each branch of the network makes its predictions independently. The training of the parameters is performed independently for each individual branch of the network for the same task, that is, predicting the 24-h forecast track. The three networks are then integrated into a fused network and the parameters are fine tuned. The different steps will be outlined in the following sections.

7.4.2.2. *Convolutional neural network for reanalysis*

Two similar CNN networks were developed for the wind and the pressure fields. They were separated into two networks because the information to extract is very different and thus different learning rates were needed. The data were stacked over height (pressure level) and time, such that the inputs of the CNNs consist of multiple 2-D (longitude, latitude) frames or channels. The Pressure CNN has six input channels (each one of size 25 $\times$ 25), while the Wind CNN input consists of 12 channels (u and v are stacked). A typical CNN architecture was used, alternating convolutional layers (Conv layer) and max-pooling layers, with FC layers at the end (Simonyan and Zisserman 2014). All hidden layers are equipped with the rectification (ReLU) nonlinearity and batch normalization.

An evaluation on the performance by adding historical features from past time steps was performed. In addition to t and $t - 6h$, no noticeable improvement was noticed by including more previous time steps. Only the times t and $t - 6h$ were kept.

7.4.2.3. *Past tracks + metadata neural network*

Another important source of information are the previous displacements and the other IBTrACS features (see section 7.4.1). All of the available features, which can be treated as a size-9 vector of 0D components, were used. A small NN of two FC layers was developed (the green branch in Figure 7.6) to learn the future track from the 0D features. Two past displacements were used: from $t - 12h$ to $t - 6h$ and from $t - 6h$ to t.

7.4.2.4. *Fused neural network for wind, altitude, and tracks*

In order to combine the different inputs, the three individual branches of the network were first trained separately. Then, their two last layers were concatenated and a layer at the end of the network was added (see Figure 7.6). The new concatenated layers consist of the same weights as before in each branch, plus new connections from each branch to the other ones, which were initialized to zero. In this way, the function computed is (at the start) the same as previously. The whole fused network is then re-trained by allowing every weight to be re-optimized. The number of fused layers (here two) was determined by comparing four different configurations on the validation set. The loss function was taken as the root mean square error (RMSE) between the forecast and the true storm location at $t + 24h$. The Haversine formula was used, giving great-circle distances between two points on a sphere from their longitudes and latitudes. Every evaluation was repeated three times and an average score was computed in order to assess the robustness to the random weights initialization.

7.4.3. *Experimental evaluation*

7.4.3.1. *Results on the whole dataset (all basins)*

We can compare the fused network, fusing all three branches, with the three single branches of the networks. Figure 7.8 shows the 24-h forecast results on the test set, which was 14,256 time steps in total, in absolute distance error. We can see the improvement of fusing networks (mean error: 130 km) with respect to the Wind CNN (mean: 148.9 km), the Pressure CNN (mean: 172.7 km) and the Past tracks + meta NN (mean: 186.6 km) alone. We can also see the importance of separately pre-training the three networks before the fusion, as it improves the mean result by 5 km. A persistence forecast baseline was also calculated: a 24-h prediction that is four times the storm's last displacement from $t - 6h$ to t. The mean error of this baseline on the test set is 196 km, which is more than 60 km higher than the fused model.

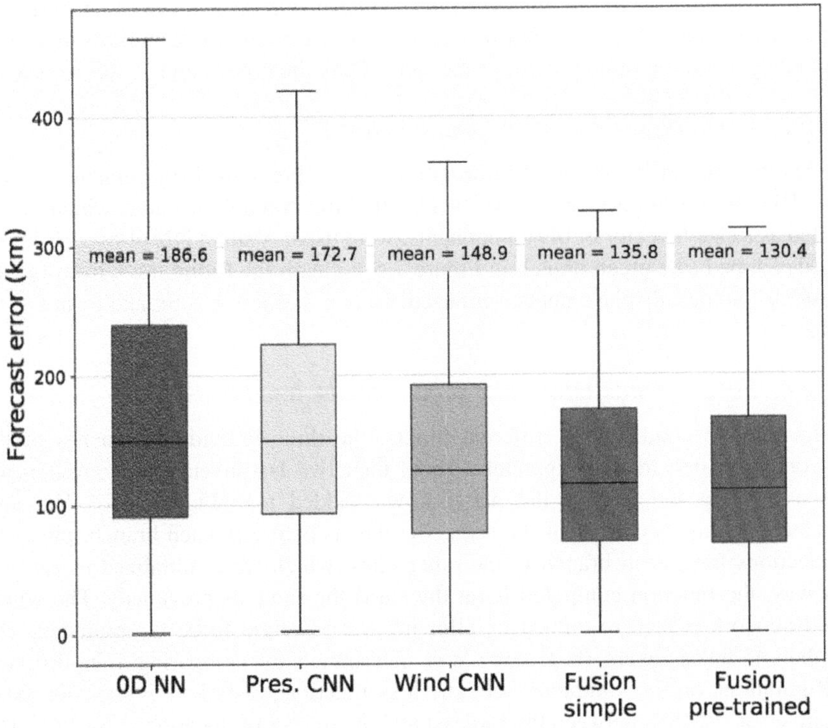

Figure 7.8. *Comparison between the three simple networks (the 0D neural network, the Pressure CNN and the Wind CNN), the fused network without separate pre-training (gray) and the fused network with pre-training (red, proposed method). Note that 24-h forecast results on the test set (storms coming from all oceanic basins), in distance between predicted and real locations. For a color version of this figure, see www.iste.co.uk/mallet/multitemporal.zip*

Moreover, if we only test on tropical cyclone time steps excluding depressions, which are storms of lower intensity, the mean prediction error drops from 130 km to

109.3 km. We can see in Table 7.2 the results on the test set for the different regions or basins: the best results are in the North Atlantic with a mean error of 130.2 km, corresponding to 26.5% of the 24-h displacement mean distance. The larger error is found in the South Pacific basin, but it is also the basin having the smaller number of samples. It is also worth noticing that the IBTrACS database is made of different source agencies depending on the location of the storm, and their reliability are not equal: this may also explain why some basins show higher errors.

Basin	Mean error (km)	Rel. to mean disp. (%)	Num. time points
North Atlantic	130.2	26.5	2,413
West Pacific	136.1	27.9	4,080
East Pacific	106.9	29.3	2,142
South Pacific	161.7	41.6	693
North Indian	138.9	51.3	2,286
South Indian	136.1	41.2	3,050

Table 7.2. *Note that 24-h forecast results for the different regions (basins), on the test set. Mean error in km and relative mean error w.r.t. the mean 24-h displacement distance*

7.4.3.2. *Comparison with statistical/consensus forecasts methods*

Two existing forecasting models were used to compare the fusion model results: CLP5[6], a statistical model which is often used to benchmark other TC forecasting methods, and OFCL, the National Hurricane Center official forecast (consensus of dynamical models)[7]. The CLP5 prediction results of years 1989–2016 were extracted in the Atlantic and Eastern Pacific basins. In Table 7.3, the fused network is compared with the statistical CLP5 on the test TC time instants at which both methods provided a forecast. This means that the comparison is made only when there is a one-to-one correspondence, which means 4,349 time steps from 258 storms. On both basins, the fused network performs better than the CLP5 model on average and in standard deviation. Moreover, the frequency of forecast errors larger to 200 km, or *bursts*, is also lower for our method, especially in the Atlantic (10% compared to 18%). Such comparison is not possible with the OFCL as this model is modified every year and they only provide forecasts of the version N of the model for the year N. The performance of the recent OFCL models on previous years is unknown and it would

6 Best track decay, combination of CLIPER5 (Climatology and Persistence model 5 day) and Decay-SHIFOR (Decay Statistical Hurricane Intensity Forecast).

7 National Hurricane Center Forecast Verification: https://www.nhc.noaa.gov/verification/verify6.shtml [Accessed 31 July 2018].

be unfair for them to compare with old results that were potentially obtained with earlier, less efficient models.

Model	Atlantic errors (km)				East Pacific errors (km)			
	Mean	SD	Bursts $> 200km$	Bursts $> 250km$	Mean	SD	Bursts $> 200km$	Bursts $> 250km$
CLP5	125	90	18.3%	9.8%	112	78	4.4%	2.3%
Fusion	112	71	10.3%	4.5%	88	52	4.1%	1.9%

Table 7.3. *Mean and standard deviation 24h-forecast errors for the Atlantic and Pacific basins on the subset of the test set where both predictions were available (total = 4,349 time steps). Bursts correspond to the ratio of track errors exceeding 200km (and 250km)*

Yet, the *yearly* results of the fused network performance with the two models on the same subset of the test set was computed. These results (mean and standard deviation) per year are shown in Figure 7.9. We can see that the fused network behaves better than the statistical approach CLP5 in most of the years. Moreover, it performs better than the OFCL forecast until year 2010 for the Pacific basin (2006 for the Atlantic). During the 2010s, the OFCL method improved and its mean errors per year are smaller (between 50 and 90 km for the Pacific, and between 60 and 100 km for the Atlantic).

Finally, the predictions are qualitatively compared for recent storms of the test set, such as TC Hermine in 2016 (Figure 7.10). The small bars connect each pair of predicted and ground truth location after 24 h. The longer the length, the larger the error. The official OFCL model has globally smaller forecast errors. Yet, on some time points the DL model outperforms the OFCL. It seems that the DL still perform poorly on the land: a future improvement could be to add the sea/land map as additional feature. Moreover, the three forecasts often have different directions. A neural network model could thus help the current forecast modelers by providing a complementary prediction that could be integrated in a consensus method, as their mistakes are different.

7.5. Applications using recurrent neural networks–convolutional neural networks

Following these two applications for intensity and track forecasting, we will now investigate how new studies incorporate differently the temporal dimension in their spatiotemporal learning framework: by using coupled RNNs and CNNs.

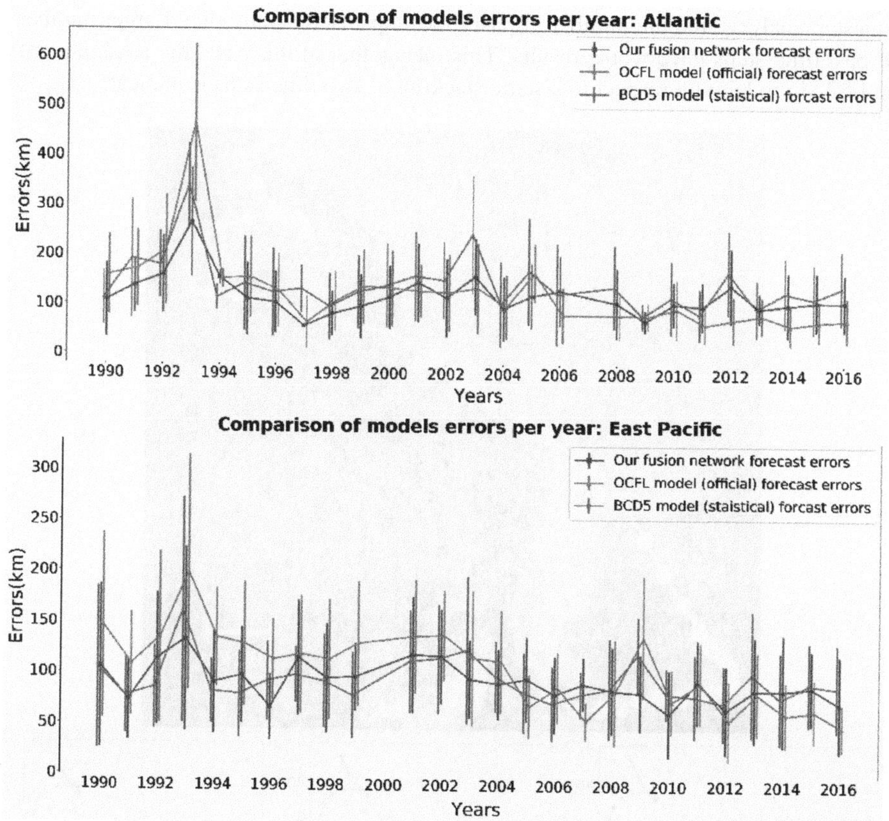

Figure 7.9. *Yearly average of 24-h storm track forecasting errors (km) and standard deviation on the test set (top figure for storms in Atlantic, bottom figure for storms in East Pacific) for our fused network forecasts (blue), the CLP5 model forecasts (green) and the official NHC forecasts (red), 1989–2016. For a color version of this figure, see www.iste.co.uk/mallet/multitemporal.zip*

7.5.1. *A hybrid CNN-LSTM model*

One of the first studies to apply a hybrid recurrent-convolutional NN to TC forecasting was published in 2019 by Chen et al. (2019). They proposed to use the LSTM recurrent architecture, coupled with a CNN, in order to learn from the 3-D time series reanalysis. The targets are twofold: the 24-h forecasted intensity and a binary 24-h formation forecasting (either the storm will develop in a TC or not). As in previous sections, they use time series with a time step of 6 h. Their results are successful, yet they have to be taken carefully as their comparison with traditional forecasts are possibly made on different datasets. We can also notice in their LSTM timestep analysis that the number of past time steps used is only two. Larger number of past time steps gave worse results. This means that in this case, the advantage of an LSTM model with respect to a static stacking of two time steps is unclear.

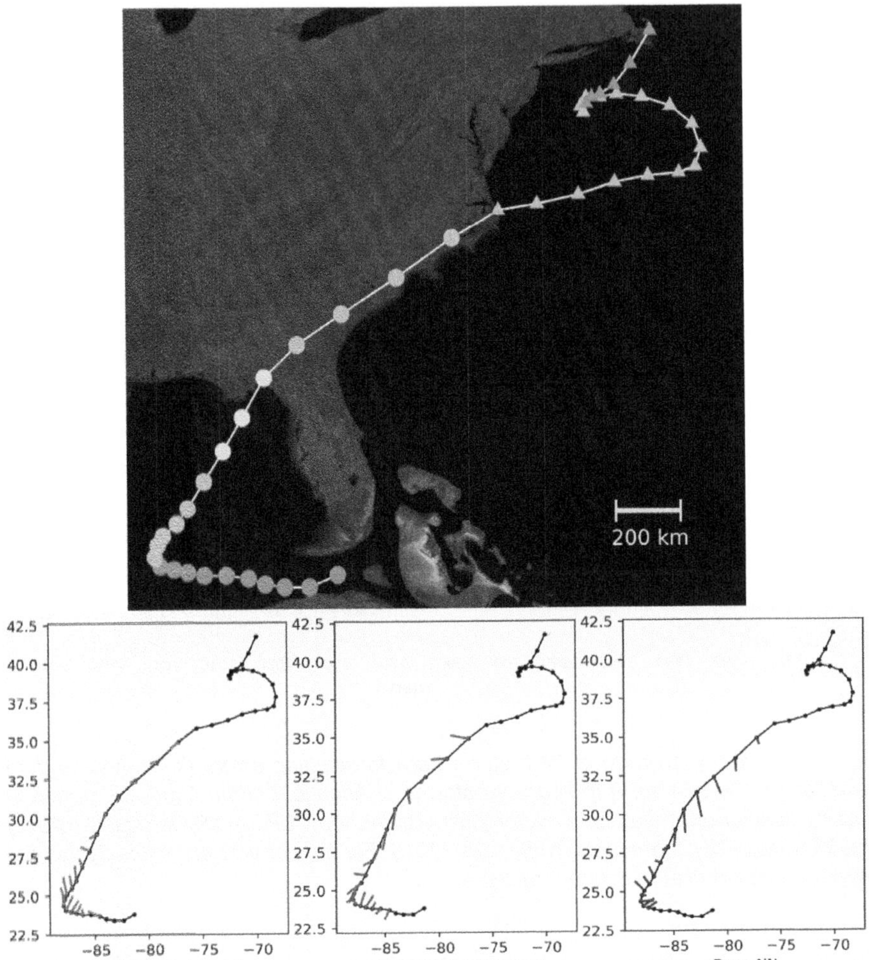

Figure 7.10. *24-h forecast errors (4 time steps ahead) on TC Hermine in 2016. The bars connect each pair of predicted and ground truth location. The longer the length, the larger the error. At the beginning and at the end of the track, the forecasts were not always available (a complete absence of an error bar means no forecast). For a color version of this figure, see www.iste.co.uk/mallet/multitemporal.zip*

7.5.2. *An attention recurrent CNN model*

As LSTM are known to be difficult to train and do not perform well in sequence-to-sequence tasks, other recurrent techniques emerged in the recent years, such as GRU or Transformer based on attention mechanisms on time series (see section 7.1.4.3). In Boussioux et al. (2022), the method combines these state-of-the-art recurrent methods with CNNs in order to forecast trajectory and intensity at a 24-h time lag (both GRU and Transformer are evaluated). The idea is to use a CNN to first encode the 3-D reanalysis data of each of the eight last time steps in a 1D vector. Secondly, these new embeddings are then used as input of GRU cells or Transformer before concatenating all time steps together in FC layers (see Figure 7.11). Their results show that the use of these recurrent networks helps to reduce significantly the errors on both intensity and track tasks. Compared to a tensor decomposition method, which does not use any recurrent model, the tracking results are always 20–30% better when using GRU or Transformer, able to take into account eight past time steps.

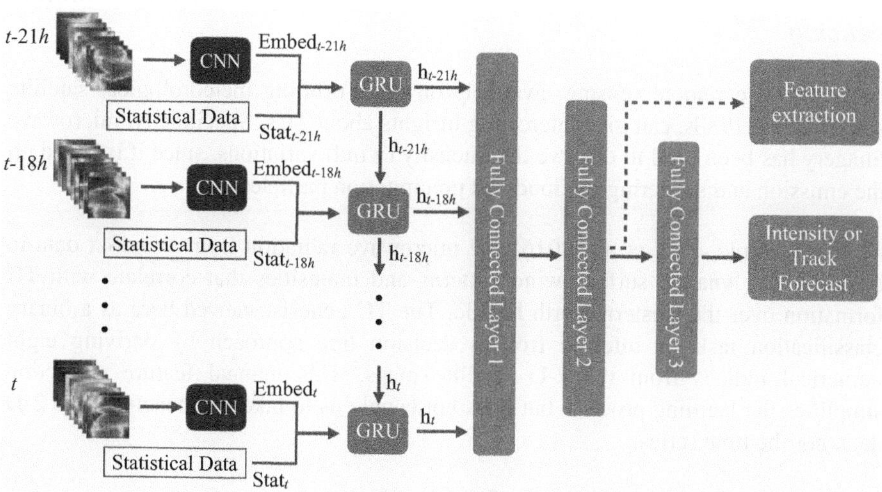

Figure 7.11. *Recurrent CNN model proposed by Boussioux et al. (2022), using GRU units on embeddings from 8 past time steps. This model can forecast intensity or track as well as providing a low-size embedding of the input data (feature extraction). Courtesy of L. Boussioux. For a color version of this figure, see www.iste.co.uk/mallet/multitemporal.zip*

The rest of the methodology follows closely the one presented in section 7.4, with some notable improvements. First, the use of 3 h time steps instead of 6 h and the use of the new ERA5 reanalysis database, which leads to more precise dynamical information. Second, the fusion between the different data types is performed at the end by a gradient boosted tree model XGBoost (Chen and Guestrin 2016) instead of fusing neural networks. This enables to easily train different XGBoost models for different tasks (intensity and track in this case). Last, once trained for a specific task, the encoding of a new reanalysis map into a low-size embedding (the *feature extraction* of Figure 7.11) can be easily calculated. This vector of size 128 captures important spatiotemporal features, and it can be directly used for other TC forecasting tasks using for example transfer learning methods.

7.6. Applications using remote sensing data

The methods developed in the previous sections are using reanalysis data, yet some ML TC forecasting research using remote sensing data have also began. Two types of satellite information have been investigated so far: microwave and infrared imaging.

7.6.1. *Deep learning intensity forecasting using microwave remote sensing*

Microwave remote sensing, available on polar-orbiting meteorological satellite since the last 1980s, can give interesting insights about TCs. Specifically, microwave imagery has been used to observe the intensity (wind) variations, since it is based on the emission and scattering of cloud and precipitation particles.

For example, Park et al. (2016) use microwave radiation from WindSat data to quantify the dynamic surface wind patterns and intensities that correlate with TC formation over the western North Pacific. The TC genesis, viewed here as a binary classification task, is inferred from a decision tree approach by deriving eight numerical indices from the 2-D satellite maps. This manual feature extraction simplifies the learning process, but does not enable us to take full benefit of the 2-D data, nor the time series.

In Wimmers et al. (2019), the TC intensity nowcasting (i.e. current or very short-term estimation) and 6-h forecasting is performed by DeepMicroNet, a CNN model from microwave satellite imaging bands 85-92 GHz and 37 GHz. While the results are hardly comparable to reanalysis methods (because reanalysis are not usable for nowcasting or 6-h forecasting as they take few hours to compute), this framework is competitive with more traditional automated satellite-based intensity estimation methods. The main advantage in developing a CNN model from satellite microwave data is its ability to provide an estimation of the current intensity in near to real time: the raw satellite imaging does not require any model extrapolation, like reanalysis, and the CNN inference can usually be performed in only few seconds.

7.6.2. *Deep learning intensity and wind structure forecasting using infrared remote sensing*

Longwave infrared satellite imaging are available by Geostationary Operational Environmental Satellites (GEOS) at a high temporal and spatial resolution. They provide an estimate of cloud-top temperature, which is a proxy of regions of deeper convection.

In Zhang et al. (2017), the infrared satellite images are used to derive the TC innercore surface wind structure, which is very important for forecasting TC intensity. First, the authors estimate some key structural TC elements, such as the eye size, using active contour segmentation techniques. Then, some standard ML methods, such as support vector machines and radial basis function neural networks, are used to estimate the main parameters of the 2-D surface wind field.

In a more recent work (Lee et al. 2020), the current TC intensity is estimated from multi-spectral infrared remote sensing, using 2-D and 3-D CNN architectures. Specifically, four channels are used: infrared 1 (IR1, 10.8 μm), infrared 2 (IR2, 12.0 μm), water vapor channel (WV, 6.7 μm) and shortwave infrared (SWIR, 3.7 μm). While only the current time step is used (without time series), it is interesting to see that using a 3-D input image, by viewing the sensor wavelength as the third dimension, does not give better results than stacking the four images as four channels.

Using infrared remote sensing as a complex spatiotemporal input for TC intensity forecasting has not yet been investigated. Nonetheless, as can be seen in the following *proposal paper* (McNeely et al. 2020), the infrared imaging could be used in a recurrent convolutional DL model in order to derive forecasted infrared imagery as well as forecasted structural features, which could lead to (i) forecast the TC intensity and (ii) provide structural summaries that are interpretable for operational forecasters.

7.7. Conclusion, current limitations and open problems

We have seen in this chapter some of the most recent advances in ML TC forecasting for tracking, intensity or formation tasks. Only a selection of works was presented, and we invite the reader to look at the review by Chen et al. (2020) on the subject for a complete overview. We have seen that the main difficulties lie in the complexity of the phenomena and on the spatiotemporal input data: 2-D or 3-D time series, coming either from reanalyis or remote sensing, to which 0D or 1D data can be added. While a consensus on using convolutional neural networks for handling the spatial inputs seems to be found, the temporal dimension is still handled in many different ways. Lastly, the combination of the different input types remains an open question too and many solutions are proposed. By analyzing the current state of the art, what is also lacking is a reference database that would be used as a benchmark by all of the future studies.

All of the presented ML methods need to be trained only once. After training, a few seconds are usually sufficient to give a forecast for a new TC. This is a significant time improvement over dynamical models, whose bottleneck is the computing speed.

The works presented above are in the large majority some proof-of-concept for 24-h forecasting or less. Yet, more long-term forecasts are needed in the real-world. I conjecture that for very long forecasts, larger than 25 x 25 degree images may be needed. Moreover, one of the key elements of current forecasts is the *track forecast cone* representing the circles of probable future locations: ML models will need to output a probability map and produce a *forecast cone* in order to be used in practice. If we want to estimate the model uncertainty, we can think of Bayesian DL approaches where all quantities are represented as probability distributions rather than point estimates, such as Monte Carlo dropout sampling (Gal and Ghahramani 2016).

For the reanalysis fields, a more refined selection of features could increase the overall performance of the models. Moreover, as it was shown that the storms close to the land or on the land have higher forecast errors, some improvements can be made by adding the land-sea mask image to the 2-D features. We could also represent the wind field by streamfunction and velocity potential as opposed to u- and v-wind components, which may help to have less correlated features. Moreover, while the ML algorithm could learn the differences of flow direction between north and south, a future improvement could be to flip the fields north-south and to change the sign of the v-wind component.

Lastly, as we have seen, most of the current methods are using current and/or past reanalysis fields. While they are usually quickly calculated, within few hours, it does increase the total forecast time accordingly. The methods relying on remote sensing data are thus faster to obtain (by supposing that the satellite data are not pre-processed and the acquisition time is short). The use of ML on "raw" data (as opposed to reanalysis), and especially on remote sensing, has a huge potential and will need to be investigated in more depth in the near future.

7.8. References

Arefin, M.R., Michalski, V., St-Charles, P.-L., Kalaitzis, A., Kim, S., Kahou, S.E., Bengio, Y. (2020). Multi-image super-resolution for remote sensing using deep recurrent networks. In *Proceedings of the IEEE/CVF Conference on Computer Vision and Pattern Recognition Workshops*, The Computer Vision Foundation.

Bai, S., Kolter, J.Z., Koltun, V. (2018). An empirical evaluation of generic convolutional and recurrent networks for sequence modeling. *arXiv:1803.01271*.

Boussioux, L., Zeng, C., Guénais, T., Bertsimas, D. (2022). Hurricane forecasting: A novel multimodal machine learning framework. *Weather and Forecasting*, 37(6), 817–831.

Chen, T. and Guestrin, C. (2016). Xgboost: A scalable tree boosting system. In *Proceedings of the 22nd ACM SIGKDD International Conference on Knowledge Discovery and Data Mining*, Association for Computing Machinery, New York.

Chen, R., Wang, X., Zhang, W., Zhu, X., Li, A., Yang, C. (2019). A hybrid CNN-LSTM model for typhoon formation forecasting. *Geoinformatica*, 23(3), 375–396.

Chen, R., Zhang, W., Wang, X. (2020). Machine learning in tropical cyclone forecast modeling: A review. *Atmosphere*, 11(7), 676.

Chung, J., Gulcehre, C., Cho, K., Bengio, Y. (2014). Empirical evaluation of gated recurrent neural networks on sequence modeling. *arXiv:1412.3555*.

Dee, D.P., Uppala, S.M., Simmons, A., Berrisford, P., Poli, P., Kobayashi, S., Andrae, U., Balmaseda, M., Balsamo, G., Bauer, D.P. et al. (2011). The ERA-interim reanalysis: Configuration and performance of the data assimilation system. *Quarterly Journal of the Royal Meteorological Society*, 137(656), 553–597.

DeMaria, M., Mainelli, M., Shay, L.K., Knaff, J.A., Kaplan, J. (2005). Further improvements to the statistical hurricane intensity prediction scheme (SHIPS). *Weather and Forecasting*, 20(4), 531–543.

Gal, Y. and Ghahramani, Z. (2016). Dropout as a Bayesian approximation: Representing model uncertainty in deep learning. In *International Conference on Machine Learning, PMLR*, International Machine Learning Society, Sydney.

Gao, S., Zhao, P., Pan, B., Li, Y., Zhou, M., Xu, J., Zhong, S., Shi, Z. (2018). A nowcasting model for the prediction of typhoon tracks based on a long short term memory neural network. *Acta Oceanologica Sinica*, 37(5), 8–12.

Giffard-Roisin, S., Gagne, D., Boucaud, A., Kégl, B., Yang, M., Charpiat, G., Monteleoni, C. (2018a). The 2018 climate informatics hackathon: Hurricane intensity forecast. In *8th International Workshop on Climate Informatics*, Boulder.

Giffard-Roisin, S., Yang, M., Charpiat, G., Kégl, B., Monteleoni, C. (2018b). Fused deep learning for hurricane track forecast from reanalysis data. In *Climate Informatics Workshop Proceedings 2018*, Boulder.

Giffard-Roisin, S., Yang, M., Charpiat, G., Kumler Bonfanti, C., Kégl, B., Monteleoni, C. (2020). Tropical cyclone track forecasting using fused deep learning from aligned reanalysis data. *Frontiers in Big Data*, 3, 1.

Hersbach, H., Bell, B., Berrisford, P., Hirahara, S., Horányi, A., Muñoz-Sabater, J., Nicolas, J., Peubey, C., Radu, R., Schepers, D. et al. (2020). The era5 global reanalysis. *Quarterly Journal of the Royal Meteorological Society*, 146(730), 1999–2049.

Jiang, G.-Q., Xu, J., Wei, J. (2018). A deep learning algorithm of neural network for the parameterization of typhoon-ocean feedback in typhoon forecast models. *Geophysical Research Letters*, 45(8), 3706–3716.

Kégl, B., Boucaud, A., Cherti, M., Kazakci, A., Gramfort, A., Lemaitre, G., Van den Bossche, J., Benbouzid, D., Marini, C. (2018). The ramp framework: From reproducibility to transparency in the design and optimization of scientific workflows. In *International Conference on Machine Learning*, Stockholm. Association for Computing Machinery, New York.

Kerns, B.W. and Chen, S.S. (2013). Cloud clusters and tropical cyclogenesis: Developing and nondeveloping systems and their large-scale environment. *Monthly Weather Review*, 141(1), 192–210.

Kim, S., Kim, H., Lee, J., Yoon, S., Kahou, S.E., Kashinath, K., Prabhat, M. (2019). Deep-hurricane-tracker: Tracking and forecasting extreme climate events. In *2019 IEEE Winter Conference on Applications of Computer Vision (WACV)*, Waikoloa, 1761–1769.

Knapp, K.R., Kruk, M.C., Levinson, D.H., Diamond, H.J., Neumann, C.J. (2010). The international best track archive for climate stewardship (IBTrACS) unifying tropical cyclone data. *Bulletin of the American Meteorological Society*, 91(3), 363–376.

Krizhevsky, A., Sutskever, I., Hinton, G.E. (2012). Imagenet classification with deep convolutional neural networks. *Advances in Neural Information Processing Systems*, 25(2).

Lee, J., Im, J., Cha, D.-H., Park, H., Sim, S. (2020). Tropical cyclone intensity estimation using multi-dimensional convolutional neural networks from geostationary satellite data. *Remote Sensing*, 12(1), 108.

Leidner, S.M., Isaksen, L., Hoffman, R.N. (2003). Impact of NSCAT winds on tropical cyclones in the ECMWF 4DVAR assimilation system. *Monthly Weather Review*, 131(1), 3–26.

Loridan, T., Crompton, R.P., Dubossarsky, E. (2017). A machine learning approach to modeling tropical cyclone wind field uncertainty. *Monthly Weather Review*, 145(8), 3203–3221.

McNeely, T., Dalmasso, N., Wood, K.M., Lee, A.B. (2020). Structural forecasting for tropical cyclone intensity prediction: Providing insight with deep learning. *arXiv:2010.05783*.

Milletari, F., Navab, N., Ahmadi, S.-A. (2016). V-net: Fully convolutional neural networks for volumetric medical image segmentation. In *2016 Fourth International Conference on 3D Vision (3DV)*, Stanford, CA, 565–571.

Moradi Kordmahalleh, M., Gorji Sefidmazgi, M., Homaifar, A. (2016). A sparse recurrent neural network for trajectory prediction of Atlantic hurricanes. In *Proceedings of the Genetic and Evolutionary Computation Conference 2016*, Association for Computing Machinery, New York, 957–964.

Mudigonda, M., Kim, S., Mahesh, A., Kahou, S., Kashinath, K., Williams, D., Michalski, V., O'Brien, T., Prabhat, M. (2017). Segmenting and tracking extreme climate events using neural networks. In *31st Conference on Neural Information Processing Systems (NIPS 2017)*.

Park, M.-S., Kim, M., Lee, M.-I., Im, J., Park, S. (2016). Detection of tropical cyclone genesis via quantitative satellite ocean surface wind pattern and intensity analyses using decision trees. *Remote Sensing of Environment*, 183, 205–214.

Piñeros, M.F., Ritchie, E.A., Tyo, J.S. (2008). Objective measures of tropical cyclone structure and intensity change from remotely sensed infrared image data. *IEEE Transactions on Geoscience and Remote Sensing*, 46(11), 3574–3580.

Rußwurm, M. and Körner, M. (2020). Self-attention for raw optical satellite time series classification. *ISPRS Journal of Photogrammetry and Remote Sensing*, 169, 421–435.

Rüttgers, M., Lee, S., You, D. (2018). Prediction of typhoon tracks using a generative adversarial network with observational and meteorological data. *arXiv:1812.01943*.

Simonyan, K. and Zisserman, A. (2014). Very deep convolutional networks for large-scale image recognition. *arXiv:1409.1556*.

Vaswani, A., Shazeer, N., Parmar, N., Uszkoreit, J., Jones, L., Gomez, A.N., Kaiser, Ł., Polosukhin, I. (2017). Attention is all you need. In *31st Conference on Neural Information Processing Systems (NIPS 2017)*, Long Beach.

Wimmers, A., Velden, C., Cossuth, J.H. (2019). Using deep learning to estimate tropical cyclone intensity from satellite passive microwave imagery. *Monthly Weather Review*, 147(6), 2261–2282.

Zhang, C., Dai, L., Ma, L., Qian, J., Yang, B. (2017). Objective estimation of tropical cyclone innercore surface wind structure using infrared satellite images. *Journal of Applied Remote Sensing*, 11(4), 046030.

List of Authors

Jérôme AMMANN
CNRS UMR 6538 LGO
IUEM Plouzané
Université de Bretagne Occidentale
France

Nicolas BAGHDADI
INRAE
UMR TETIS
Montpellier
France

Romina Vanessa BARBOSA
CNRS UMR 6539 LEMAR
UBO
IRD
Ifremer
IUEM Plouzané
France

Hassan BAZZI
Atos France
Technical Services
Bezons
France

Maxime BEAUCHAMP
IMT Atlantique Bretagne
Pays-de-la-Loire
Technopôle Brest-Iroise
France

Santiago BELDA
University of Alicante
Applied Mathematics
and
Space Geodesy Group
Alicante
Spain

Sean BEMIS
Department of Geosciences
Virginia Tech
Blacksburg
USA

Francesca BOVOLO
Remote Sensing for Digital Earth
Fondazione Bruno Kessler
Trento
Italy

Nesrine CHEHATA
Bordeaux INP – ENSEGID
France

Guillaume COULOUMA
LISAH
INRAE
IRD
Institut Agro
University of Montpellier
France

Stefano D'ARONCO
EcoVision Lab
ETH Zurich and University of Zurich
Switzerland

Ronan FABLET
IMT Atlantique Bretagne
Pays-de-la-Loire
Technopôle Brest-Iroise
France

Denis FEURER
LISAH
INRAE
IRD
Institut Agro
University of Montpellier
France

Sophie GIFFARD-ROISIN
Université Grenoble Alpes
and
Université Savoie Mont Blanc
CNRS
IRD
IFSTTAR
ISTerre
Chambéry
France

Zacharias KANDYLAKIS
Remote Sensing Laboratory
National Technical University of Athens
Greece

Christina KARAKIZI
Remote Sensing Laboratory
National Technical University of Athens
Greece

Konstantinos KARANTZALOS
Remote Sensing Laboratory
National Technical University of Athens
Greece

Hatem MABROUK
Institut National Agronomique
de Tunisie (INAT)
Tunis
Tunisia

Clément MALLET
Université Gustave Eiffel
IGN
ENSG
Saint-Mandé
France

Sylvain MASSUEL
G-EAU
AgroParisTech, Cirad
INRAE
IRD
Institut Agro
University of Montpellier
France

Mehmet OZGUR TURKOGLU
EcoVision Lab
ETH Zurich and University of Zurich
Switzerland

Luca PIPIA
Institut Cartogràfic iGeològic de
Catalunya (ICGC)
Barcelona
Spain

Konrad SCHINDLER
EcoVision Lab
ETH Zurich and University of Zurich
Switzerland

Yoann THOMAS
CNRS UMR 6539 LEMAR
UBO
IRD
Ifremer
IUEM Plouzané
France

Jochem VERRELST
Image Processing Laboratory (IPL)
Universitat de València
Spain

Fabrice VINATIER
LISAH
INRAE
IRD
Institut Agro
University of Montpellier
France

Jan Dirk WEGNER
EcoVision Lab
ETH Zurich and University of Zurich
Switzerland

Index

A, B, C

D, F, G

H, I, L

M, N, O

P, Q, R

S, T

U, V, W

Printed and bound by CPI Group (UK) Ltd, Croydon, CR0 4YY

13/09/2026

14961070-0004